D0232319

Apostolos Syropoulos
Antonis Tsolomitis
Nick Sofroniou

DIGITAL TYPOGRAPHY
USING LATEX

With 68 Illustrations

Extra
Materials
extras.springer.com

Springer

Apostolos Syropoulos
366, 28th October St.
GR-671 00 Xanthi
GREECE
apostolo@ocean1.ee.duth.gr

Antonis Tsolomitis
Dept. of Mathematics
University of the Aegean
GR-832 00 Karlobasi, Samos
GREECE
atsol@iris.math.aegean.gr

Nick Sofroniou
Educational Research Centre
St. Patrick's College
Drumcondra, Dublin 9
IRELAND
nick.sofroniou@erc.ie

Library of Congress Cataloging-in-Publication Data
Syropoulos, Apostolos.
    Digital typography using LaTeX / Apostolos Syropoulos, Antonis Tsolomitis, Nick Sofroniou.
        p. cm.
    Includes bibliographical references and indexes.
    Additional material to this book can be downloaded from http://extras.springer.com.
    ISBN 0-387-95217-9 (acid-free paper)
    1. LaTeX (Computer file) 2. Computerized typesetting. I. Tsolomitis, Antonis. II.
Sofroniou, Nick. III. Title.
    Z253.4.L38 S97 2002
    686.2′2544—dc21   2002070557

ACM Computing Classification (1998): H.5.2, I.7.2, I.7.4, K.8.1

ISBN 0-387-95217-9 (alk. paper)   Printed on acid-free paper.

Printed on acid-free paper.

Printed in the United States of America.

9 8 7 6 5 4 3 2 1   SPIN 10791970

Typesetting: Pages created by the authors using LaTeX

www.springer-ny.com

Springer-Verlag   New York Berlin Heidelberg
A member of BertelsmannSpringer Science+Business Media GmbH

Dedicated to the fond memory of Mikhail Syropoulos,
my beloved brother,
to my parents,
Georgios and Vassiliki,
and to my son,
Demetrios-Georgios.
— A.S.

◆

To my parents,
Panagiotis and Evangelia,
and to my wife,
Angeliki.
— A.T.

◆

To my father,
Andreas Sofroniou,
who introduced me to computers
when they were few and far between.
— N.S.

CONTENTS

FOREWORD

This book explores a great number of concepts, methods, technologies, and tools–in one word resources–that apply to various domains of typesetting. These resources have been developed and are used by the members of a very special community of people, which is also a community of very special people: the TEX community. To understand the motivation that led these special people to develop and use these resources, I believe it is necessary to make a short flashback. Since it is true that the past (uniquely?) determines the present and the future, I decided to divide this foreword into three parts: *The Past*, *The Present*, and *The Future*.

At this point, I am asking the readers to excuse my tendency of sometimes becoming autobiographic. This is very hard to avoid when talking about people and events important to one's life, and, after all, avoiding it could mean betraying the subject I would like to talk about.

The Past

Back in the 1980s, when I started working on my Ph.D. thesis, people in my department at the time (the Math Department, University of Lille, Northern France) were using a piece of software called "ChiWriter." This DOS program produced a very ugly low-resolution output of text and mathematical formulas. Others preferred to use IBM's Selectric II typewriter machines, spending hours and hours switching balls between Roman, Italic, and Symbol characters. Then came the day when the department finally bought a Macintosh Plus (with 1 MB of RAM and a 20 MB external hard drive!) and we installed *Textures* (a Macintosh implementation of TEX) on it. That day, my thesis advisor gave me a photocopy of the *TEXbook*, which I spent the whole night reading.

The last appendix chapter of that book was called "Joining the TEX community" and talked about TUG (the TEX Users Group), *TUGboat* (the newsletter of TUG) and so on. But the reader must realize that at that time things were quite different from today: computers were of course unfriendly, expensive, and slow, but the main difference was that there was as yet no Internet. Without the Internet, distances were more real than today, and for people like me who had not yet traveled to the States, places such as

"Stanford" or "Princeton" were infinitely far away and seemed to exist only for the privileged few. This is probably hard to understand today, but at that time, imagining the "TEX community" for me was like seeing a Star Trek episode or an old Hollywood movie: it was about people knowing and communicating with each other and acting together, but in a totally different place, time, and context—there could *de facto* be no interaction between them and myself.

That was in 1986, and then came the day when, during a stay at the Freie Universität Berlin, two things happened: I met and became friends with Klaus Thull (one of the European TEX veterans), and I opened my first *TUGboat*. By a coincidence so strong that one would be tempted to consider it as paranormal, the first *TUGboat* page I read was exactly page 22 of volume 9 (1), namely the one containing Silvio Levy's examples of Kazantzaki's text typeset in Silvio's Computer Modern Greek. Here is a translation of that text, reminiscent of the storm in Beethoven's sixth symphony:

> "At this moment I understand how heavy the mystery of confession is. Until now no one knows how I spent my two years at Mount Athos. My friends think I went there to see Byzantine icons, or because of a secret longing to live a bygone era. And now, look, I feel embarrassed to speak.
>
> How shall I put it? I remember a late afternoon in the spring, when a storm overtook me as I was coming down Mount Taygetos, near Pentavli. The whirlwind was so fierce I fell flat on the ground so I wouldn't be blown off the mountain. Lightning encircled me from everywhere and I closed my eyes to keep from being blinded and waited, face down, on the bare earth. The whole towering mountain shook and two fir trees next to me snapped in the middle and crashed to the ground. I felt the thunderbolt's brimstone in the air, and suddenly the deluge broke, the wind died down, and thick warm drops of rain struck the trees and soil. It pelted the thyme, oregano, and sage, and they shook off their odors and scented the whole earth."

Goethe (and Beethoven) wanted to communicate "von Herzen zu Herzen"; well, this is exactly what happened to me: altogether, the marvelous inebriating contents of this text which I had not read before, its appearance (which at that time I also found marvelous), and its context were quite a shock. That same day, I was able to communicate with Silvio (at that time still at Princeton) through e-mail. A few days later, Klaus and I had written our first joint *TUGboat* paper and submitted it to Barbara Beeton, again through e-mail. Suddenly, there were no frontiers anymore: the TEX community was quite real, and a new world opened in front of me. It is obvious that without traveling to Freie Universität Berlin, without Klaus, without e-mail, without *TUGboat*, none of these would happen.

In the summer of 1990, just a month after I defended my Ph.D. thesis, Tereza (who later became my wife) and I went to the TEX Users Group meeting in Cork, Ireland, and we had the chance to meet there all those mythical people who made TEX–the pioneers of the TEX community–except Donald Knuth himself, whom I met two years later, in Stockholm, in the pure Bergmanian atmosphere of the late Roswitha Graham's house. The occasion was the ceremony where Donald Knuth was conferred an honorary doctor's degree at the Kungl Tekniska Högskolan. Roswitha cashed in on that opportunity and organized a small but very interesting Nordic TUG meeting.

In the late 1980s and early 1990s many wonderful things happened (to name only one: the fall of the Berlin wall while Klaus spent the whole night cycling from East to West Berlin and back). At the same time, using communication tools such as mailing lists and ftp, the TEX community was able to communicate more and more and became wider and more powerful.

But who were these people and where did they come from? The twenty-first century reader should realize that in the 1980s and early 1990s, when Linux was in the mind of its creator and GNU software was not widely known, public domain software did not have the same degree of popularity and reputation as it has today. On the other hand, computers and commercial software were horribly expensive. The psychology of computer users was different as well: there was a tremendous psychological gap between "users" and "programmers"; especially, Macintosh and Windows users would be shocked if they had to type something that even vaguely looked like programming code, and writing TEX was indeed "programming," even if learning TEX was far more pleasant than learning, for example, Fortran IV or 8086 Assembler–not to mention the frightening task of implementing TEX on different platforms, which was, at that time, sometimes still unavoidable for people who simply wanted to use TEX for their documents. In France, in the early 1980s, there were Ph.D.s written on the process of implementing TEX on specific platforms.

It is not surprising that most members of the TEX community were students or scientists from computer science, mathematics, or physics departments. Because they had a reason to use TEX (writing their reports and publications), and because they had the means to communicate with each other, many of them contributed to TEX by writing code, and surprisingly enough, the TEX code that they wrote was very often not connected to the subject of their studies and research. Some projects were linguistic (extending TEX's capabilities to other languages and scripts), others typographical (facing the challenges of book typesetting), others artistic, ludic, or educational. In fact, what happened was, on a smaller scale, the same phenomenon as with Web pages some years later: students and scientists suddenly had the possibility to include their private life and hobbies in their work context and to share them with the community. The human dimension of TEX (and later of the Web) was flexible enough to allow input from various areas of human activities and interests. *TUGboat* was a wonderful mirror of that activity.

There were also the human needs of creativity and commitment: many TEX users wrote some code for their own needs, realized then that such code could be useful to others, extended it and wrapped it into a package with documentation and examples, and finally committed themselves to supporting it. By doing that, others became interested and communicated with them to express gratitude and suggestions for further development, which in turn resulted in reinforcing that commitment even more, and so on. Years before the widespread use of the Internet, the TEX community was already what we now call a *virtual community*, providing a positive and creative identity to people.

That identity was—and still is—one of the most charming aspects of TEX.

The Present

In the years that followed, the emergence of the Web brought big changes to the TEX community and to the perception of TEX by computer users in general. Thanks to HTML, it is quite natural today for everybody to be able to read and write "code." On the other hand, Adobe's PDF file format has bridged the gap between TEX output and electronic documents (and there is indeed a version of TEX producing PDF output directly). DVI was defined as a "device independent" and "typographically correct" file format: it was abstract enough to be usable on any platform and at the same time precise enough to be able to describe a printed page without loss of information. This was, more or less, also the case for the PDF format, which has the enormous advantage of being self-contained in the sense that it contains all resources (images, fonts, etc.) necessary for displaying and printing the document.

Finally, thanks to Linux and GNU, public domain software is nowadays very well-reputed, and, quite naturally, TEX is still part of every public domain operating system. That is why it gained popularity among computer gurus who used it to prepare their documents with other tools.

For every new TEX user, the contact with the TEX community (which has been such a big deal for me) has become instantaneous, since nowadays almost everybody is connected to the Web. TEX code can be distributed to the whole community—and this includes people in places unimaginable ten years ago—in a few minutes or hours. Even better, collaborative development tools such as sourceforge.net allow people to work simultaneously on an arbitrary number of different versions of the same software, however extensive and complicated this software may be.

The Web was very profitable for TEX for a number of reasons. Besides providing the TEX community with the means to be a true virtual community, it also made the principle of the dual nature of a document (source code versus compiled result) to become completely natural: when you write HTML code and preview it in your browser, you see two different representations of the same document. In other words, the "WYSIWYG" principle (which in the 1980s was quite an annoyance to TEX) has, at last, lost its supremacy.

Also, thanks to the Web and to political changes, there are no frontiers anymore, and standards such as Unicode have emerged to allow communication in all languages. TEX has always been a pioneer in multilingual typesetting, a feature that becomes more and more important today. As we will see in a while, a successor to TEX is one of the few (if not the only) software packages nowadays allowing true multilingual typesetting.

But are all things really well in the best of all possible worlds?

Talking of free software, let us return to one of the biggest achievements in the public domain, namely the Linux operating system, developed by hundreds of people all around the world. The obvious question to ask is: can TEX be compared to Linux? Unfortunately *not*, for several reasons.

First of all, is the absence of a Linus Torvalds for TEX: in fact, the author of TEX, Donald Knuth, one of the biggest computer scientists of the twentieth century and indeed a fabulous person with interests far beyond computer science, unfortunately decided to stop working on TEX once a certain number of goals were achieved. This happened in 1992, when version 3 of TEX was released. New versions after that were just bug fix releases. There are some small groups of people working on specific TEX-related projects (such as the LATEX group, the Ω group, the $\mathcal{NTS}$ group, etc.) and some institutions maintaining specific TEX packages (such as the $\mathcal{AMS}$). But outside of these, there is no coordination of the individual programming efforts.

Secondly, the goal to be reached in further developing TEX is not quite clear. TEX is a program dedicated to *typography*, a craft that very few people actually have studied, some people have learned by themselves—mainly by actually making books—and most people are generally unaware of. To continue our comparison with Linux, the latter is an operating system and hence deals with the global use of the computer: it is easy to imagine improvements, and if you lack imagination, you can always look into commercial operating systems to get ideas. TEX is the *only* piece of software dedicated to typography, and it does a *very* good job. Some people even believe that TEX *is* already perfect and hence there is no need for further improvement. But what *is* the ultimate goal of TEX, its *raison d'être*?

For years now, pessimists have been predicting TEX's extinction, but TEX is still alive and kicking! Maybe the most important reason for that is that TEX bridges the gap between the cultural heritage of the precomputer era and us today. Typography is both a craft and an art 500 years old, and Donald Knuth actually learned it and encoded his knowledge to TEX so that TEX is a "typographer-in-your-machine." Using just standard LATEX, people unaware of typography can produce decent documents by including in their text some markup reminiscent of XML. With a little more effort, and using a little more than standard LATEX, people aware of typography can produce brilliant documents. This degree of proficiency at attaining the sublime is cruelly missing from contemporary commercial software where the goal is not really commitment to our cultural heritage. TEX is a craftsman's tool like in the good old days: using such a tool, a novice can produce decent results and a master can make works of art. And, as always with Donald Knuth, a work of art in the context of TEX is both beautiful typesetting and efficient programming.

This book presents some of the achievements of the TEX community in the last two decades. For reasons inherent to the TEX users community, the tools presented are of various degrees of quality, efficiency and compatibility. There are so many tools (or packages, in LATEX parlance) available from the Comprehensive TEX Archive Network that there are strong chances you will find a package for any of your potential needs.

But how efficient will that package be, or how compatible with other packages written by other authors? This is an important question because improvements or resolutions of conflicts require a good knowledge of LaTeX. Often, there is a high level of support by the author of the package. But what happens when the author is hard to reach, or even unknown? Others in the TeX community may help you, but, as always in the public domain, there is no guarantee that you will get the help you need precisely when you need it.

This situation may seem frightening to people who expect absolute efficiency and immediate compatibility from software they use. There is a working scheme that is better fit to TeX and LaTeX, namely that of small groups of people sharing the same computer resources and being assisted by a "system administrator" (or "guru"). The "guru" is supposed to know TeX and LaTeX sufficiently well and to have the necessary time and energy to solve problems for the rest of the group, which can then smoothly use the software. Unfortunately, this organizational scheme does not fit individual personal computer users, who have to be simultaneously users and administrators.

So, how does one deal with problems in LaTeX packages? Well, experience shows that if you are a convinced LaTeX/TeX user, then you always manage to get by the problems, either by searching in literature (and books such as this one are very important for that very reason) by diving into the code and trying to "make it work," or, finally, by contacting other members in the community, even if the developers of the package are unreachable. A combination of these three methods actually works best. What is important is to realize that you are extremely lucky to be able to do all three: you have valuable books (such as this one and others), you can indeed dive into the code since it is open and freely distributed, and you can indeed contact others since there is a virtual—and furthermore friendly and united—community. Commercial software does not offer these opportunities.

The reader may have noticed that this book often mentions Ω and Λ. Where do these mysterious names come from and how do they fit in the "TeX and friends" context?

Ω, one of the major current TeX projects, is an effort by two people (John Plaice and myself) to develop a successor to TeX. It started two years after Donald Knuth's decision to freeze TeX. The philosophy of Ω is to take TeX as a starting point and to progressively add techniques and tools allowing the resolution of specific typesetting problems one at a time. The first major goal was to achieve typesetting in all languages of the world in the most natural and efficient way. In particular, one of the tasks that Ω seeks to accomplish is Unicode compliance (as explained in the book, Unicode is a standard 21-bit encoding for information interchange).

But Ω has other goals as well and is in fact an open platform for enhancements and additions to TeX. The name Ω has been chosen because traditionally the last letter of the Greek alphabet stands for ultimacy, "the ultimate tool," and also probably because 50% of Ω's development team is Greek. Finally, because choosing a Greek letter as the invariable and nontranslatable name and logo of a program is an additional argument for using the Unicode encoding (just as the fact of lowering the letter 'E' in the TEX logo was a very clever way to show the absolute need of using TEX to typeset even its own name).

Contrarily to Ω, which is existing, and quite extensive software, Λ is just a nickname, a kind of parody of the LATEX name: In fact, the "La" in LATEX comes from "Lamport", as in Leslie Lamport, the author of pre-1992 LATEX. The word "Lambda" also starts with "La", but has no relationship whatsoever with "Lamport" and is a Greek letter just like "Omega." Λ stands (as explained in this book) for the current LATEX (an achievement of the LATEX team, headed by Frank Mittelbach) when used in conjunction with the Ω engine.

It is quite probable that future versions of LATEX (for instance, version 3) will either be entirely written for Ω or at least have parts dedicated to Ω, in which case the Λ nickname will be useless. Also, due to the fact that the greatest part of Ω resources has not yet been released publicly, and that the Ω team still has to make a certain number of important global decisions, some information on Ω contained in this book may undergo minor changes in the future. In particular, there is (at the time this text is being written in March 2002) still no standard user-level LATEX interface for Ω.

Nevertheless, the basics of Ω will not change, and this book has the merit of being the first one to describe some of the very fundamental aspects of Ω, such as Ω translation processes, Ω virtual property lists, and so on and to illustrate them by examples.

The Future

The "future of TEX" (including the question of whether there is a future for it at all) has been a popular discussion subject for years in the TEX community. In fact, TEX is the sum of a big variety of different things, and for each one of them one can more or less predict its destiny, but one can hardly do this for the sum of them.

For example, TEX is both a programming language and a program (a "compiler" for that language): one could imagine that the program survives (for example as a typesetting or "rendering" engine inside a bigger system, and rumors circulate that this is already the case in Adobe InDesign); on the other hand, one could imagine Ω or some other successor to TEX becoming more and more different from TEX but—for reasons of upward compatibility—keeping the same programming language for input.

Besides being a programming language and a program, TEX is also a popular notation for mathematical formulas: mathematicians worldwide use TEX notation when writing formulas in, for example, e-mail messages: x^2 + y^2 < 1 with or without dollars is a natural choice for expressing the formula $x^2 + y^2 < 1$ in a text-only context. For writing mathematical formulas, TEX is exhaustive, clear, unambiguous, and short enough–all of the qualities of a good notation.

In recent years, the computer industry has become more and more involved in typesetting engine projects: the context in which source code of some kind has to produce more or less rigid formatted output becomes more and more important. After the first enthusiastic years of explosion of the Web, people realized that HTML (even combined with CCS) was definitely *not* sufficient for formatting documents. XML provided the necessary standard for structuring documents in an arbitrarily fine way, but still there was no "standard" way to *represent* an XML document. In October 2001, a new standard filled that gap: XSL-FO. The tools provided by XSL-FO for formatting documents are a quite serious challenge, and a new generation of XSL-FO-compliant typesetting engines is slowly emerging.

More generally, the current trend is to use XML as the basis of every kind of file format. For example, the SVG standard is, in some sense, an "XML-ized version of PostScript." One could very well imagine all file formats involved in TEX becoming XML-compliant: the input file could be pure XML "processing instructions" for including code in the TEX language the DVI file format could be replaced by SVG, the font metrics could be expressed in XML, illustrations could be in SVG instead of EPS, and so on. In that case, TEX (or Ω, or some other successor to TEX) would simply transform one XML document into another one. The fact that XML document transformation is nowadays an increasingly popular and important concept is by no means a coincidence.

Another area where Ω can be applied to revolutionize the electronic document is that of adaptive documents. A research project in that area deals with *vario-documents*, namely documents that contain a big number of page descriptions and display the right one according to context parameters, just as HTML browsers reflow text when their display window is resized. Only here each page description of the document has been compiled in advance by a "super-Ω," always with the same high typesetting quality standards.

Yet another area of drastic improvement of Ω's capabilities would be an on-the-fly interaction between typesetting and dynamic fonts. Already, in VectorTEX (a commercial TEX for Windows platform), Dimitri Vulis has included METAFONT capabilities into TEX. By using more modern font formats, such as OpenType, one could obtain a dialog between the font and TEX's typesetting engine so that each one instructs the other on constraints and context parameters and so that the final result is optimal for both.

There is also the more global, operating system-oriented point of view: Ω could very well become a server, and arbitrary client applications could send requests with text extracts and macros or parameters and receive in return small parts of page descriptions.

All of these "mutation" scenarios could be compared with the common skeleton of many science-fiction stories, where humans mutate to become less and less organic. Usually sci-fi authors want to express the fact that despite and beyond the changes of the human body (including an artificial brain), a core of *humanity* will always emerge as a fundamental quality of mankind. This is exactly the case for TEX: I am convinced that however drastically TEX (and its successors) will change in the future, its fundamental quality, which is the love of one man—and not just any man!—for good typography and good programming will always prevail and will always be the ultimate guarantee for the survival of this magnificent tool.

If this book succeeds in transmitting the fundamentally human quality of TEX and its successors, due to the love, sweat, and tears of Don Knuth and the hundreds of members of the active TEX community, then it will have reached its goal. I sincerely hope it does.

Yannis Haralambous
Brest, France

❖

March, 2002

Preface

What Is This Book About?

Our era is characterized as the "information era" mainly because computers (i.e., machines that manipulate information) are used in virtually all aspects of human life. One particularly interesting aspect of this phenomenon is that computers are used in areas where people traditionally thought that these machines had no use. One such area is fine arts (music, typography, painting, etc.).

Strictly speaking, typography is both an art and a craft. Typography is an art because it exists to honor content, and consequently, it can be deliberately misused. On the other hand, it is a craft, by which the meaning of a text (or its absence of meaning) can be clarified, honored, and shared, or knowingly disguised.

Many computer programs provide the means by which one is able to produce printed matter (books, leaflets, etc.). Most of them strive to provide a user-friendly interface that sometimes tries to guess the writer's intentions. However, it is a fact that all of these systems fail to produce the result that a traditional typographer would produce. There are many reasons for this serious drawback. For example, when the writer uses a friendly user interface, he or she is provided with a quite limited set of formatting tools that cannot handle all possible cases. This is quite evident when it comes to the typesetting of mathematical text, which is very demanding.

However, if one is provided with a programming notation specifically designed for typesetting purposes, then one loses the friendly user interface, but this is usually compensated by the output quality. In this book we make every possible effort to show that it is worthwhile to go to the trouble of learning such a programming notation. The programming notations we present are LaTeX (and its variant, pdfLaTeX) and Λ. They are markup languages specifically designed to ease the creation of scientific and nonscientific documents alike. Currently, the only evident difference between LaTeX and Λ is the fact that LaTeX operates on top of the TeX typesetting engine and Λ on top of the Ω typesetting engine. Otherwise, there is no obvious difference between the two notations. Virtually any document produced with LaTeX can be produced with Λ.

Reading the Book

Who Should Read It?

TEX in general and LATEX in particular are programming notations, and many newcomers wonder whether they can master the basics of the systems easily. Regarding LATEX, the answer is yes! LATEX has been designed so that even uninitiated people can produce excellent documents with the least possible effort, and this is exactly one of the goals of this book: to teach the novice all that is necessary so that he or she can be able to create high quality documents quickly with the tools described in this book.

 This book contains many text blocks that are marked with the symbol that marks this paragraph and are narrower than the usual text. These text blocks go into the details of the various typesetting tools and describe ways that allow users to customize them. Consequently, they should be read only by readers who have a good understanding of LATEX basics. Naturally, all novice readers will reach this level of understanding once they carefully study the rest of the text and try to do all the exercises (solutions to all exercises are provided at the end of the book).

So, this book is for novice as well as advanced LATEX users. Therefore, the book is suitable for everyone who wants to learn to use the system and its variations. Although LATEX and Λ are excellent typesetting tools for all sorts of documents, many people still think that they are the tools of choice only for mathematical typesetting. By presenting the multilingual capabilities and the other capabilities of these systems, we hope to make clear that these tools are just the best typesetting tools for all kinds of documents and all kinds of users!

The Book in Detail

Let us now describe the contents of each chapter.

The first chapter explains what LATEX/Λ is in general. We discuss the advantages of the logical document preparation versus the visual document preparation. Next, we provide information regarding the document preparation cycle and the various tools that are involved. The chapter concludes with general information regarding the programming notation.

In the second chapter we discuss various things that are essential for the preparation of even the simplest document. More specifically, we present the various characters that have a predefined meaning and the sectioning commands. We also discuss how one can prepare the title or the title page of a document. Next, we explain how one produces the various logos (e.g., how one can get the LATEX logo). Then, we discuss the preparation of articles, letters, and proceedings articles. We conclude by presenting a tool that allows us to combine many different documents into a single one.

In the third chapter we discuss various issues related to fonts, such as font shapes, series, and families. We continue with the presentation of the various font selection commands as well as the various symbol access commands. Also, we present ways that one can get important symbols such as the € symbol, the letters of the phonetic alphabets, astronomical symbols, and more, and since accented letters are found in most languages, we conclude the chapter by presenting tools that facilitate the placement of accents over letters.

The fourth chapter presents tools that can be used to typeset lists and catalogs, as well as poems, quotations, and more. In addition, we give all of the details that are necessary for the customization of these tools.

In chapter five we describe how one can typeset mathematical content using LaTeX. We present the available symbols and the symbol access commands. In addition, we present the necessary tools that the creation of complete mathematical texts. The last two-thirds of this chapter are for those who will use this chapter for reference for demanding mathematical text, and it can safely be skipped on first reading. The chapter concludes with a presentation of how one can generate MathML content from Λ sources. In addition, we discuss how it is possible to generate hypertext content from Λ sources.

Chapter six presents all of the core LaTeX features that have not been described in the previous five chapters. Topics covered in this chapter include references and hyperreferences, commands that generate white space, floats, page styles, and layout, slide preparation, and the definition of new commands and environments.

The seventh chapter presents a number of very useful packages (i.e., "systems" that extend the functionality of LaTeX) and do not comfortably fit in any other place.

Chapter eight shows how we can prepare the bibliography and the index of a document. We also show how we can prepare multilingual bibliographies and how we can create a simple package that can assist us in the generation of glossaries.

In chapter nine, we present a number of tools that allow LaTeX users to create simple drawings. These tools include the `picture` environment, the PiCTeX package, and METAPOST. We also discuss ways to include images in LaTeX and pdfLaTeX files, and since color and graphics are two closely related issues, we also discuss how we can create colorful documents.

Not many years ago, the English language dominated scientific writing, and this was reflected in most books on LaTeX; these books assumed that their readers would typeset their documents in English. However, this situation has changed, and nowadays most people prefer to use their mother tongue in their writings. Naturally, all of these people need typesetting tools to prepare their documents in their native languages. The tenth chapter describes all of the currently available tools for typesetting documents in a variety of languages. The first part of the chapter is devoted to the description of the typesetting tools, while the second part presents the typesetting facilities that are available for around forty languages or groups of languages.

To err is human, and this is the subject of the eleventh chapter, where we present common errors and error recovery strategies.

Chapter twelve is devoted to a description of the steps necessary for successfully installing new fonts (particularly scalable fonts) in an existing TEX installation.

The book concludes with five appendices that describe the generation of PostScript files from LATEX files, visual editing with XDVI and EMACS, the typesetting of XML files with LATEX, the transformation of LATEX files to HTML files, and the new features that will be introduced to the the the Ω typesetting engine.

The bibliography mentions only material published in some journal, periodical, or newsletter or as a book. Program manuals and "system" documentation usually accompany the corresponding software and in general are available from the CTAN (see page 12). There are two indexes: a name index and a subject index. In the subject index, a boldfaced page number denotes the page where the subject is discussed in detail (or defined). If for some subject there is no such page number, this means that the subject is considered well-known stuff.

The TEXLive CD-ROM that is included with this book offers a complete TEX system for Linux, Solaris 8 x86/SPARC, and Win32 platforms. This encompasses programs for typesetting and printing of LATEX/Λ documents, all of the packages described in this book, plus many other useful packages and extensive font libraries. The CD-ROM includes a large amount of general documentation about TEX, as well as the documents that accompany specific software packages. In addition, the CD-ROM contains all the book examples plus a number of selected exercises in the directory omegabook. The CD-ROM was compiled by Sebastian Rahtz.

Typographic Conventions

For most programs we use their respective logos when we are referring to them in the text. In case there is no such logo, we use small caps to write the program name (e.g., DVIPS). But the reader is warned to enter the program name with lowercase letters when attempting to use them. So, for example, the reader must type latex and dvips in order to use LATEX and DVIPS.

Acknowledgments

In this book, we present formatting tools for very many languages, and naturally we do not speak most of them. So we had to ask for help from native speakers (or fluent speakers, in the worst case) to verify the linguistic accuracy of the corresponding sections. We thank the following people for providing us with comments and suggestions that substantially improved the corresponding language sections: Takanori

Uchiyama (Japanese language), Jazier Bezos (Spanish language), Jin-Hwan Cho (Korean language), Serguei Dachian (Armenian language), Oliver Corff (Mongolian language), and Chakkapas Visavakul (Thai language).

We also would like to thank the following people for their help, suggestions, and constructive comments: Ichiro Matsuda, Norbert Preining, Koaunghi Un, Nguyen Duc Kinh, Olaf Kummer, Denis Girou, Andrea Tomkins, Sivan Toledo, Georgios Tsapogas, Vassilis Metaftsis, Harald H. Soleng, and Sebastian Rahtz for his excellent work on the T_EXLive CD-ROM.

Special thanks go to the Data Analysis Lab of the Department of Electrical Engineering of the Democritus University of Thrace and to the Department of Mathematics of the University of the Ægæan for providing the necessary resources for the creation of this book. Also, the first author of this book wishes to thank Sotirios Kontogiannis, Osman Osmanoglou, Georgios Toptsidis, and Kostantinos Sotiriadis for many stimulating and thought-provoking late-night discussions! The third author wishes to thank the Educational Research Centre at Saint Patrick's College for enabling him to contribute to this project. We also thank the anonymous reviewers who helped us to substantially improve the text of the book; and John Plaice for sharing with us his vision for Ω. Last but not least, we thank Wayne Yuhasz, executive editor of Springer-Verlag N.Y.; his assistant, Wayne Wheeler; Frank Ganz, the Springer T_EX evaluations manager for his help with some PostScript Type 1 fonts; Hal Henglein, the copyeditor; and Lesley Poliner the Spinger production editor.

The writing of a book is not an easy task at all, and of course this book is no exception. But in certain cases it is far easier if there is a starting point. For this book we used many ideas and the presentation style of [23]. The present book contains references to many web sites, but since it is a fact that web sites change web hosts rather frequently, we provide a web page with all the Web links of this book. The page also contains some other information regarding this book and it is located at http://ocean1.ee.duth.gr/LaTeXBook/ and mirrored at http://iris.math.aegean.gr/LaTeXBook/.

Apostolos Syropoulos
Xanthi, Greece

❖

Antonis Tsolomitis
Samos, Greece

❖

Nick Sofroniou
Dublin, Ireland

❖

June, 2002

1

INTRODUCTION

Computer Science is a fast growing discipline that rapidly engulfs exciting new disciplines such as Digital Typography and Mathematical Typesetting. Indeed, today Digital Typography is an active research field of Computer Science. In this chapter we introduce the fundamental concepts related to digital typesetting with TeX. We briefly present all of the relevant ideas that are necessary for the rest of this book.

1.1   What Is TeX?

The term "Digital Typography" refers to the preparation of printed matter by using only electronic computers and electronic printing devices, such as laser-jet printers. Since electronic printing devices are widely available, one often needs a digital typesetting system. TeX is a digital typesetting system designed by Donald E. Knuth. He designed TeX [19] mainly because, as he was struggling to finish the books of *The Art of Computer Programming*, he became disappointed with the computer technology available at the time.

   According to its creator, the idea for TeX was actually born on February 1, 1977, when Knuth accidentally saw the output of a high-resolution typesetting machine [16] (this article has been reprinted in [17]). He was told that this fine typography was produced by entirely digital methods (unfortunately, we are not aware of these methods), yet he could see no difference between the digital type and "real" type. At that moment he realized that the central aspect of printing had been reduced to bit manipulation. By February 13, he had changed his plan to spend the next year in South America; instead of traveling to some exotic place and working on Volume 4 of *The Art of Computer Programming*, he decided to stay at Stanford and work on digital typography. It is interesting to note that the 4th Volume of *The Art of Computer Programming* has not been published yet. By August 14, 1979, Knuth felt that TeX was essentially complete and fairly stable. In the meantime, he worked also on METAFONT [18], the companion program of TeX that he used to create the Computer Modern typefaces [15] that are now the standard font for TeX. Later on, he rewrote both TeX and METAFONT using the *literate* programming methodology that he also developed [17]. The product of this work was a system that is now known as TEX82. Knuth further developed his systems, and both of them are now frozen, in the sense that no further improvements will be done by him apart from some bug fixes. Since Knuth wants people to help him to find all possible remaining errors in his programs, he is offering the amount $327.68 to anyone who finds a bug. For more information on this offer, we suggest you to read the first few lines of the files tex.web[1] and mf.web[2] that contain the source code of both systems. The present version of TEX is 3.14159 and that of METAFONT is 2.718. Readers with a mathematical background will realize that the version numbers are identical to the first few digits of the numbers π (i.e., the circumference of a circle whose diameter is one) and e (i.e., the base of the natural logarithms). It is Knuth's wish to name the final version of TEX the version π and the final version of METAFONT version e by the day he dies. Although TEX and METAFONT are free software, they are trademarks of the American Mathematical Society (or $\mathcal{AMS}$ for short) and of Addison–Wesley Publishing Company, respectively.

Since TEX and METAFONT are frozen, one is not allowed to extend these systems and call them TEX and METAFONT, respectively. However, Knuth has encouraged researchers to extend his systems and to produce new systems. So, we now have many systems that have evolved from the original work by Knuth. The most notable TEX extensions are Ω, pdfTEX, ε-TEX, and $\mathcal{NTS}$ ($\mathcal{NTS}$ stands for New Typesetting System). Ω is a Unicode version of TEX that provides all of the necessary tools for real multilingual typesetting and has been developed by Yannis Haralambous and John Plaice. The program pdfTEX [26], a version of TEX capable of directly producing PDF output, originally developed by Hàn Thế Thành, is currently being further developed by its original developer, Hans Hagen and Sebastian Rahtz. ε-TEX [25], a TEX extension that can handle languages written from left to right and languages written from right to left, has been developed by the team that now develops $\mathcal{NTS}$, a TEX extension currently written in Java that will one day replace TEX (at least that is what the designers hope) and is being developed by Karel Skoupý with assistance by Phil Taylor. On the other hand, METAPOST by John Hobby is a reimplementation of METAFONT that produces PostScript output instead of bitmaps, which METAFONT produces.

TEX is a typesetting language (i.e., a programming language specifically designed to ease the generation of beautiful documents). The language has a wide range of commands that allow users to take into account every possible detail of the generated document. However, even expert computer programmers would have a really hard time if they were to produce even a simple document without additional help. Since TEX is a programming language, it offers the ability to define macros (i.e., to define new keywords that will have the combined effect of *primitive* commands when used). Moreover, TEX is designed in such a way that one can create a collection of macros designed to facilitate the document preparation process. Such macro collections are

---

1. Available from ftp://ftp.dante.de/pub/tex/systems/knuth/tex.
2. Available from ftp://ftp.dante.de/pub/tex/systems/knuth/mf.

known as *formats*. Knuth himself has designed the plain format, which was quite popular for some time.

Although the plain format is quite useful, there are many things that the casual user has to master in order to write even simple documents. This remark and the fact that the casual user wants to write a letter, a simple article or report, or even a simple book led Leslie Lamport to create the LATEX format. LATEX allows its user to write very quickly a letter, an article, a report, or even a book. Moreover, when compared to usual word-processing systems, LATEX has many other advantages, which are the subject of the next section. The present version of LATEX is called LATEX 2_ε and it is the one that we will present in this book. LATEX 2_ε has been developed by a team lead by Frank Mittelbach. When one uses Ω, LATEX becomes Λ (pronounced *lambda*), while when one uses pdfTEX it becomes pdfLATEX. Unlike TEX, LATEX is not frozen and is the subject of continuous development. The next version of LATEX will be called LATEX3 and will be a substantial improvement of the current version. The main advantages over its predecessor include the unified approach to multilingual typesetting, the simplification of the font access process, and more. For more information regarding the LATEX3 project, the interested reader should consult the LATEX project Web page at http://www.latex-project.org.

The reader may wonder why the name of the TEX system is written in this way and, moreover, how one should pronounce the name of the system. First of all the system's name is written this way to avoid confusion with TEX, an editor that was very popular by the time TEX was developed. Second, the letters that make up the TEX logo are the first three letters of the common root of the Greek words τέχνη (art, craft) and τεχνολογία (technology). Consequently, TEX should be pronounced "tekh," where the "kh" is pronounced as in the name Mikhail, and LATEX might be pronounced "latekh." The letter ε in the LATEX 2_ε logo comes from the word ἔκδοση (edition), so the logo actually means LATEX second edition. The "La" part in the LATEX logo comes from the last name of its creator: La(mport)TEX.

1.2 Logical versus Visual Design

Contrary to common belief, the preparation of a good document is a difficult task. By using an ordinary document preparation system, one is forced to make important decisions about the layout and the structure of the document. Thus, one has to decide on the page format and its general appearance and, at the same time, the text must be organized so that readers will not have any difficulty understanding it. Most common document preparation systems force their users to work on both aspects of the document preparation process. Certainly, this is not a severe restriction when it comes to the preparation of a nondemanding text. But, if someone has to prepare either a long document or a really demanding document, then this document preparation process may become a nightmare! Hence, it is extremely important for a document preparation system to assist its users in at least the visual design of their documents. In this way, the writer will concentrate on the logical design of the document and will let the document preparation system do the visual design. The advantage of this approach is that the visual design reflects the logical structure of the document. Systems that have this property are called markup languages. LaTeX is a system that pays more attention to the logical design than to the visual design, so it is a markup language. We will now give a simple example by which we hope things will become clearer.

Suppose that Michael wants to write an article about mathematics that will contain formulas and proofs based on these formulas. It is common practice in mathematical text to put a unique number at the end of each equation and to refer to it by this number. If Michael uses an ordinary document preparation system, then he has to manually enter the number for each equation since these systems treat equation numbers as an ordinary piece of text and nothing more. On the other hand, LaTeX assigns to each equation a number by incrementing the value of a *counter* (i.e., a computer storage location). Moreover, it provides a facility by which one can easily refer to any number that has been assigned to an equation, a page, and so forth. So, if Michael has the following equation in his article

$$e^{i\pi} + 1 = 0 \tag{1.1}$$

and for some reason he decides to insert another equation before it, LaTeX will automatically renumber all equations and, more importantly, it will produce the correct references in his text. Of course, if he had opted to use an ordinary document preparation system, he would have to manually change all references, something that is really error-prone. But things can get even worse. Suppose that Michael submits his article for publication to some journal and they accept it but want him to number equations with Latin numerals. Then he would have to manually change everything, and it is obvious what that means. But if he had opted to use LaTeX, he could have made the change by adding just a couple of lines of code.

By emphasizing the logical design of the document preparation process, LaTeX makes its users more productive and, consequently, allows them to concentrate on their real work (i.e., the writing of their text). Moreover, since TeX, as well as all typesetting engines based on TeX, is free software and available for virtually any computing system, LaTeX gives its users the rare chance to be able to switch between computing systems without any problem.

Readers who want to learn more on the subject of this section should consult the Web page http://ricardo.ecn.wfu.edu/~cottrell/wp.html. This Web page is maintained by Allin Cottrell.

1.3 Preparing a Document with LaTeX

The preparation of a document with LaTeX is usually done in two steps. The first involves the use of a text editor by which the user types a manuscript. This usually disappoints newcomers, who are accustomed to the so-called WYSIWYG (What You See Is What

You Get) document preparation systems (i.e., systems where the user directly types the text into a so-called *graphical user interface*, or GUI for short). But, as we have already explained, this has the big drawback that it does not allow the users to easily do what they really want to do. However, let us continue with the description of the document preparation process with LaTeX. Since LaTeX is a markup language, one has to type not only text but also commands, or "tags," that will assist LaTeX in the formatting process. It is important to note that our text must be saved in a plain text file; in other words, the resulting file must contain only the characters that we have typed and nothing more. Thus, users can use even a fancy word processing system to type their text and not just a simple text editor, perhaps because they want to use its spell-checking capabilities, but they must always remember to save their text in a plain text file. Once we have created a text file that contains the LaTeX source of our document, we are ready to feed it to the TeX typesetting engine with the LaTeX format preloaded. If there are no errors in our input file, then TeX will generate a DVI (DeVice Independent) file, which will contain all of the information that is necessary to either print or view, on our computer screen, the resulting formatted document. However, since this file does not contain the fonts necessary to print or view the document, one has to use a driver program. This program will automatically use the font information contained in the DVI file to correctly produce the formatted output. The viewing program is not standard and depends on the particular TeX installation. For example, on Unix, people usually view DVI files with a program called xdvi, originally developed by Eric Cooper and modified for X by Bob Scheifler, for X11 by Mark Eichin, and currently being maintained by Paul Vojta. On the other hand, many TeX installations provide their users with a printing program, but it is common practice to transform the DVI file into PostScript, by using the program dvips by Tomas Rokicki, and to print it either on a PostScript printer or on any printer using a PostScript driver such as Ghostscript by L. Peter Deutsch. The following diagram presents the basic document preparation cycle with LaTeX:

The diagram above omits various aspects of the document preparation cycle. For example, it does not present the bibliography generation as well as the index and glossary generation. Moreover, it does not present the generation of the various font-associated files. TeX uses the so-called TeX Metric Files (or TFM for short), files that contain the dimensions of each glyph as well as kerning and ligature information for a font, in order to correctly typeset a TeX source file. On the other hand, when one wants to view or print a file, the driver must either generate the so-called packed bitmap files (or PK for short), which contain resolution-dependent bitmaps of each glyph, or include the font outlines. (There will be more on fonts in the relevant chapters.)

In the case where the typesetting engine is pdfLATEX, the output file can be either a DVI or a PDF file. If it is a PDF file, this means that we can print or view it directly with Acrobat Reader by Adobe, Inc. Moreover, one can also use Ghostscript since this program can handle PDF files as well. But, now it is time to pass from theory into practice.

On most computing systems, a filename consists of two parts—the main filename and the filename extension. Usually, these two parts are separated by a period (for example, text.doc or text.txt). When one creates a text file that contains LATEX markup, it is customary to have tex as the filename extension. This way, the user does not have to type the complete filename when the file is fed to TEX. Now, we are ready to create our first LATEX file.

Using your favorite text editor, create a text file that will contain the following four lines:

```
\documentclass{article}
\begin{document}
Hello from \LaTeXe!
\end{document}
```

For the moment, you should not pay any attention to what you have typed. Now, suppose that the resulting text file is called example.tex. If we enter the following command at the prompt (e.g., an MS-DOS prompt of Microsoft Windows or a Unix xterm), LATEX will process our file and it will generate, among others, a DVI file:

```
$ latex example
This is TeX, Version 3.14159 (Web2C 7.3.1)
(example.tex
LaTeX2e <2000/06/01>
Babel <v3.6k> and hyphenation patterns for american, english,
greek, loaded.
(/usr/local/teTeX/share/texmf/tex/latex/base/article.cls
Document Class: article 2000/05/19 v1.4b Standard LaTeX
document class
(/usr/local/teTeX/share/texmf/tex/latex/base/size10.clo))
(example.aux)
[1] (example.aux) )
Output written on example.dvi (1 page, 368 bytes).
Transcript written on example.log.
```

Note that the $ sign indicates the system prompt; for example, in MicroSoft Windows this might be C:\. So what follows this sign, on the same line, is what the user enters. Moreover, the program output has been slightly modified so that it can fit the page, and this applies to all of the program output that follows. In the program output above, we can easily identify the versions of both TEX and LATEX that we are using. Furthermore, the system lets us know that it has created three files with main filename example and filename extensions aux, dvi, and log. The aux file contains auxiliary information that can be used for the creation of the table of contents, among other things. The dvi file is the DVI file that TEX has just generated, and the log file contains log information that is useful for debugging purposes in case there is an error in our LATEX source file. TEX indicates its progress by printing a left square bracket and the number of the page that it will start to process. When the page is shipped out to the DVI file, it prints a right square bracket. The total number of pages successfully processed as well as the total size of the DVI file appear at the end.

Since we have managed to successfully generate the DVI file, it is now possible to create a PostScript file from it by using the DVIPS driver:

```
$ dvips example
This is dvips(k) 5.86 Copyright 1999 Radical Eye Software
(www.radicaleye.com)
' TeX output 2000.10.08:0100' -> example.ps
<texc.pro>. [1]
```

In cases where the DVIPS driver cannot find the necessary PK files, it will try to generate them:

```
$ dvips example
This is dvips(k) 5.86 Copyright 1999 Radical Eye Software
(www.radicaleye.com)
' TeX output 2000.10.10:1241' -> example.ps
kpathsea: Running mktexpk --mfmode ljfour --bdpi 600
--mag 1+0/600 --dpi 600 cmr10
mktexpk: Running mf \mode:=ljfour; mag:=1+0/600; nonstopmode;
input cmr10
This is METAFONT, Version 2.7182 (Web2C 7.3.1)

(/usr/local/teTeX/share/texmf/fonts/source/public/cm/cmr10.mf
(/usr/local/teTeX/share/texmf/fonts/source/public/cm/cmbase.mf)
(/usr/local/teTeX/share/texmf/fonts/source/public/cm/roman.mf
(/usr/local/teTeX/share/texmf/fonts/source/public/cm/romanu.mf [65]
[66] [67] [68] [69] [70] [71] [72] [73] [74] [75] [76] [77] [78]
[79] [80] [81][82] [83] [84] [85] [86] [87] [88] [89] [90])
(/usr/local/teTeX/share/texmf/fonts/source/public/cm/romanl.mf [97]
[98] [99] [100] [101] [102] [103] [104] [105] [106] [107] [108] [109]
[110] [111] [112] [113] [114] [115] [116] [117] [118] [119] [120]
[121] [122])
(/usr/local/teTeX/share/texmf/fonts/source/public/cm/greeku.mf [0]
[1] [2])
(/usr/local/teTeX/share/texmf/fonts/source/public/cm/romand.mf [48]
[49] [50] [51] [52] [53] [54] [55] [56] [57])
```

```
(/usr/local/teTeX/share/texmf/fonts/source/public/cm/romanp.mf [36]
[38] [63] [62])
(/usr/local/teTeX/share/texmf/fonts/source/public/cm/romspl.mf
[16] [17] [25] [26] [27] [28])
(/usr/local/teTeX/share/texmf/fonts/source/public/cm/romspu.mf [29]
[30] [31])
(/usr/local/teTeX/share/texmf/fonts/source/public/cm/punct.mf
[33] [60] [35] [37] [39] [40] [41] [42] [43] [44] [46] [47] [58] [59]
[61] [64] [91] [93] [96])
(/usr/local/teTeX/share/texmf/fonts/source/public/cm/accent.mf
[18] [19] [20] [21] [22] [23] [24] [32] [94] [95] [125] [126] [127])
(/usr/local/teTeX/share/texmf/fonts/source/public/cm/romlig.mf [11]
[12] [13] [14] [15])
(/usr/local/teTeX/share/texmf/fonts/source/public/cm/comlig.mf
[34] [45] [92] [123] [124]) ) )
Font metrics written on cmr10.tfm.
Output written on cmr10.600gf (128 characters, 24244 bytes).
mktexpk: /var/tmp/texfonts/pk/ljfour/public/cm/cmr10.600pk:
successfully generated.
<texc.pro>. [1]
```

As we see from the program screen output, DVIPS could not find the PK at the requested resolution for the font cmr10. So, DVIPS calls METAFONT to generate the missing font. Once the PK file is successfully generated, DVIPS resumes and generates the final PostScript file. In case we want to generate a resolution-independent PostScript file, we have to configure the file psfonts.map so that the DVIPS will embed the outline font files into the final PostScript file (details will be discussed later):

```
$ dvips example
This is dvips(k) 5.86 Copyright 1999 Radical Eye Software
(www.radicaleye.com)
' TeX output 2000.10.10:1241' -> example.ps
<texc.pro><texps.pro>. <cmmi10.pfb><cmr7.pfb><cmr10.pfb>[1]
```

The PFB file is a binary PostScript outline font file. The corresponding nonbinary or ASCII files are called PFA files. Sometimes, the driver fails to embed the outline font files, although it has been configured to do so and the files are part of our TeX installation. In this case, the -j0 switch for DVIPS usually resolves the problem. If we want to view a DVI file that uses PostScript fonts, then xDVI calls GSFTOPK by Paul Vojta to generate PK files from the font outlines since xDVI can handle only PK files. Note that the latest versions of xDVI are capable of rendering PostScript fonts directly without using GSFTOPK.

If we had opted to use ε-LATEX, the resulting DVI file would have been identical to the one produced by LATEX since ε-TEX operates identically to TEX if we do not use its extended capabilities:

```
$ elatex example.tex
This is e-TeX, Version 3.14159-2.1 (Web2C 7.3.1)
(example.tex
LaTeX2e <2000/06/01>
Babel <v3.6k> and hyphenation patterns for american, english,
greek, loaded.
(/usr/local/teTeX/share/texmf/tex/latex/base/article.cls
Document Class: article 2000/05/19 v1.4b Standard LaTeX
document class
(/usr/local/teTeX/share/texmf/tex/latex/base/size10.clo))
No file example.aux.
[1] (example.aux) )
Output written on example.dvi (1 page, 368 bytes).
Transcript written on example.log.
```

If we had opted to use pdfLATEX, the output would be a PDF file:

```
$ pdflatex example
This is pdfTeX, Version 3.14159-13d (Web2C 7.3.1)
(example.tex
[/usr/local/teTeX/share/texmf/tex/pdftex/base/pdftex.cfg]
LaTeX2e <2000/06/01>
Babel <v3.6k> and hyphenation patterns for american, english,
greek, loaded.
Configured for pdftex use [1997/11/26]
(/usr/local/teTeX/share/texmf/tex/latex/base/article.cls
Document Class: article 2000/05/19 v1.4a Standard LaTeX document
class
(/usr/local/teTeX/share/texmf/tex/latex/base/size10.clo))
(example.aux)
[1[/usr/local/teTeX/share/texmf/tex/pdftex/base/standard.map]]
(example.aux) )<cmmi10.pfb><cmr7.pfb><cmr10.pfb>
Output written on example.pdf (1 page, 15680 bytes).
Transcript written on example.log.
```

Since pdfLATEX embeds the necessary fonts into the resulting PDF file, the screen output lets us know which fonts pdfLATEX has embedded into the PDF file. Of course, it is possible to create PDF files from PostScript files directly by using the program PS2PDF. This program is actually an application of Ghostscript and can only be used on a command line.

In the case where we are using Λ, the source file can be a Unicode file and not just an extended ASCII file. In any extended ASCII file, we are allowed to type up to 256 different characters, while in a Unicode file we are allowed to type up to 65,536 different characters. So, we can directly type text in any possible language. We will elaborate on this subject in Chapter 10, which presents the multilingual capabilities of LaTeX/Λ. Let us see now what the screen output will be when we use Λ:

```
$ lambda example
This is Omega, Version 3.14159--1.8 (Web2C 7.3.1)
Copyright (c) 1994--1999 John Plaice and Yannis Haralambous
(example.tex
LaTeX2e <2000/06/01>
Hyphenation patterns for american, english, greek, loaded.
(/usr/local/teTeX/share/texmf/tex/latex/base/article.cls
Document Class: article 2000/05/19 v1.4b Standard LaTeX
document class
(/usr/local/teTeX/share/texmf/tex/latex/base/size10.clo))
(example.aux)
[1] (example.aux) )
Output written on example.dvi (1 page, 392 bytes).
Transcript written on example.log.
```

Although the output file is called example.dvi, it is not a DVI file but rather an ΩDVI file. This new file format is actually an extended DVI file in which Ω can store information regarding Unicode fonts, writing directions, and so on. Because of this fact, one needs special drivers to handle the resulting ΩDVI files. To generate a PostScript file, one has to use the odvips driver:

```
$ odvips example
This is (Omega) odvips(k) 5.86 Copyright 1999 Radical Eye Software
(www.radicaleye.com)
'Omega output, Version 3.14159--1.8, 2000.10.08:1227' -> example.ps
<texc.pro>. [1]
```

On the other hand, if we want to view an ΩDVI file we have to use the oxdvi driver.

1.4 How Does TeX Typeset?

A typesetting system has to perform many operations in order to yield excellent output. One of its chief duties is to take a long sequence of words and break it up into individual lines of the appropriate size. In order to do this successfully, the system has to find the best breakpoints. TeX initially takes a paragraph and tries to find these breakpoints without employing the hyphenation mechanism that is available. If this is not possible, then it hyphenates all words according to the hyphenation patterns that are built into a particular format file and then tries to find the breakpoints, which of course in some cases will be in the middle of a word. Of course, we can instruct TEX to avoid breaking a line at specific points. In certain situations, TEX fails to produce a line of the appropriate size. If the line is longer than this size, we have an *overfull* box. On the other hand, if the line is shorter, we have an *underfull* box.

Things are even more difficult for page breaks. TEX usually guesses what would be the ideal breakpoint. This is mainly related to the fact that when TEX was designed, computer memory was an expensive resource and of very limited size. Most certainly, new typesetting systems could deal with this drawback, but since TEX decides page breakpoints in a very reasonable way, there has not been any significant progress on the matter.

Another interesting aspect of TEX's functionality is that it treats each character as a little box that can be virtually placed everywhere on the page (see page 39). This way, one can achieve interesting results such as the following alternative dollar symbol $, which is not usually available in most widely available fonts.

1.5 More Information and Resources

TEX is a typesetting system that has attracted the attention of many people. Moreover, since it is an extremely flexible system, many people work on the creation of TEX extensions and the development of new macros or formats that aim at facilitating the document preparation process. This fact had led a group of people to create the TEX Users Group (TUG for short), a nonprofit organization dedicated to the promotion and further development of TEX and its descendants. TUG publishes the quarterly newsletter *TUGboat*, which features refereed articles on various aspects of digital typography with TEX. More information on TUG can be found at their Internet site: http://www.tug.org. Since TEX is also heavily used by non-English-speaking people, there are many LUGs (i.e., local TEX users groups) that are dedicated to the promotion of digital typography with TEX in their respective countries and the development of tools that facilitate the preparation of documents in their respective languages. More information on these groups can be found at http://www.tug.org/lugs.html. Most of these groups publish newsletters similar to *TUGboat*; for example the Greek TEX Friends publish the semi-annual newsletter Εὐτυπον, NTG, the Dutch group, publishes the semiannual newsletters MAPS, and GUST, the Polish group, publishes a semi-annual bulletin. It is worthwhile considering becoming a member of your local TEX users group and/or of TUG.

Apart from these resources, one can download TEX installations for virtually any computing system, TEX packages, and fonts from either ftp://ftp.dante.de/tex-archive (maintained by DANTE, the German group), ftp://ftp.tex.ac.uk/tex-archive (maintained by UKTUG, the UK group), or ftp://ctan.tug.org/

tex-archive (maintained by TUG). These three sites constitute what is commonly known as the "Comprehensive TeX Archive Network," or CTAN, for short. Moreover, most TeX groups have mailing lists where people can ask questions regarding anything related to TeX. The Usenet newsgroup comp.text.tex is the official TeX forum for advanced and novice users. However, before sending any question to this group, you are strongly advised to consult the TeX Frequently Asked Questions Web page at http://www.tex.ac.uk/cgi-bin/texfaq2html. Finally, we suggest that you might like to have a look at the LaTeX Navigator site at http://tex.loria.fr/tex.

2

THE FILE STRUCTURE

In this chapter, we describe the general structure of a LATEX/Λ file. Since a LATEX/Λ file is composed of characters, we elaborate on the characters that one is allowed to type into a valid file and present some special characters with a predefined meaning. Next, we present the concept of a document class, the standard LATEX classes, and the classes provided by the American Mathematical Society. Furthermore, we discuss how one can create the title of a document and a title page. Next, we present how one can get some of the standard logos that are frequently used in the TEX world. We continue by presenting a real-world LATEX file and conclude with the presentation of a package that allows the combination of several LATEX files into a single document.

2.1 The Characters We Type

A user communicates with a computer by either typing in letters, digits, or symbols or by using some pointing device (e.g., a mouse). In the first case, these letters, digits, and symbols are collectively called characters. Each character is internally encoded as a sequence of binary digits (i.e., the digits "0" and "1") of a fixed length. This means that each character is equal to some number and, consequently, one can compare characters. Early computing systems provided only uppercase English letters, digits, a few symbols, and some special characters, such as the newline character, the end of file character, and so on. This limitation was imposed mainly because computers at that time had limited memory. Soon, people realized that they could not type in an ordinary English text with this limited character set, so, as computer technology advanced, computer manufacturers proposed new, larger character sets. The ASCII (American Standard Code for Information Interchange) character set was the one adopted by most computer manufacturers. ASCII contains 128 characters and includes all English letters in both cases, the ten digits, all symbols that are on a common keyboard, and 32 control characters. However, as computers became available to non-English-speaking people, there was a need to provide extended character sets so that non-English-speaking people could type in texts in their own languages. This fact led the various national standards organizations to define extensions of ASCII that contained at most 256 characters. These extended ASCII character sets were approved by the International Standards Organization, and now each of them has a unique name. For example, ISO-8859-7 is the name of the extended ASCII used in Greece. Similarly, ISO-8859-9 is the one used in Turkey, ISO-8859-1 the one used in Western Europe, and ISO-8859-5 is the default character set in countries that use the Cyrillic alphabet. Although people can write texts in their own language, it is still difficult to exchange files containing characters belonging to some extended ASCII. The main reason is that characters above 127 (i.e., the numbers that represent these characters are greater than 127) are not the same in two different extended ASCIIs, so it was necessary to define a new character set that would contain all possible letters, symbols, ideograms, and so on, in order to allow data exchanges without any problem. This necessity led to the definition of the Unicode character set. Unicode does contain all of the necessary characters to correctly type in a text in any language currently in use but also many mathematical symbols, characters not presently in use, such as the accented vowels of polytonic Greek, and many symbols that are in common use such as the symbol ®. Of course, one is also allowed to have characters from different languages in the same file (e.g., it is possible to have Japanese, Greek and Arabic text in the same file). Unicode provides for two encoding forms: a default 16-bit form called UCS-2 and a byte-oriented form called UTF-8. The Unicode standard version 3.1 is code-for-code identical with International Standard ISO/IEC 10646. If we use the 16-bit form, we can encode more than 65000 characters, while if we use the UTF-16 extension mechanism, we can encode as many as 1 million additional characters. The reader interested in learning more about Unicode may consult the relevant Web page at http://www.unicode.org.

TEX is a typesetting engine that can handle only files that contain characters belonging to some extended ASCII character set. For this reason, it is not particularly well-suited for multilingual document preparation, especially when it comes to languages that do not use the Latin alphabet. On the other hand, Ω is a typesetting engine that can handle Unicode files, so it is particularly well-suited for multilingual document preparation.

Although a LATEX file can contain ASCII characters and a Λ file can contain Unicode characters, there are a few characters that cannot be typed in directly as they have a predefined meaning. These characters are the following ones:

$$\# \quad \$ \quad \% \quad \& \quad \tilde{} \quad \_ \quad \hat{} \quad \backslash \quad \{ \quad \}$$

Let us now explain the special meaning of each of these characters. The character # (called *sharp*) is used to name the parameters of a parametric macro. However, this mechanism is primarily used in plain TEX and by people who create new formats and packages. The character $ (called *dollar*) is used to designate that one wants to write mathematical formulas. The same symbol is used to designate the end of mathematical text. The character % (called *percent*) is used to write comments (i.e., a sequence of characters that is completely ignored by LATEX). When we place the % character in a line,

LATEX ignores this character and everything to the right up to the end of the current line. Moreover, in certain cases, it prevents the typesetting engine from putting in some unwanted white space. The character & (called *ampersand*) is used in the construction of tables. The character ˜ (called *tilde*) usually stands for an unbreakable space; that is, if we put it between two character sequences without any space before or after it (e.g., Figure˜1) TEX will not attempt to put these two sequences on different lines or pages. However, in certain cases, it does not act like an unbreakable space. Such a case occurs when one prepares a manuscript in polytonic Greek. The characters _ (called *underscore*) and ˆ (called *circumflex*) are used to enter subscripts and superscripts in mathematical formulas, respectively. The characters { (called *left brace*) and } (called *right brace*) are used to define what is called in Computer Science a *local scope* (i.e., a place where all changes are local and do not affect the rest of the code). Readers familiar with C, Java, or Perl programming will identify this mechanism with the block structure provided by these languages. The character \ (called *backslash*) is the *escape* character (i.e., a character that makes special characters nonspecial and vice versa). For example, when it is in front of a word, the word is treated as a command. Certainly, there are some things that may not be clear at the moment, but they will become clear as we proceed. Now, since these characters are special and one is not allowed to type them in directly, the question is: "How can we type in these characters in a LATEX file"? The answer is given by the following table:

Symbol	Command	Symbol	Command
#	\#	$	\$
%	\%	&	\&
^	\textasciicircum	_	_
˜	\textasciitilde	\	\textbackslash
{	\{	}	\}

Thus, in order to get 40% off, we have to type in the characters 40\% off.

▶ **Exercise 2.1** What are the characters that one has to type in to get the following sentence:

You have a 30% discount and so the price is $13.

□

In any ordinary text, one has to be able to write paragraphs. In LATEX, paragraphs are really easy: we just put a blank line between the two paragraphs. In other words, at the end of a paragraph, we simply press the enter key two times.

▶ **Exercise 2.2** Now that you know how to create paragraphs, create a complete LATEX file that will typeset the following text:

The characters { and } are special. They are used to create a local scope.

Comments are introduced with the character % and extend to the end of the line.

☐

On rare occasions where one cannot type in a blank line, an alternative solution is to type in the command \par.

➤ **Exercise 2.3** Redo the previous exercise using \par. ☐

The discussion that follows may be skipped on first reading.

From the discussion above one may conclude that the characters above are like the reserved words of usual programming languages; that is, symbols or keywords that have a predefined meaning that cannot be changed in any way. Fortunately, this is not the case. TeX assigns to each character a *category code* (i.e., a number that characterizes its functionality) but this is something that can be changed either locally or globally. TeX provides the primitive command \catcode that allows its users to change the category code of any character. If we want to change the category code of, say, the character %, then we have to enter the command

$$\text{\catcode`\%=n}$$

where n is a number from 1 to 15 representing a particular category code. Note that the construct `\% is one of the ways to refer to a particular character. Alternatively, we can use its code point expressed in decimal, hexadecimal (prefixed by "), or octal (prefixed by '). It is even possible to refer to a particular character with the construct `^^hh (or `^^^^hhhh if we are using Ω), where hh (or hhhh) is the character's code point expressed in (lowercase) hexadecimal. Moreover, if we omit the ` symbol, we get a method to refer to any character of the character set we use. The complete list of the available category codes follows.

Category	Description	Category	Description
0	Escape character	1	Beginning of group
2	End of group	3	Math shift
4	Alignment tab	5	End of file
6	Parameter	7	Superscript
8	Subscript	9	Ignored character
10	Space	11	Letter
12	Other character	13	Active character
14	Comment character	15	Invalid character

Some explanations are in order. When we start using any format, all characters that are not assigned a particular category code will become *other* characters. For example, the character @ has no particular usefulness so it becomes an other character. An active character is one that can be used as a macro (i.e., one can instruct TeX to perform certain operations whenever the next input character is that particular active character). Initially, the term letter refers only to all of the uppercase and lowercase letters.

Now, if one wants to make the character * the comment character, then the command \catcode'*=14 achieves the desired effect. However, one must be very careful when changing category codes, as this may have unpredictable effects. Moreover, if one wishes to change the category codes of the characters { and }, we strongly advise the use of a local scope defined by the commands \begingroup and \endgroup. These commands define a local scope exactly like the two braces do. For example, the following code fragment makes the characters { and } behave like letters and assigns to the character [the category code 1 and to the character] the category code 2 in a safe way:

```
\begingroup
\catcode'\{=11  \catcode'\}=11
\catcode'\[=1   \catcode'\]=2
. . . . . . . . . . . . . . . . . . . . . . . . . . . .
\endgroup
```

2.2 Document Classes and Packages

A document class specifies the general layout of our document as well as the various sectioning commands that are available for the particular document that one is preparing. For example, the book document class must provide commands for chapters, while for the article document class this is meaningless. LATEX comes with a number of standard classes available for general use, which are shown in the following table.

Document Class	Description
article	An article, useful for the preparation of papers
book	For book preparation
report	For report preparation
letter	For letter preparation
slides	For slide preparation
proc	An article in conference proceedings

Apart from these document classes, there are a few more standard document classes with a special usage:

ltxdoc Used only for the typesetting of the LATEX kernel and LATEX packages.
ltxguide Used for the typesetting of the various documents that accompany each release of LATEX 2_ε, such as LATEX for Authors, and so on.
ltnews Used for the typesetting of the LATEX News, a leaflet that briefly describes what is new to each LATEX 2_ε release.
minimal A minimal document class that is mostly used for debugging purposes.

But how can we tell LATEX which document class we want to use? Since LATEX is a markup language, there must be a command by which one specifies the document class to be used. Not surprisingly, this command is called \documentclass. This command takes at least one argument (i.e., the name of a valid document class and some optional arguments). The general form of this command is shown below.

\documentclass[*optional args*]{*doc class*}

Of course, one may choose not to specify any optional argument. In this case, the user can either omit the square brackets altogether or just leave them without specifying anything in between them.

With the optional arguments, one can specify the default paper size on which the document will be printed, whether the document will be printed two-sided. The supported paper sizes are shown in the following table:

Paper Size	Page Dimensions
letterpaper	8.5 in × 11 in[†]
legalpaper	8.5in × 14 in
executivepaper	7.25 in × 10.5 in
a4paper	210 mm × 297 mm
a5paper	148 mm × 210 mm
b5paper	176 mm × 250 mm

[†] 1 in = 72.27 pt and 1 in = 2.54 cm.

The default paper size is letterpaper. Note that the text width and the text length are both predefined for each particular paper size. Certainly, one can easily change the paper size and the text dimensions, but we will come back to this issue in Chapter 6. The other available optional document parameters are shown in the following table.

Parameter	Brief description
10pt	Normal letter size at 10 pt
11pt	Normal letter size at 11 pt
12pt	Normal letter size at 12 pt
twoside	Two-sided printing
oneside	One-sided printing
twocolumn	Document is typeset in two columns
landscape	Document is typeset in landscape mode
titlepage	Forces LaTeX to generate a separate page for the document header and abstract
leqno	Equation numbers appear at the left end of the page
fleqn	Equations are being typeset at the left margin of the page
draft	Forces LaTeX to print a line overflow indicator at the end of a line that is longer than the predefined size and does not include external graphics files

The default font size is 10 pt (i.e., by omitting this option we inform LaTeX that we want to typeset our document in 10 pt). If the draft parameter is in use, then TeX prints the symbol ■ at the end of each line that is longer than the predefined line length. Moreover, it prints a rectangle in the case where we are including some graphics file prepared, for instance, with a drawing tool. If we want to specify two or more optional arguments, we separate them with commas (e.g., draft, 10pt). Note that the optional arguments can be specified in any order.

➤ **Exercise 2.4** Suppose now that one wants to prepare an article at 11 pt for two-sided printing. Write down the corresponding \documentclass command.   □

The American Mathematical Society has created a number of classes suitable for their publications. These classes provide additional functionality and are particularly well-suited for mathematical text. The complete list of these classes is shown in the following table.

Document Class	Description
amsart	The article class of the AMS
amsbook	The book class of the AMS
amsproc	For proceedings preparation
amsdtx	Used to typeset the source code of packages
amsldoc	Used to typeset documentation

More information on these classes is provided in Section 5.5.

A package is a LaTeX file that provides additional functionality to the functionality already provided by the LaTeX kernel. Nowadays, packages are written in a form of literate programming suitable for LaTeX. Any package comes with at least one .dtx file and one .ins file. The .dtx file(s) contain the documentation as well as the source code of the package. One can directly typeset each .DTX file by feeding it to LaTeX. On the other hand, a user can extract the source code of the package by feeding the .INS file to LaTeX. This procedure will yield one or more .STY files. Each .sty file contains the source code of a LaTeX package. After generating the .sty files, one has to install them (i.e., to put them in a directory directly searchable by the TeX typesetting engine). The exact location where we put the resulting files is system-dependent. Note that all of the packages described in this book are part of any modern TeX installation. Yet how can we inform LaTeX that we want to use a particular package? Before answering this question, we have to present the general structure of a LaTeX file:

```
\documentclass[optional-args]{doc-class}
Preamble
\begin{document}
Text
\end{document}
```

The preamble is the place where we put the necessary commands to use any packages that we choose. Moreover, this is the place where we put any user-defined macro definitions. The text area is the place where we actually write our text. In order to use a package, we have to add the \usepackage command. Since many packages provide options, one can specify which options one wants to use with a particular package. On the other hand, if one or more packages do not provide any options, one can directly specify all of the required packages as arguments of the \usepackage command. So, there are two forms of the \usepackage command,

\usepackage[*option(s)*]{*package*}
\usepackage{*package(s)*}

where *option(s)* and *package(s)* is either the name of one option, a package, or a list of comma-separated options and package names, respectively. For instance, if one wants to use the graphicx package with the dvips option, then the appropriate package inclusion command is

\usepackage[dvips]{graphicx}

On the other hand, if one wants to use the packages ulem and letterspace, then the correct package inclusion command is

\usepackage{ulem,letterspace}

2.3 Sectioning Commands

Even the simplest document is partitioned into text chunks for at least two reasons. The first reason is that the document can be created more easily. The second reason is that the reader can be guided through the text by consulting the table of contents, which, in turn, is based on the title of each text chunk. Every LATEX class provides a number of sectioning commands that are suitable for that particular document class. For example, the book document class provides, among other things, sectioning commands for parts and chapters. The general format of the LATEX sectioning commands is as follows:

\Section{*Section title*}   or   \Section*{*Section title*}

Here, \Section is the name of the particular sectioning command, and, of course, what goes inside the curly brackets is the actual title of the \Section. LATEX uses the arguments of the various sectioning commands to construct the table of contents, if we ask LATEX to generate it. Sectioning commands that have a trailing asterisk are not included in the table of contents and also generate headings without numbers. This is very useful since, for example, it may make no sense to put an acknowledgment section into the table of contents. LATEX provides its users with a plethora of sectioning commands:

\part   \chapter   \section
\subsection   \subsubsection   \paragraph
\subparagraph

The command \part is used to partition a book or report into *parts*. The command \chapter is used to partition a document into *chapters*, but it can be used only when preparing a book or a report. The other commands can be used in any document class and are suitable for sections, subsections, and so on.

▶ **Exercise 2.5** Write the command that is necessary to create the section heading of this section. □

The argument of a sectioning command is also used to typeset the so-called *running heads* (i.e., the text we see at the top of the page). In many cases, this argument can be quite long so the running head does not fit well into the page. The obvious solution is to have a shorter text as a running head. For this reason, every sectioning command can have an optional argument, which must be enclosed in square brackets and is placed just after the name of the sectioning command; for example

$$\text{\\section[\textit{Short title}]\{\textit{Long title}\}}$$

so the *Short title* will be used as a running head. Moreover, LATEX will insert this title into the table of contents, if we ask it to generate it. Of course, in the case where we have a starred sectioning command, for example

$$\text{\\section*[\textit{Short title}]\{\textit{Long title}\}}$$

the *Short title* will be used to typeset the running head. It is important to note that the sectioning commands are commands where one has to omit the square brackets altogether when it is not necessary to specify an optional argument.

If one is not satisfied with the predefined sectioning command, it is easy to define one's own commands in a separate package file. In order to do this, be aware of the fact that most sectioning commands are defined in terms of the command \@startsection, while the sectioning commands for parts and chapters are defined in terms of the command \secdef. The command \@startsection has six required arguments, optionally followed by a *, an optional argument, and a required argument:

\@startsection{*name*}{*level*}{*indent*}{*beforeskip*}
                {*afterskip*}{*style*}
                optional * [*altheading*]{*heading*}

The meanings of the various arguments are explained below:

name The name of the user-level command (e.g., section).

level A number denoting the depth of the section (e.g., chapter has level equal to one, and so on).

indent The indentation of the heading from the left margin (see Figure 2.1).

beforeskip The white space that is left above the heading (see Figure 2.1). This argument is actually what we call a glue (i.e., a length that can shrink and stretch). In general, a glue is a length followed by the keyword plus and the maximum length that the glue can stretch and/or the keyword minus and the maximum length that the glue can shrink.

afterskip The glue that is left after the heading (see Figure 2.1). In case it is a negative glue, it just leaves a horizontal space.

style Command to set the appearance of the heading.

altheading An alternative heading to use in the table of contents and in the running heads.

heading The heading used in the table of contents and in the running heads.

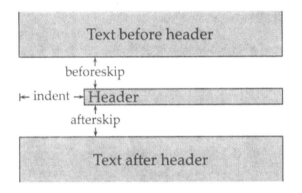

Figure 2.1: Graphical representation of the arguments indent, beforeskip, and afterskip of the \@startsection command.

A typical example definition is

```
\newcommand{\section}{\@startsection {section}{1}{0pt}%
                      {-3.5ex plus -1ex minus -.2ex}%
                      {2.3ex plus .2ex}%
                      {\normalfont\Large\bfseries}}
```

The command \secdef can be used to define (demanding) sectioning commands. It has two arguments

$$\text{\secdef}\{unstarcmds\}\{starcmd\}$$

where *unstarcmds* is used for the normal form of the sectioning command and *starcmds* for the starred form of the sectioning command. Here is how one can use this command:

```
\newcommand{\firstarg}{}
\newcommand{\chapter}{...\secdef{\cmda}{\cmdb}}
\newcommand{\cmda}[2][]{\renewcommand{\firstarg}{#1}%
```

```
                              \ifx\firstarg\empty
                                 \relax ...
                              \else ... \fi}
           \newcommand{\cmdb}[1]{...}
```

The \ifx construct is used to check whether the first argument is empty or not. In case it is, we put the code that processes the second argument after the command \relax, otherwise we put the code that processes both arguments after the command \else. The \relax command prevents TEX from consuming more tokens than is necessary, and it actually does nothing.

In case one wants to change the appearance of the sectioning commands, the command can be redefineed \@seccntformat can be redefined. For example, if we would like old style numerals, then the following redefinition achieves the desired effect:

```
      \renewcommand{\@seccntformat[1]}{%
         \oldstylenums{\csname the#1\endcsname}\quad}
```

The command \oldstylenums should be used to typeset a number using old style numerals. The construct \csname *string*\endcsname makes up a command of the *string*. To turn a command into a string, use the \string command.

The table of contents of a document can be constructed by issuing the command \tableofcontents. This command can be put in any place that the user feels is appropriate. However, it is common practice to have the table of contents in either the beginning of the document, usually after the title or title page, or at the end, just before the index and the bibliography.

In some cases, one may want to be able to manually add something to the table of contents. In cases such as this, one can use the command

\addcontentsline{*table*}{*type*}{*entry*}

where *table* is the file that contains the table of contents, *type* is the type of the sectioning unit, and *entry* is the actual text that will be written to the table of contents. For example, the command

\addcontentsline{toc}{chapter}{Preface}

adds to the .toc file (i.e., the table of contents) the chapter entry Preface. In case we want to add an entry to the list of tables or figures, the table is called lot or lof, respectively (see Section 6.5).

The command \addcontentsline is defined in terms of the commands \addcontents and \contentsline. The first command has two arguments: the *table* and a *text*; this command adds the *text* to the *table* file, with no page number. The second command has three arguments: the *type*, the *entry*, and the *page*, which can be either a number or a command that yields the current page number; the command produces a *type* entry in the *table*. For example, the entry for Section 2.5 of this book in the table of contents is produced by the command

`\contentsline{section}{\numberline{2.5}Basic Logos}{26}`

The command `\numberline` puts its argument flush left in a box whose width is stored in the internal length variable `\@tempdima`. Now, we give the definition of `\addcontentsline`:

```
\newcommand\addcontentsline[3]{%
  \addtocontents{#1}{%
    \protect\contentsline{#2}{#3}{\thepage}}}
```

➤ **Exercise 2.6** Create a new sectioning command that will behave like the `\section` command and in addition will put an asterisk in front of the section number in the table of contents. [Hint: Use the `\let` command (see Section 4.5.1)] □

Every document class defines for each sectioning command an `\l@`*type* command. The general form of this command is

`\l@type{entry}{page}`

These commands are needed for making an *entry* of type *type* in a table of contents, or elsewhere. Most of the `\l@`*type* commands are applications of the command

`\@dottedtocline{level}{indent}{numwidth}{title}{page}`

where *indent* is the total indentation from the left margin, *numwidth* the width of the box that contains the section number if the *title* (i.e., the contents of the entry) has a `\numberline` command, and *page* the page number. Here is an example:

`\newcommand*\l@section{\@dottedtocline{1}{1.5em}{2.6em}}`

Note that the `\@dottedtocline` produces a dotted line. If we do not like this effect, then a good solution is to try to delete the part that produces the leaders.

When creating a large document, one may need to have appendices. LaTeX provides the command `\appendix`, which resets the numbering of the various sectioning commands and, depending on the document class, forces the top sectioning command to produce alphabetic numbers instead of Arabic numbers. So, if we are preparing a book, the chapter numbers in the appendix appear as letters.

If we are preparing a book, we usually want to have a preface and/or a prologue. Moreover, it is customary for the page numbers of this part of our document to be typeset using the Roman numbering system (i.e., i, ii, iii, etc.). The book document class provides the commands `\frontmatter`, `\mainmatter`, and `\backmatter`, which can be used to divide a book into three (logical) sections. The effect of the first command is to start Roman page numbering and to turn off chapter numbering. The second

```
\documentclass{book}        \chapter{The Language}
\usepackage{.....}          ......................
\begin{document}            \appendix
\frontmatter                \chapter{More on Nothing}
\chapter{Preface}           ......................
................            \backmatter
\tableofcontents            \chapter{Solutions }
\mainmatter                 ......................
\chapter{History}           %Bibliography & Index
................            \end{document}
```

Figure 2.2: A skeleton LaTeX file suitable for book preparation.

command resets the page numbering, starts Arabic page numbers, and turns on chapter numbering. The third command just turns off the chapter numbering. In Figure 2.2, we give a skeleton LaTeX file that can be used in the preparation of a complete book. The bibliography and index parts are commented out since we have not yet discussed the creation of these particular parts.

The appendix package by Peter Wilson provides some commands that can be used in addition to the \appendix command. It also provides a new environment that can be used instead of the \appendix command. The \appendixpage command will typeset a heading in the style of a \part heading for the document class. The name of the heading is given by the value of \appendixpagename. The \add-appheadtotoc command inserts an entry into the table of contents. The text to be inserted is stored in the variable \appendixtocname. These commands can be used in conjunction with the \appendix command. We can change the predefined value of the commands \appendixtocname, \appendixpagename, and \appendixname by a simple redefinition, such as

```
\renewcommand{\appendixtocname}{List of Appendices}
```

The appendices environment can be used instead of the \appendix command. It provides more functionality than is possible by the commands listed above. This additional functionality is accessible through options that the package provides:

toc Puts a header (e.g., "Appendices") into the table of contents before listing the appendices. The header used is stored in \addappheadtotoc.

page Puts a title (e.g., "Appendices") into the document at the point where the environment begins. The title used is stored in \appendixpagename.

title Puts a name (e.g., "Appendix") before each appendix title in the body of the document. The name used is stored in \appendixname.

titletoc Puts a name (e.g., "Appendix") before each appendix listed in the table of contents. The name used is stored in \appendixname.

`header` Puts a name (e.g., "Appendix") before each appendix in the page headers. The name used is stored in \appendixname. This is default behavior for document classes that have chapters.

The `subappendices` environment can be used to add appendices at the end of a main document division as an integral part of the division. This environment recognizes only the `title` and `titletoc` options.

2.4   The Document Title

In any document, we want to have either a compact title part or a separate title page. In any case, we want to include the title of the document, the list of authors, and their affiliations. In addition, we often need to include an abstract, which will briefly describe the contents of the document. Another important piece of information that one may want to include is the time stamp of the document. This time stamp can be either the date that the document was processed by LaTeX or any other date (e.g., in the case where it is a program manual, the time stamp may be the actual date when the program was released). There are two ways to construct the title of a document prepared with LaTeX: either by using the command \maketitle and its associated commands or by using the `titlepage` *environment*. An environment creates a local scope with a predefined functionality. The local scope is delimited by \begin{*environment*} and \end{*environment*}. An environment involves the execution of the command \environment, the typesetting of the *body* of the environment according to the rules set by the command \environment, and the execution of the command \endenvironment. If there is no command \endenvironment, LaTeX simply ends the local scope. We now present in turn the two ways that one can prepare the title of a document.

The title of a document must be specified with the \title command:

\title{*Title*}

The argument of the command is the actual title of the document. The authors, their respective affiliations, and the date of the document can be specified as follows:

```
\author{
          First & Last name\\
          Organization\\
          Street address\\
          Postal Code City, Country\\
          . . . . . . . . . . . . . . . . . .
          \and
          First & Last name\thanks{...}
          . . . . . . . . . . . . . . . . . .
       }
```

```
\date{Date}
\maketitle
```

As is obvious, the details of each author are separated by the command \\, which forces LaTeX to change lines. The command \and is used when a document has more than one author. In the example above, we specified two authors. However, if we want to specify one more author, we have to write one more \and command after the details of the second author. The command \thanks is used when an author wants to express gratitude to people or organizations that have provided assistance in the course of the preparation of the work presented in the document. The argument of the \date command will be used to typeset the time stamp of the document. If the argument is empty (i.e., \date{ }) LaTeX will not typeset a time stamp. However, if we want LaTeX to use the current date as the document's time stamp, we simply do not specify the \date command. In this case, LaTeX prints the date using the \today command. Once we have finished with the author information, we include the \maketitle command to force LaTeX to typeset the title. This means that if we do not write down the \maketitle command, LaTeX will not produce the title of the document. In Figure 2.3, you can see an example of a typical article title.

TrueType Fonts and LaTeX

Apostolos Syropoulos	Antonis Tsolomitis
Department of Civil Engineering,	Department of Mathematics
Democritus University of Thrace,	University of the Aegean
GR-671 00 Xanthi, GREECE	Karlovasi
E-mail: `apostolo@obelix.ee.duth.gr`	GR-832 00 Samos, GREECE
	E-mail: `atsol@iris.math.aegean.gr`

November 1, 2000

Figure 2.3: A typical article title.

In a typical scientific document, there is usually an *abstract* that contains a brief description of the text that follows. One can put the abstract just before or after the \maketitle command. The environment abstract is used to write down the abstract of a document.

```
\begin{abstract}
Text of the abstract.
. . . . . . . . . . . . . . . .
\end{abstract}
```

Note that a book is not allowed to have an abstract, as this makes no sense.

➤ **Exercise 2.7** Write a small document and write the abstract before and after the command \maketitle. What's the difference? ☐

Since what LATEX provides to its users may not satisfy everybody, the system allows its users to reprogram the \maketitle command and its associated commands or just use the titlepage. This environment generates a new page where the author is free to put whatever suits the needs of the particular document being prepared. Of course, at this moment, our knowledge of LATEX is too limited so there are only a few things we can do. As the reader proceeds with this the book, new commands to create an excellent title page will be discovered.

2.5 Basic Logos

A logo is a name, symbol, or trademark designed for easy and definite recognitionm, so the word TEX is a logo. The natural question is: "How can one produce this and the other TEX-related logos?" The answer is in the following table.

Logo	Command
TEX	\TeX
LATEX	\LaTeX
LATEX 2$_\varepsilon$	\LaTeXe

All of these commands have a peculiar behavior, which is demonstrated by the following example (note that the character ␣ is a visual representation of the space character):

Plain TEXis easy	Plain␣\TeX␣␣␣␣␣␣is␣easy
but LATEX is easier!	but␣\LaTeX\␣is␣easier!

It is obvious that in the first example, regardless of the number of spaces that follow the command \TeX, TEX is setting no space between the logo and the next word. This happens simply because these commands *consume* all of the white space that follows them. However, in the second example, we see that the use of the command \␣ produces the intended result. The effect of this command is to force TEX to produce a reasonable interword space.

➤ **Exercise 2.8** According to the regulations set by the Greek Postal Services, when one affixes a printed address label on a postal objèct, the first three digits of the postal code must be separated by a single space from the remaining two digits. Moreover, we have to add two spaces between the postal code and the name of the town that immediately follows. Write down the necessary commands to typeset the following address according to the regulations of the Greek Postal Services:

Basil Papadopoulos
33, Andrianoupoleos Str.
671 00 Xanthi

(Hint: use the command \\ to force LaTeX to change lines.) ☐

The package mflogo by Ulrik Vieth provides two commands that generate the following logos:

Logo	Command
METAFONT	\MF
METAPOST	\MP

In addition, there are many other TeX-related logos that we will encounter in subsequent chapters.

2.6   Article Preparation

We are now in a position to write a complete real LaTeX document. Using your favorite text editor, type the following LaTeX code into a file.[1] Save the file and call it article.tex.

```
\documentclass[a4paper,11pt]{article}
\begin{document}
\title{Ways to the Moon?}
\author{R. Biesbroek\\
        JAQAR Space Engineering\\
        Den Haag\\
        The Netherlands
        \and
        G. Janin\\
        Mission Analysis Section\\
        ESOC\\
        Darmstadt\\
        Germany}
\date{August 2000}
\maketitle
\begin{abstract}
ESA has conducted several studies on missions to the Moon
in recent years\ldots
\end{abstract}
```

---

1. This is part of an article published in the ESA Bulletin, number 103, August 2000, pp. 92–99.

```
\tableofcontents
\section{Previous lunar missions}

Lunar exploration began on 1 February 1959 when the Soviet
satellite Luna-1 flew past the Moon.\ldots

\section{Going to the Moon now}

The Moon is the Earth's only known natural satellite\ldots

\subsection{The direct way: fast but expensive}

The ''classical'' lunar mission begins from a so-called
''parking orbit'' around the Earth\ldots

\section*{Acknowledgments}
The support received from Dr. W. Ockels in the preparation
of this article is gratefully acknowledged.
\end{document}
```

Before you feed the file to LATEX, we must explain what the command \ldots is doing. This command simply produces an *ellipsis* (i.e., three spaced-out dots) since the file contains part of the complete document. Now, we feed the file to TEX or to any of the related typesetting engines:

```
$latex article
This is TeX, Version 3.14159 (Web2C 7.3.1)
(article.tex
LaTeX2e <2000/06/01>
Babel <v3.6k> and hyphenation patterns for american, english,
greek, loaded.
(/usr/local/teTeX/share/texmf/tex/latex/base/article.cls
Document Class: article 2000/05/19 v1.4b Standard LaTeX
document class
(/usr/local/teTeX/share/texmf/tex/latex/base/size11.clo))
No file article.aux.
No file article.toc.
[1] (article.aux) )
Output written on article.dvi (1 page, 1604 bytes).
Transcript written on article.log.
```

As is obvious from the program output, there are two missing files called article.toc and article.aux. The first file is supposed to contain the table of contents of the document, while the second file contains auxiliary data useful for the correct label resolution, among other things. When the \tableofcontents is used, LATEX searches for a file that has the same name as the input file but with file extension is .toc. This file is used to store the generated table of contents while processing the input file. The file is (re-)generated every time our file is processed by LATEX, just in case the contents of the file have changed. Now, we have to reprocess our file with LATEX to get the table of contents:

```
$ latex article.tex
This is TeX, Version 3.14159 (Web2C 7.3.1)
(article.tex
LaTeX2e <2000/06/01>
Babel <v3.6k> and hyphenation patterns for american, english,
greek, loaded.
(/usr/local/teTeX/share/texmf/tex/latex/base/article.cls
Document Class: article 2000/05/19 v1.4b Standard LaTeX
document class
(/usr/local/teTeX/share/texmf/tex/latex/base/size11.clo))
(article.aux)
(article.toc) [1] (article.aux) )
Output written on article.dvi (1 page, 1984 bytes).
Transcript written on article.log.
```

Finally, we can view the resulting output file. In Figure 2.4, the reader can see the formatted output.

It is interesting to note that the first word of the first paragraph of each new section is not indented, whereas in all subsequent paragraphs the first word is indented. This is common practice in American typography, but in many continental European countries, the typographic style demands that even the first paragraph must be indented. This effect can be achieved in two ways. The most simple one is to use the package indentfirst. Alternatively, one can put the command \indent just before the first word of the paragraph. However, the drawback of this approach is that it is not an efficient solution for a large document. Note that there is also a command that has the opposite effect of the \indent command; this is the \noindent command.

2.7 Letter Preparation

The letter document class offers a special environment for writing letters. A LATEX file that uses this document class can contain an arbitrary number of such environments so one can create many letters with a single LATEX file. Initially, one has to specify the sender's address with the \address command. Moreover, one has to specify the name of the sender with the \signature command. Once we have specified this information, we can start writing our letters. The text of each letter is actually written in the body

Ways to the Moon?

R. Biesbroek

JAQAR Space Engineering

Den Haag

The Netherlands

G. Janin

Mission Analysis Section

ESOC

Darmstadt

Germany

August 2000

Abstract

ESA has conducted several studies on missions to the Moon in
recent years...

Contents

1 Previous lunar missions

Lunar exploration began on 1 February 1959 when the Soviet satellite Luna-1 flew past the Moon.....

2 Going to the Moon now

The Moon is the Earth's only known natural satellite...

2.1 The direct way: fast but expensive

The 'classical' lunar mission begins from a so-called 'parking orbit' around the Earth...

Acknowledgments

The support received from Dr. W. Ockels in the preparation of this article is gratefully acknowledged.

Figure 2.4: The formatted output of the LaTeX file on page 29.

of the letter. The environment has one argument, which is the recipient address. Inside the letter environment, we can specify how the letter will open and how it will close by giving the appropriate text as arguments to the commands \opening and \closing. The \cc command can be used after the closing command to list the names of people to whom we are sending copies. There is a similar \encl command for the list of enclosures. A postscript can be generated with the \ps command. Note that this command does not generate any text; consequently, one has to type the "PS" oneself. Here is the code of a complete example:

```
\documentclass[a4paper,12pt]{letter}
  \address{%
          DigiSetter Inc.\\
          1, \TeX\ Drive\\
          Sparta}
  \signature{%
            Apostolos Syropoulos\\
            Sales Manager}
\begin{document}
 \begin{letter}{%
  Dr. Nikolaos Sofroniou \\
  Department of Digital Typography\\
  3, \LaTeX\ Str.\\
  Thebes}
  \opening{Dear Sir,}
  Please find enclosed the information you have requested
  regarding our digital setter.
  \closing{Yours very truly,}
  \ps{PS. All prices are in Spartan Talanta.}
  \cc{Dr. Antonis Tsolomitis}
  \encl{A prospectus of DSET-100.}
 \end{letter}
\end{document}
```

The output of this example can be seen in Figure 2.5.

2.8 Producing Proceedings Articles

Proceedings articles are typeset in two columns, and actually any article can be immediately converted into a proceedings article. The corresponding document class provides the command \copyrightspace, which is used to produce blank space in the first column, where a copyright notice belongs. Note that this command should appear after any footnotes (see Section 4.5).

DigiSetter Inc.
1, TEX Drive
Sparta

December 4, 2007

Dr. Nikolaos Sofroniou
Department of Digital Typography
3, LATEX Str.
Thebes

Dear Sir,

Please find enclosed the information you have requested regarding our digital setter.

Yours very truly,

Apostolos Syropoulos
Sales manager

PS. All prices are in Spartan Talanta
cc: Dr. Antonis Tsolomitis
encl: A prospectus of DSET-100.

Figure 2.5: A complete letter.

2.9   Combining Individual LATEX Files

The problem of combining a set of individual LATEX files into a single LATEX file occurs very
frequently. For example, one faces this problem in the preparation of the proceedings
of a workshop or conference. The combine document class, by Peter Wilson, provides a
solution for this problem. A master file that uses the combine document class imports a
set of individual LATEX files, which use the same document class, and when fed to LATEX
generates a single DVI file. Sectioning, cross-referencing, bibliographies, and so on, are
local to each imported file. Here is a simple example file that might be used to typeset
conference proceedings, among other things:

```
\documentclass[colclass=hms,packages]{combine}
\begin{document}
\title{Proceedings of the...}
\author{George Taylor\thanks{Support...}}
\maketitle
\tableofcontents
\section*{Preface}
This is ...
In the paper by Chris Andersson...
\begin{papers}
\coltoctitle{On the use of the hms class}
\coltocauthor{John Smith}
\import{hms}
\newpage
\end{papers}
\begin{papers}
\coltoctitle{On the use of the combine class}
\coltocauthor{Peter Wilson}
\import{combine}
\end{papers}
\section*{Acknowledgment}
We would like to thank...
\end{document}
```

The combine class offers many options apart from providing all of the class options appropriate for the class of the individual documents. Here, we present the most important options.

book, report, and letter By default, the article document class is assumed for both the main and the imported files. These options change the class to book, report, or letter, respectively.

colclass=SomeClass This option makes SomeClass the class that is used to typeset the whole document.

packages By default, all \usepackage commands in imported files are ignored. If this option is specified, the use of all \usepackage commands will be enabled. However, we must stress that only the first occurrence of a package is actually used, and it is not available to any file imported later.

classes This option enables the imported documents to be of different classes. However, this option is error-prone and best avoided.

layout This option enables the imported documents to have their own page layouts. We feel that this option is redundant since it makes no sense to have a document with parts that have different page layouts.

folios The page numbers are sequential throughout the document. It is a good idea to avoid using this option with the plain page style.

notoc Disables the inclusion of the table of contents in any imported document.

nolof Disables the inclusion of the list of figures in any imported document.

nolot Disables the inclusion of the list of tables in any imported document.

maintoc Adds the table of contents, list of figures, and list of tables of all imported documents to the corresponding table and lists of the main document.

notitle Disables title printing by any \maketitle in any imported document.

noauthor Disables author printing by any \maketitle in any imported document.

date This option allows dates generated by \maketitle commands to be printed.

nomaketitle Disables all output generated by \maketitle.

nopubindoc Disables the printing of the \published information within an imported document.

nopubintoc Disables the printing of the \published information with the main table of contents.

The combine document class provides a few additional commands and environments that facilitate the creation of the single document. The environment papers provides a wrapper around imported file(s). This environment may have an optional argument that can be specified in square brackets immediately after the beginning of the environment, that is executed immediately at the start of the environment, and its default action is to force TeX to skip the current page and to start its typesetting business at the next odd-numbered page. To avoid the default action, one must specify an empty option argument (i.e., \begin{papers}[]).

The command \import{TeXfile} is used to import the individual LaTeX files and should be used only within a papers environment. TeXfile is the name of a LaTeX file without the default .tex filename extension. Moreover, the TeXfile must be a complete LaTeX file.

The commands \maintitlefont, \postmaintitle, \mainauthorfont, \postmain-author, \maindatefont, and \postmaindate control the typesetting of the main document's \maketitle command. Each part of the main title is typeset as if the corresponding main and post commands surround this particular element, for example

{\maintitlefont \title \postmaintitle}

Here is a simple example:

```
\renewcommand{\maintitlefont}{\hrule\begin{center}%
    \Large\bfseries}
\renewcommand{\postmaintitle}{\end{center}\hrule%
    \vspace{0.5em}}
\renewcommand{\mainauthorfont}{\begin{center}\large%
    \begin{tabular}[t]{c}}
\renewcommand{\postmainauthor}{\end{tabular}\par%
    \end{center}}
```

The \title and \author commands in each individual file are typeset by \maketitle commands in the imported files. Their typesetting is controlled by the commands \importtitlefont, \postimporttitle, \importauthorfont, \postimportauthor, \importdatefont, and \postimportdate. These commands are used by the combine document class exactly like the \main and \post commands. Here is a real-world example that the first author has used in the preparation of a mathematics journal:

```
\renewcommand{\importtitlefont}{\vspace*{0.75in}\begin{center}
\large\bfseries \artTitle}
\makeatother
\renewcommand{\postimporttitle}{\par\end{center}\vskip 3.0em}
\renewcommand{\importauthorfont}{\begin{center}
  \large\scshape \lineskip .5em%
  \begin{tabular}[t]{c}}
\makeatletter
\renewcommand{\postimportauthor}{\end{tabular}\par\vskip1em%
{\@received \@revised}\end{center}}
```

The \bodytitle[*short title*]{*long title*} command is similar to a \chapter or a \section command, depending on the document class of the document. It may be used to enter a numbered title heading into the main document or table of contents for the following imported file. The starred version produces an unnumbered title heading and makes no entry in the table of contents. The commands \coltoctitle and \coltocauthor have one argument, the title and the author(s) of an imported file. The command \published[*short*]{*long*} can be used to put the *long* text into the body of the main document. If the optional argument is not used, then *long* is also added in the main table of contents. The command \pubfont controls the appearance of the text of a \published command.

3

Fonts and Their Use

A very common complaint from beginners to the TEX system and its derivatives is "where are the fonts?" or "am I restricted to the default font?". In this chapter we will see first how to access different font families and within them special symbols. Moreover, we will see size-changing commands and also deal with typographical issues, which the wealth of fonts available to TEX makes necessary.

However, what is a font? Looking it up in a dictionary, you will find something like this "a complete assortment of types of one sort, with all that is necessary for printing in that kind of letter." We will use the word *glyph* to talk about the shape of each letter or character. A font is a file that contains the description of glyphs, usually–for example, PostScript Type 1 fonts or METAFONT fonts–in a mathematical manner; that is, the curves that make up the glyph's design are described as the graphs of functions with certain characteristics. For each glyph, the font file sets up a name for it and then the mathematical description is given, the same for the next glyph, and so on. We will come back to this internal structure of the font files when we deal with the procedure for installing new fonts.

3.1 Classification of Fonts

Generally speaking, there are two main categories of fonts: serifed and sans serif (also known as *gothic*). The serifed fonts are those that have *serifs*, or little decorative lines at the end of their strokes, like the default font of this text. The second category is the sans serif font, which does not have these decorations. The following example makes things clear:

Computer Modern serifed type	Computer Modern sans serif type
Lucida Serifed type	Lucida sans serif type
Lucida Serifed Ελληνικά στοι-	Lucida sans serif Ελληνικά στοι-
χεία (greek)	χεία (greek)

The serifs are useful because they serve as a guide to the eye, making reading more comfortable. That is why they are usually preferred for text work.

Another parameter that categorizes fonts is whether they are of variable widths or fixed widths. Fixed-width fonts are fonts where each glyph has the same width as all others, that is, the letter i, for example, has the same width as the letter m. These fonts often remind us of the typewriter and are usually used for computer program listings:

radii   in variable-width serifed font

`radii`   in fixed-width font

radii   in variable-width sans serif font

These categories—the serifed, the sans serif, and the typewriter fonts—usually provide a full set of glyphs—enough to typeset most articles or books. In LaTeX, we call them *families*, and in order to use them, we use the commands

```
{\rmfamily text}     {\sffamily text}     {\ttfamily text}
```

or alternatively

```
\textrm{text}     \textsf{text}     \texttt{text}
```

for serifed (also called "Roman"), sans serif and typewriter, respectively.

Inside each of these families, other parameters lead to further classification. Each of these families is now divided into *series*. Series have different *weights*. Weight is the width of the curves that draw each glyph. Let us first note that the width of these lines is usually modulated. In some fonts, the modulation is heavy, resulting in high-contrast fonts, whereas for others the modulation is moderate or sometimes nonexistent.

heavy modulation, moderate modulation and no modulation

However, font families are usually accompanied by a set of glyphs that have *overall* heavier lines and **are the bold series of the font**. The bold series are used for emphasis, but their use must be kept to a minimum, as they make the "page color" look unbalanced. We make use of the bold series with the commands

```
\textbf{text}   or   {\bfseries text}
```

Now, inside each series, we have different shapes, such as the italic shape. In typography, it is very common to want to emphasize something. It may be a single word or even a full paragraph (such as the statement of a theorem in mathematics). For this task, we use the italic shape of the type. This is done with the commands

```
\textit{text}   or   {\itshape text}
```

For example, *this is written in italic shape*. No font family can be considered complete if it does not come with a companion italic shape. Another way to emphasize, which is also popular, is with the slanted shape. *This is the slanted shape*, and it becomes available with the commands

$$\texttt{\textbackslash textsl\{text\}} \quad \text{or} \quad \texttt{\{\textbackslash slshape text\}}$$

Note that the slanted shape is the same as normal upright shape but is printed with a slope other than 90 degrees. The italic shape is essentially a different design. Slanted printing is a simple mathematical transformation. That is why one should not really pay for slanted type, as TEX can easily print with slope from the upright version of the font. This will be explained in Chapter 12, where we will discuss techniques of font installation. The italic shape is something to look for in a (modern) font family. It is something to help you check how much the designer really thought about typographical perfection for his font creations. Bold series usually have their own italic shape, but it is rarely used. We must add here that all of the preceding commands of series and shape changes can be combined as in

$\texttt{\{\textbackslash bfseries\textbackslash itshape text\}}$ producing *text*.

Table 3.1: Family series and shape-changing commands.

Command	Corresponds to ...	Example Text
\textrm{...}	{\rmfamily ...}	Roman Family
\textsf{...}	{\sffamily ...}	Sans Serif Family
\texttt{...}	{\ttfamily ...}	Typewriter Family
\textmd{...}	{\mdseries ...}	Medium Series
\textbf{...}	{\bfseries ...}	**Bold Series**
\textup{...}	{\upshape ...}	Upright Shape
\textit{...}	{\itshape ...}	*Italic Shape*
\textsl{...}	{\slshape ...}	*Slanted Shape*
\textsc{...}	{\scshape ...}	SMALLCAPS SHAPE
\emph{...}	{\em ...}	*Emphasized*
\textnormal{...}	{\normalfont ...}	Normal document font

The commands in the second column are actually *declarations*. Any declaration should be used in a local scope unless we want it to globally affect the typesetting of our document. Given a \declaration, we can create a local scope in the following way:

\begin{declaration}

. . . .

\end{declaration}

that is, we can turn it into an environment.

We will pause for a moment to talk about a technique of emphasizing that is considered poor practice. Because of typewriters and the usually incomplete fonts that come with word processors, many people have learned to emphasize by underlining the text. That is really poor, as it disturbs the balance of the page color and makes it look unprofessional. Even worse, word processors underline without any regard to letters such as p, q, and others that extend below the writing line (also called "baseline"). For example, many times you can hardly distinguish an underlined j from an underlined i. Although LATEX can produce better underlining if instructed properly, this is not a good technique for emphasizing. If the user insists on using underline, either the command \underline or the package ulem by Donald Arseneau may be used. The following example shows both ways. Note that the ulem package provides additional commands such as the \uwave command.

This is a test of the package ulem. Although underlining is not suggested for emphasizing, the user may find this useful in certain applications.

```
This is a \uline{test
of the package \textsf{ulem}.
Although underlining is}
\uwave{not} \underline{suggested}
for emphasizing, the user may
find this \uuline{useful} in
certain applications.
```

A very special shape inside a series is the SMALL CAPITALS SHAPE. Small capitals are capitals of reduced size usually around the size of the lowercase type. They are a sign of typographical perfection for the following reasons. First, capitals are hard to read, and that is why they are avoided (Figure 3.1). The second reason is technical. When we say reduced size we do not mean plain scaling. Plain scaling will reduce the width of the lines that make up the glyphs. Thus, the result will have color problems (see Figure 3.2). True small capitals are scaled in width and height, but the line width is not scaled by the same factor as the other dimensions. This is why the small capitals must come as an additional shape of a well-designed font, since the user cannot create them with simple scaling. Small capitals are used exclusively when we want to set capitals inside text. For example, we write: SPAIN and GREECE are both members of EU.

Many commercial fonts come with other shapes as well, such as swash capitals or other ornamental shapes:

THESE ARE SWASH CAPITALS

THIS BOOK IS THE FIRST BOOK THAT PROVIDES A DETAILED DESCRIPTION OF MULTILINGUAL ISSUES AND USEFUL INFORMATION ABOUT THE OMEGA AND LAMBDA TYPESETTING ENGINES.

THIS BOOK IS THE FIRST BOOK THAT PROVIDES A DETAILED DESCRIPTION OF MULTILINGUAL ISSUES AND USEFUL INFORMATION ABOUT THE OMEGA AND LAMBDA TYPESETTING ENGINES.

Figure 3.1: All capitals is hard to read.

TRUE SMALL CAPITALS   FAKE SMALL CAPITALS

Figure 3.2: True and fake small capitals.

Thus, in order to instruct LATEX to use a specific type, it is imperative to tell the program exactly what you want. A command with the following arguments must be available:

$\{family\}\{series\}\{shape\}$

Such a command of course exists, and it has one more argument that we will discuss now. This argument is called *encoding*. TEX constructs the page that outputs by choosing the glyphs from a 16 × 16 matrix, so when you type on your keyboard the letter A, TEX consults a specific entry of this matrix for information about the letter A. The font itself contains only the information about glyphs in a nonmatrix manner. Thus, a matrix from the set of the glyphs of the font is constructed during the font installation procedure. The way the glyphs are laid out in this matrix is called the *encoding vector*, or *encoding* for short. This is a powerful technique that gives us the opportunity to use any number of glyphs. We only have to switch encoding and a full 16 × 16 matrix of new glyphs becomes available. The most standard encoding vector is called OT1. In Figure 3.3 we

	'0	'1	'2	'3	'4	'5	'6	'7	
'00x		Δ	Θ	Λ	Ξ	Π	Σ	Υ	"0x
'01x	Φ	Ψ	Ω	ff	fi	fl	ffi	ffl	
'02x	1	J	`	´	˘	ˇ	¯	˚	"1x
'03x	ˌ	ß	æ	œ	ø	Æ	Œ	Ø	
'04x	˜	!	”	#	$	%	&	’	"2x
'05x	(	)	*	+	,	-	.	/	
'06x	0	1	2	3	4	5	6	7	"3x
'07x	8	9	:	;	¡	=	¿	?	
'10x	@	A	B	C	D	E	F	G	"4x
'11x	H	I	J	K	L	M	N	O	
'12x	P	Q	R	S	T	U	V	W	"5x
'13x	X	Y	Z	[	“	]	^	·	
'14x	`	a	b	c	d	e	f	g	"6x
'15x	h	i	j	k	l	m	n	o	
'16x	p	q	r	s	t	u	v	w	"7x
'17x	x	y	z	–	—	˝	˜	¨	
	"8	"9	"A	"B	"C	"D	"E	"F	

Figure 3.3: Computer Modern in OT1 encoding.

see the matrix for the Computer Modern font. We printed the matrix in eight columns in order to fit the page. This font has no glyphs (in this encoding) after the 128th entry of the matrix. Now, we present the font selection command. Suppose that we know that the Palatino font is named "ppl." Then, to select Palatino in medium (not bold) series in italic shape with the OT1 encoding vector, we issue the command

\usefont{OT1}{ppl}{m}{it}

This command can be broken into simpler commands, each of them setting the value for encoding, family, and so forth separately. This enables us to change only one of the characteristics of the type that we want to use. The commands are

\fontencoding{OT1}, \fontfamily{ppl}, \fontseries{m}, \fontshape{it}.

These commands must be followed by the \selectfont command in order to take effect. Of course, if, for example, one sets the font family by, say, \fontfamily{ppl}, then to select the italic shape one can just use the \itshape or \textit commands discussed earlier. These commands are considered the low-level interface for selecting fonts and font attributes, as opposed to the high-level commands such as \textit. Most of the time, the user just needs to load the appropriate package and can forget about these commands. They are necessary though for difficult tasks when we need several fonts (such as for the creation of this book). Font packages that are usually available in a modern installation are described in the table 3.2.

Table 3.2: Standard font packages in modern installations.

Package	Serif Font	Sans Serif Font	Typewriter Font
avant	Default (cmr)	AvantGarde, pag	Default (cmtt)
avantgar	AvantGarde, pag	Default (cmss)	Default (cmtt)
bookman	Bookman, pbk	AvantGarde, pag	Courier, pcr
chancery	Zapf Chancery, pzc	Default (cmss)	Default (cmtt)
charter	Charter, bch	Default (cmss)	Default (cmtt)
concrete	Concrete, ccr + EulerMath	Default (cmss)	Default (cmtt)
courier	Default (cmr)	Default (cmss)	Courier, pcr
helvet	Default (cmr)	Helvetica, phv	Default (cmtt)
ncntrsbk	New Century Schoolbook-Roman, pnc	Default (cmss)	Default (cmtt)
palatcm	Palatino, ppl + cmr-Math	Default (cmss)	Default (cmtt)
palatino	Palatino, ppl	Helvetica, phv	Courier, pcr
pandora	Pandora, panr	Pandora Sans, pss	Default (cmtt)
times	Times, ptm	Helvetica, phv	Courier, pcr
utopia	Utopia, put	Default (cmss)	Default (cmtt)

Although these packages provide a wealth of fonts, more fonts are available for special purposes and are usually installed by default. One such case is the fonts provided by the oldgerm package. This package provides the commands \gothfamily, \frakfamily, and \swabfamily that give access to the old German fonts 𝕲𝖔𝖙𝖍𝖎𝖘𝖈𝖍, 𝖆𝖑𝖘𝖔 𝖈𝖆𝖑𝖑𝖊𝖉 𝕿𝖊𝖗𝖙𝖚𝖗 (Gothisch, also called Textur), 𝔉𝔯𝔞𝔨𝔱𝔲𝔯 (Fraktur), and 𝔖𝔠𝔥𝔴𝔞𝔟𝔞𝔠𝔥𝔢𝔯 (Schwabacher) designed by Yannis Haralambous. For using the fonts locally (such as here), it also provides the commands \textgoth, \textfrak, and \textswab. If any of the fonts above are not available in your installation, you can find them in the CTAN or in many Internet mirrors around the globe. If we really need the low-level interface, then it is useful to know the parameters necessary, given in Table 3.3. Note that not all series or shapes are always available for every font. Moreover, some fonts have extra shapes (such as swash), in which case we need to customize our installation (see Chapter 12). Note also that some font designers consider outline not just a shape but a family.

➤ **Exercise 3.1** Typeset the following paragraph:

> Because of typewriters and the usually **incomplete** fonts that come with WORD PROCESSORS, many people have learned to *emphasize* by *underlining* the text. That is **really** *poor* as it disturbs the *balance* of the *"page color"* and makes it look unprofessional.

☐

➤ **Exercise 3.2** An old technique of emphasizing is by spacing out the text to be emphasized. In older times or in countries where italic shapes were not available, the typographer still preferred to avoid underlining and preferred to space out the text. This is still in use, but mainly for stylistic reasons. Since it obviously affects the page color, it should be used with caution. Probably the easiest way of doing this is by using the letterspace package by Phil Taylor. The package provides the command

\letterspace to *size* {text to be spaced out by *size*}

Table 3.3: Font attribute parameters.

Parameter	Possible Values
\fontencoding	OT1 (standard Latin), T1 (extended Latin), OT2 (Cyrillic), "custom" (such as LGR that selects Greek, LHE that selects Hebrew, and others)
\fontseries	ul (ultra light), el (extra light) l (light), sl (semi light), m (medium=normal), sb (semi bold), b or bx (bold), eb (extra bold), ub (ultrabold)
\fontshape	n (upright=normal), it (italic), sl (slanted), sc (small caps), ui (upright italic), ol (outline shape, e.g.,)

where *size* is a length followed by a length unit (see Section 4.1). Use this package to typeset the following:

> **Theorem 1** Every natural number bigger than 1 has a prime divisor.
>
> S k e t c h o f P r o o f : We use induction. For the number 2, the result is correct since 2 is a prime itself. Let *n* be a natural number. If *n* is prime, then we are done. If not, then assume that *k* is a divisor different from 1 and *n*. Then, *k* is less than *n* and consequently (*by induction*) it has a prime divisor, say *p*. Then, *p* is a divisor of *n* as well.

[Hint: The \letterspace command and the text to be spaced out must be in a box (put the whole thing in a \mbox{}, see Section 6.10); otherwise, you will get a new line after the end of the spaced-out text.] The letterspace package can also be used for big type sizes. If setting in big sizes, then spacing out the text a little bit makes it more readable. Let us close this exercise by mentioning that the letterspace package defines the parameters \naturalwidth and \linewidth, which are the natural width of the text to be spaced out and the width of the line, respectively. These parameters make possible commands such as

```
\letterspace to 1.7\naturalwidth {Sketch of Proof:}
```

□

➤ **Exercise 3.3** Typeset the next paragraph. Select fonts according to the following phrases (consult Table 3.2).

> This is Avant Garde, Bookman Old Style, *Chancery italic.* Also available in modern installations are the Charter font, the Concrete font, the Courier font, the Helvetica font, the New Century Schoolbook Roman font, the Palatino font, the Pandora font, the Times font, and the Utopia font. If the oldgerm package is available, then you can write with 𝔉𝔯𝔞𝔨𝔱𝔲𝔯, 𝔊𝔬𝔱𝔦𝔩𝔠𝔥, or 𝔖𝔠𝔥𝔴𝔞𝔟.

□

3.2 Accessing more Glyphs

As we discussed in the previous section, the standard way of accessing more glyphs is to change the encoding vector. With this technique, it would seem that Unicode is irrelevant to TeX since one may argue that if we need additional glyphs, we change the encoding and that is all. Unfortunately, this is not true. It would be true if we always needed less than 256 glyphs for the typesetting of, say, a paragraph, as we cannot expect the user to switch encodings all the time within sentences. This problem appears with languages such as Arabic, Ethiopian, or even Chinese, where we need to access many glyphs simultaneously. But TeX cannot use bigger font matrices, and this need led to the birth of Ω. Both Ω and Λ can access Unicode fonts, making the need for encodings obsolete. For this reason, they also greatly simplify the writing even of languages that use the Latin script. Since these tools are not yet widely used, we will continue with the discussion of common encoding vectors.

To access glyphs as in "Hölder's rôle in España" one has two possible ways. Either use commands such as \"o, \^o, and \~n and let the encoding switching be done automatically by the command or switch the encoding to T1 (by issuing, for example, \fontencoding{T1}\selectfont) and have a way through the keyboard to type characters such as these directly. The latter is the preferred way when the main typesetting language has such special letters as those found in Spanish, German, or French. Of course, we cannot expect a user from these countries to type commands for accessing simple characters such as ö. For this reason, keyboard drivers exist in these countries that are used so that people directly punch the appropriate keys on their keyboards. Consequently, the right approach for them is the T1 encoding.

In the English literature, however, such characters are not accessed very frequently, and the mechanism of access through commands is the preferred one. Table 3.4 contains the most common special commands for accessing these glyphs.

Table 3.4: Examples of special characters. The characters of the last two rows of the first table and all of the characters of the second table are available in the T1 encoding. All other characters are available in the OT1 encoding. The \k command adds the character *ogonek* to its argument.

\`e	è	\'e	é	\^o	ô	\"e	ë
\~n	ñ	\=o	ō	\.o	ȯ	\u{o}	ŏ
\v{o}	ǒ	\H{o}	ő	\t{oo}	o͡o	\c{c}	ç
\d{o}	ọ	\b{o}	o̲	\r{a}	å	\i	ı
\j	ȷ	\AE	Æ	\ae	æ	\ss	ß
\OE	Œ	\oe	œ	\O	Ø	\o	ø
\L	Ł	\l	ł	!`	¡	?`	¿
\DH	Đ	\dh	ð	\DJ	Đ	\dj	đ
\NG	Ŋ	\ng	ŋ	\TH	Þ	\th	þ

\guillemotleft	«	\guillemotright	»	\guilsinglleft	‹
\guilsinglright	›	\quotedblbase	„	\quotesinglbase	‚
\textquotedbl	"	\textsterling	£	\textsection	§
\k{a}	ą				

Apart from the special letters that we saw, there are some special symbols that we usually want to access. The commands given in Table 3.5 work with the standard OT1 encoding and do not work with the T1 encoding. However, the symbols may be accessible in both encodings with different commands. For example, in the T1 encoding, we use \textsterling in order to get £, but in OT1 encoding we use the \pounds command.

Table 3.5: Additional symbols for the OT1 encoding.

\dag	†	\ddag	‡	\S	§
\P	¶	\copyright	©	\pounds	£
\textregistered	®	\SS	SS	\lq	'
\texttrademark	TM	\aa	å	\AA	Å
\textvisiblespace	␣	\textcircled{s}	ⓢ	\rq	'

LaTeX also provides the command \symbol for accessing a symbol from the font matrix using not a name but its position. The syntax of this command is \symbol{*number*}, where *number* is a decimal number, an octal number, or a hexadecimal number denoting a glyph position. Octal numbers are prefixed by ' (e.g., '143) and hexadecimal numbers are prefixed by " (e.g., "1A and not "1a).

If we have a font, for example, pzdr, and we want to see the glyphs it contains so that we can also see their position in the matrix, we use the command

$ latex nfssfont.tex

The program will ask which font we want to see and then, in order to obtain the matrix, we will write \table and \bye at the next prompts (the program will guide us through these steps, see Section 12.2.2). The result for the pzdr font is given in Figure 3.4. Now, if we want to produce ✦, we can use the command \symbol{"F7} or the command \symbol{'367} (since the octal 367 is the same with F7 in the hexadecimal system) after we select the pzd family. Note that this font is encoded in the U encoding vector, so the font encoding must also be set to U.

Less frequently used symbols are provided by the package textcomp. This package provides access to symbols such as those in the following table:

Symbol	Command
℧	\textmho
℘	\textleaf
♪	\textmusicalnote
¥	\textyen

The glyphs in standard installations are provided by the font tcrm. Readers may see the full list of symbols available by running latex nfssfont.tex for the font tcrm1000 and find the commands for each glyph by looking in the file textcomp.sty of their installation.

	'0	'1	'2	'3	'4	'5	'6	'7	
'04x		☇	✄	☞	✖	☭	✆	✇	"2x
'05x	✈	☒	☛	☞	☡	☎	✎	✐	
'06x	✏	☞	☞	✓	✔	✕	✖	✗	"3x
'07x	✘	✙	✚	+	✛	†	✞	✝	
'10x	✠	✡	✢	✣	✤	✥	✦	✧	"4x
'11x	★	☆	✪	✫	✬	★	★	★	
'12x	☆	✱	✲	✳	✴	✵	✶	✷	"5x
'13x	✸	✹	✺	✻	✼	✽	✾	✿	
'14x	❀	❁	❂	❃	❄	❅	❆	❇	"6x
'15x	❈	❉	❊	❋	●	○	■	❏	
'16x	❐	❑	❒	▲	▼	◆	❖	❘	"7x
'17x	❘	❙	❚	❛	❜	❝	❞		
'24x		❡	❢	❣	♥	❥	❦	❧	"Ax
'25x	♣	♦	♥	♠	①	②	③	④	
'26x	⑤	⑥	⑦	⑧	⑨	⑩	❶	❷	"Bx
'27x	❸	❹	❺	❻	❼	❽	❾	❿	
'30x	①	②	③	④	⑤	⑥	⑦	⑧	"Cx
'31x	⑨	⑩	❶	❷	❸	❹	❺	❻	
'32x	❼	❽	❾	❿	→	→	↔	↕	"Dx
'33x	↘	➝	↗	➔	➔	→	→	→	
'34x	➡	➠	➢	➣	➤	➥	➦	➧	"Ex
'35x	➨	➩	➪	➫	➬	➭	➮	➯	
'36x		➱	➲	➳	➴	➵	➶	➷	"Fx
'37x	➸	➹	➺	➻	➼	➽	⇒		
	"8	"9	"A	"B	"C	"D	"E	"F	

Figure 3.4: Standard output of nfssfont.tex.

▶ **Exercise 3.4** Use commands to access accented letters, and typeset the following list of names or phrases:

> Hàn Thê Thành, Łatała, Livshič, Gödel, Geometriæ Dedicata, Naïve Set Theory, Åke Lundgrem är född 1951 i byn kusmark utanför Skellefleå, How many €'s is a US$, The University of the Ægean, ¿¡©1998 TV España!?, Großen Planeten, à des sociétés, Hüseyn ağabeyi ile birlilcte halâ çalışıyor.

Note that the Euro (€) is available in (at least) the LGR encoding (see also Section 3.2.1). If you have access to a European keyboard, try typing the paragraph above directly after you switch the encoding to T1. We should point out that double accents as in "Hàn Thê Thành" can be typeset using the math mode that will be discussed in chapter 5. This is because fonts do not contain such complex characters, so we have to construct them. However, these difficult cases can be handled in a much better way, discussed in Section 3.5. □

3.2.1 Euro Font

There are special packages for the Euro font. One is the eurosym package by Henrik Theiling. The package provides the official Euro symbol, following the specifications of the European Union, in METAFONT with the commands \officialeuro and \euro. It also provides the command \EUR{} for writing an amount of money. For example, \EUR{321{,}5} gives 321,5 €. (Note that we have to write the comma between curly brackets to make TeX realize that this symbol is not a punctuation mark.) The position of the € defaults to the right of the number but can be changed to the left of the number if the package is loaded with the option left (e.g., € 321,5). Note that the \EUR command inserts a small space between the number and the Euro symbol.

For creating PDF files, the preferred way is to use the package europs by Joern Clausen. This package uses the Type 1 fonts created for the Euro symbol by Adobe and distributed for free on their Internet site (http://www.adobe.com). The package provides access to the official design with the command \EURofc. It also provides access to (nonofficial) designs that fit with bold series and families such as serifed, sans serif, and typewriter. Italics are also included. The serifed family is accessed with the \EURtmcommand, the sans serif with the \EURhv command, and the typewriter with the \EURcr command. Table 3.6 shows all of the symbols privided.

Table 3.6: The Euro symbol (europs package).

	Official	EuroSerif	EuroSans	EuroMono
Normal	€	€	€	€
Oblique	€	€	€	€
Bold	€	€	€	€
Bold-oblique	€	€	€	€

3.2.2 The **wasysym** Fonts

Another package with interesting symbols is the wasysym package by Alex Kielhorn. Since the package provides many conceptually unrelated symbols, we present the symbols in groups. First, we provide the command that can be used to access the symbol and next to the command the actual symbol.

Electrical and physical symbols:

\AC	~	\HF	≈	\gluon	∿∿∿	\photon	∿∿∿	\VHF	≋

Mathematics:

\apprle	≲	\apprge	≳	\Box	□	\Diamond	◇
\iint	∬	\iiint	∭	\invneg	⌐	\Join	⋈
\leadsto	⤳	\LHD	◀	\lhd	◁	\logof	⊛
\ocircle	○	\oiint	∯	\RHD	▶	\rhd	▷
\sqsubset	⊏	\sqsupset	⊐	\varint	∫	\varoint	∮
\unlhd ⊴	\unrhd ⊵	\wasypropto	∝				

General symbols:

\agemO	℧	\ataribox	▨	\bell	♪
\Bowtie	⋈	\blacksmiley	●	\brokenvert	¦
\cent	¢	\checked	✓	\clock	◷
\currency	¤	\diameter	⌀	\DOWNarrow	▼
\frownie	☹	\female	♀	\invdiameter	⌀
\kreuz	✚	\LEFTarrow	◀	\lightning	↯
\male	♂	\permil	‰	\phone	☎
\pointer	☞	\recorder	℞	\RIGHTarrow	▶
\smiley	☺	\sun	☼	\UParrow	▲
\varangle	∢	\wasylozenge	⌑	\wasytherefore	∴

Various circles:

\CIRCLE	●	\Circle	○	\LEFTCIRCLE	◖
\LEFTcircle	◐	\Leftcircle	◖	\leftturn	↺
\RIGHTCIRCLE	◗	\RIGHTcircle	◑	\Rightcircle	◗
\rightturn	↻				

Astronomical symbols:

\ascnode	☊	\astrosun	☉	\descnode	☋	\earth	♁
\fullmoon	○	\jupiter	♃	\leftmoon	☾	\mars	♂
\mercury	☿	\neptune	♆	\newmoon	●	\pluto	♇
\rightmoon	☽	\saturn	♄	\venus	♀	\vernal	♈
\uranus	♅						

Phonetic signs:

\dh	ð	\DH	Ð	\inve	ə	\openo	ɔ	\thorn	þ	\Thorn	Þ

Astrological symbols and the zodiacal symbols:

\aries	♈	\taurus	♉	\gemini	♊	\cancer	♋
\leo	♌	\virgo	♍	\libra	♎	\scorpio	♏
\sagittarius	♐	\capricornus	♑	\aquarius	♒	\pisces	♓
\conjunction	☌	\opposition	☍				

Musical notes:

\eighthnote	♪	\fullnote	○	\halfnote	♩	\quarternote	♪
\twonotes	♫						

APL symbols (i.e, symbols used in APL programs):

\APLbox	⎕	\APLcirc{*}	⍟	\APLcomment	⍝
\APLdown	∇	\APLdownarrowbox	⍗	\APLinput	⍞
\APLinv	⌹	\APLleftarrowbox	⍇	\APLlog	⍟
\APLminus		\APLnot{*}	⍲	\APLrightarrowbox	⍈
\APLstar	⋆	\APLup		\APLuparrowbox	⍐
\APLvert{*}	⍒	\notbackslash	\	\notslash	/

Note that \APLcirc, \APLnot, and \APLvert have a required argument.

Polygons and stars:

\CheckedBox	☑	\davidsstar	✡	\hexagon	⬡	\hexstar	✶
\octagon	◯	\pentagon	⬠	\Square	☐	\varhexagon	⬡
\varhexstar	✳	\XBox	☒				

3.2.3 Phonetic Fonts

For linguistic purposes, a special phonetic font is available using the package phonetic by Emma Pease. Table 3.7 gives the commands provided by the package. Note that the oblique glyphs can be accessed by feeding the corresponding command as argument to the \textit command. The following example shows the phonetic transcriptions of Navaho (native USA) words:

dlǫ́ǫ́ʔ	(prairie dog)	tʃʼah	(hat)
łitsxʷo	(yellow-orange)	tsííł	(haste)
ʔakʼos	(neck)	xaιh	(winter)

People interested in linguistics and TEX may find more information at the following URL: http://www.ifi.uio.no/~dag/ling-tex.html.

\schwa	ə	ə	\hausab	ɓ		\hausaB	Ɓ	Ɓ
\thorn	þ	þ	\hausad	ɗ	ɗ	\hausaD	Ɗ	Ɗ
\glottal	ʔ	ʔ	\hausak	ƙ	ƙ	\hausaK	Ƙ	Ƙ
\esh	ʃ	ʃ	\varg	g		\planck	ħ	
\yogh	ʒ	ʒ	\udesc	ɥ	ɥ	\roty	ʎ	ʎ
\eth	ð	ð	\pwedge	ʌ		\rotm	ɯ	ɯ
\emgma	ɱ	ɱ	\varomega	ω	ω	\ibar	ɨ	
\engma	ŋ	ŋ	\revD	ɑ	ɑ	\vod	ʋ	
\enya	ɲ	ɲ	\fj	fj	fj	\ubar	ʉ	
\flap	ɾ	ɾ	\openo	ɔ	ɔ	\rotOmega	ʊ	ʊ
\voicedh	ɦ	ɦ	\varopeno	ɔ		\vari	ɩ	ɩ
\hookd	ɗ	ɗ	\vara	ɑ		\barj	ɟ	ɟ
\rotvara	ɒ	ɒ	\rotr	ɹ	ɹ	\rotw	ʍ	
\epsi	ɛ	ɛ	\barlambda	ƛ	ƛ	\riota		

o̞	\ut{o}	o̩	\td{o}	o̜	\syl{o}		
m͡n	\labvel{mn}	aᵇb	a\upbar b	ç	\rc{c}		
o̞	\hill{o}	o̥	\od{o}	mʰ	m\uplett{h}		
ô	\ohill{o}	ō̆	\acbar{\'}{o}	ő	\acarc{\'}{o}		

Table 3.7: Symbols available in the phonetic package.

3.3 Automated Special Glyphs Selection

This section provides information on certain features but not "how-to" knowledge. TEX has the ability to select special glyphs automatically when there is a typographical need for them. The most standard is the automatic selection of the glyphs fi fl ffi and ffl whenever TEX meets the sequences fi, fl, ffi, and ffl in the text (see Figure 3.5). Without these combined glyphs, the letter f extends so much to the right that it gets into the space of the next letter. Especially if it is followed by the letter i, the result looks poor (see Figure 3.5 again). These special combined glyphs are called *ligatures*. As we said above, the ligatures of f with i or l are the most common ones. Some font families, though, provide additional ligatures such as f with h, b, or j. For example, the kerkis font family, freely distributed by the Department of Mathematics of the University of the Aegean, Greece (see http://iris.math.aegean.gr/software/kerkis), provides the additional ligatures: fji and fjord. Other kinds of ligatures exist, but they are usually more ornamental in design and are used only when we want to create a special feeling with our text. Such ligatures are ligatures between s and t, c and t, and so on.

Another very common ligature is related to dashes. LATEX provides three dashes. The (simple) dash -, the en dash –, and the em dash —. These are produced with the ligature mechanism and were typed as -, --, and ---, respectively. Note that LATEX

$$\textit{fitness} \quad \textit{fitness}$$
$$\text{fitness} \quad \text{fitness}$$

Figure 3.5: Fitness without and with ligatures.

provides the commands \textendash and \textemdash, which produce an en dash or an em dash, respectively.

The plain dash, or hyphen, is the "joining" dash as in "David Ben-Gurion Airport." The en dash is used for ranges: "the information is on pages 17–58" and to join two proper names such as "Heine–Borel" to avoid confusion with single persons with a hyphenated surname (e.g., Ben-Gurion). The em dash is the dash used for parenthetical purposes: "she was angry —please believe me— and this is why she did not talk to us."

Ligatures have another use also. Some languages (or for ornamental reasons in Latin-derived languages) require many accents or letters to change form depending on their location in the text. This is very common in Arabic, for example. Polytonic writing is needed to typeset classical Greek texts. For example, we write θα>ρ<ρ˜ων, and the ligature mechanism combines the special characters with the following letter to produce ϑαῤῥῶν. The following text is from ΜονᾼΔΙΑ ΕΠΙ Ιογλιανᾼ. of Λιβανιος.

> Ὦ πόποι, ἡ μέγα πένθος οὐκ᾽ Ἀχαιίδα γῆν μόνον, ἀλλὰ
> καὶ πᾶσαν ὁπόσην ὁ ῾Ρωμαίων κοσμεῖ θεσμός, κατείληφε·
> μᾶλλον μὲν γὰρ ἴσως ἦν Ἕλληνες οἰκοῦσιν, ἅτε καὶ μᾶλλον
> αἰσθανομένην τοῦ κακοῦ, διήκει δ᾽ οὖν καὶ διὰ πάσης γῆς, ὡς
> ἔφην, ἡ πληγὴ τύπτουσά τε καὶ κατατέμνουσα τὰς ψυχάς, ὡς
> οὐκέτ᾽ ὂν βιωτὸν ἀνδρὶ βελτίστῳ τε καὶ ὅτῳ τοῦ εὖ ζῆν ἐπιθυμία.

To get this text, we have typed

```
>'Ω πόποι, >˜η μέγα πένθος o>υκ'' >Αχαιίδα γ˜ην μόνον,
>αλλ'α κα'ι π˜ασαν <οπόσην <o <Ρωμαίων κοσμε˜ι θεσμός,
κατείληφε; μ˜αλλον μ'εν γ'αρ >ίσως <'ην <'Ελληνες o>ι-
κο˜υσιν, <άτε και μ˜αλλον α>ισθανομένην το˜υ κακο˜υ,
διήκει δ'' o>˜υν κα'ι δι'α πάσης γ˜ης, <ως >έφην, <η
πληγ'η τύπτουσά τε κα'ι κατατέμνουσα τ'ας ψυχάς, <ως
o>υκέτ'' >'ον βιωτ'ον >ανδρ'ι βελτίστωι τε κα'ι <ότωι
το˜υ ε>˜υ ζ˜ην >επιθυμία.
```

Many commercial font families provide special letters for the English alphabet for ending or beginning a word. This practice used to be very common with Greek, and it is a necessity for some other languages. That is why D.E. Knuth created the "vargreek" letters such as $\vartheta, \theta, \varphi, \phi$, and so on, but he made them available only in math mode

(Knuth did not provide text fonts for Greek in his fonts). These letters had a special use in the past, and it was common practice to use the second form of the letter at the beginning of words. The most complete set of alternatives is provided by the kerkis font family. βϐ, ζϛ, θϑ, ϖπ, ρϱ, σς, and φϕ are provided. For example, the word ζιζάνιο will be typeset as ϛιζάνιο. It is clear from the above that the ligature mechanism is quite powerful. We will learn how to setup ligatures when we discuss font installation in Chapter 12.

Although this ligature mechanism will suffice for many tasks in English, this is not the case for languages such as Arabic, Persian, Khmer, and others. For example, Arabic needs something much more powerful since the shape of the letters changes depending on the position of the letter inside a word or in abbreviations. Although special systems have been developed (for example, ArabTEX), the correct approach seems to be Ω and, in particular, Λ. This is mainly because systems that are specially designed for one language have serious compatibility problems with other languages and with the standard LATEX packages. For example, a scholar writing an article or book about Arabic in a non-Latin and non-Arabic language would be out of luck with such systems. Ω loads the text as it reads the file in its buffer and, before the construction of the DVI file, it applies several filters, called *Omega* Translation Processes (ΩTP for short), that the user chooses to load. These procedures take place in the background. The user does not deal with them directly, but they manage to make the correct substitutions, performing contextual analysis on the text. This allows one to use even the most difficult languages written with the Arabic alphabet, such as Pashto and Sindhi (of course, other languages, such as Arabic, Berber, Farsi, or Urdu, are also supported). We thoroughly discuss ΩTPs in Section 10.2.

The user has the choice of typing in Latin letters (using a *transliteration* table) or, if an Arabic keyboard is available, directly in Arabic, informing the system of what keyboard is in use through encoding related commands. A transliteration table is a correspondence between the Latin alphabet and the letters of the language that we want to typeset. This is how one can manage to write a text, say, in Greek, without having a Greek keyboard available. The user can look up the documentation at http: //omega.cse.unsw.edu.au:8080/index.html. We shall show an example in Sindhi. Since we have no access to a Sindhi keyboard, we will use the transliteration table found in the link above. Our input here is entered in the Latin alphabet, but we shall inform the program that we want to typeset Sindhi so the program looks up the transliteration table and typesets correctly. Our input is the following text.

```
tn-hn kry AsAn khy pn-hnjy =z-hnn khy sjA=g rkh nU pUndU
||eN pn-hnjy jdUj-hd meN .=dA-hp pydA kr ny. AhU b/ m lUm
kr nU pUndU t/ sndh meN hr A yy wqt chA chA thy r-hyU
Ahy ||eN dshmn AsAn jy ||eN AsAn jy jdUj-hd jy khlAf
k-h.rA k-h.rA g-hA.t g-h.ry r-hyU Ahy.
```

Now, we pass this file through Λ and then either preview the resulting DVI file on screen with oxdvi or pass the DVI file through odvips to produce PostScript. The result follows.

تنهن كري اسان كي پنهنجي ذهنن كي سجاڳ رکڻو پوندو ۽ پنهنجي جدوجهد
م ڏاهپ پيدا كرڻي. اهو ڀ معلوم كرڻو پوندو تہ سنڌ م هر آئي وقت چا
چا ٿي رهيو آهي ۽ دشمن اسان جي ۽ اسان جي جدوجهد جي كلاف كهڙا
كهڙا گهاٽ گهڙي رهيو آهي.

There may be cases where a user wants to avoid a ligature. The easiest way is to put a space of zero width between the two letters that TEX automatically substitutes with the ligature glyph. For example, the word fitness used in Figure 3.5 without the ligature fi was produced by the input text f{}itness. We will get the same effect with f{i}tness or {f}itness.

3.4 Size-Changing Commands

In this section, we show how to change the size of fonts. The \documentclass command accepts three size-relative options. These are 10pt (default), 11pt, and 12pt (see Section 4.1). One may ask "why not accept any size?" or "how can I produce a document at 18 pt?". The answer to this is simple. The three sizes above are the usual sizes that people use to typeset their documents. Also, LATEX must load information about the relative sizes of math operators, and this restricts us to supporting the standard text sizes. For example, books are usually set at 10 or 11 pt. However, for special situations, one may want to change this. This is why LATEX provides commands such as \Large or \small. These commands are declarations and change the font size relatively to the predefined document font size. Table 3.8 shows the size-changing commands at the type size of this book. Note that these commands do not provide a fixed size but a size proportional to your basic document font size chosen at the \documentclass command. Sometimes, one may need more sizes. For example, you may be preparing a poster and then need some type at 72 pt. For such a specific task, one usually looks for a special package. For this task, the a0poster document class by Gerlinde Kettl and Matthias Weiser provides additional functionality that makes it easy to typeset a poster with LATEX and to print it on a DIN A0 printer using DVIPS. Furthermore, if one wants to typeset a document in either 8 pt, 9 pt, 14 pt, 17 pt, or 20 pt, then a reasonable solution is to use one of the "extsizes" document classes, listed below, by James Kilfiger:

Name	Corresponds to
extarticle	article
extbook	book
extletter	letter
extproc	proc
extreport	report

These document classes can have, in addition to the usual optional arguments, the following arguments:

Table 3.8: Size-changing commands.

Visual Result	Input Command
tiny size	`{\tiny tiny size}`
script size	`{\scriptsize script size}`
footnote size	`{\footnotesize footnote size}`
small size	`{\small small size}`
normal size	`{\normalsize normal size}`
large size	`{\large large size}`
Large size	`{\Large Large size}`
LARGE size	`{\LARGE LARGE size}`
huge size	`{\huge huge size}`
Huge size	`{\Huge Huge size}`

Parameter	Brief Description
8pt	Normal font size at 8 pt
9pt	Normal font size at 9 pt
14pt	Normal font size at 14 pt
17pt	Normal font size at 17 pt
20pt	Normal font size at 20 pt

➤ **Exercise 3.5** Suppose that one wants to prepare a book at 8 pt and an article at 14 pt. Write down the corresponding \documentclass command. ☐

Now, suppose that for some reason you want to typeset a document at 16.35 pt and you do not want to go to the trouble of creating a new document class. Then, you have to go to the low-level interface commands. LaTeX provides the command

\fontsize{*size*}{*skip*}

where *size* is the size of the font and *skip* is the size of the baseline skip. By "baseline skip" we mean the distance between two consecutive lines of text. Note that if we require really big sizes we must change the baseline skip so that the lines are nicely spaced apart. Sizes can be followed by units such as centimeters (cm) or inches (in). If we do not write a unit, LaTeX will assume units of points. Of course, such a low-level command must be followed by the \selectfont command:

```
{\fontsize{.5cm}{16pt}\selectfont type at 0.5cm}
{\fontsize{.5in}{20pt}\selectfont type at 0.5in}
```

However, even in this case, the user may not be happy. This is because LaTeX will choose a size close to what has been requested, but not exactly this size. As we run LaTeX, it will warn us with a message such as

```
LaTeX Font Warning: Size substitutions with differences
(Font)                 up to 11.25499pt have occurred.
```

The reason is that it does not know whether or not you will request sizes for subscripts or superscripts (for example, for math formulas), and then it has to compute the relative sizes to the size asked. This makes things difficult, and that is why LaTeX will usually make a choice close to, but not exactly, the size that you request.

In order to force a certain size, you have to use primitive commands of TeX. To do this, you also need to know the font that will be used. For example, if one wants to use cmr at 16.35 pt and Times at 14.2 pt, the commands \mycmr and \mytimes must be defined by

```
\font\mycmr=cmr10 at 16.35pt
\font\mytimes=ptmr7t at 14.2pt
```

where \font is TeX's primitive font-selection command. Note that one needs to know the internal name of the font in order to call it properly. Now, it is possible to write

```
{\mycmr The cmr-font at 16.35points} and
{\mytimes The Times-font at 14.2points}
```

and get

The cmr-font at 16.35points
The Times-font at 14.2points.

What you pay for this is that you should not expect the regular size-changing commands to cooperate with the commands that you defined above. But this may be irrelevant to your application. It is also possible to refine the LaTeX sizes so that its choices are even better. However, this means that you have to define new math sizes as well. This is done using the \DeclareMathSizes command. We will see how to use this command in Chapter 12.

➤ **Exercise 3.6** Typeset the following:

`Starting with large typewriter typeface`

GOING ON WITH LARGE SMALL CAPITALS

and ending with small sans serif bold typeface. □

➤ **Exercise 3.7** Typeset the following:

This is the Pandora font family at

13.3pt 21.5pt 34.8pt

□

3.5   Advanced Accents

TₑX's primitive command \accent is used to place accents over letters. The syntax of the command is as follows:

\accent *number* *letter*

where *number* is a decimal number, an octal number, or a hexadecimal number denoting the position of an accent in a font that will be placed on the *letter* (possibly in some other font!). However, this command has several weak points that the package accentbx by A.S. Berdnikov attempts to overcome by defining new macros similar to the \accent command. Several languages, for example the Nivh or the Saam (Lappish) languages, need multiple accents on a letter. They can be above, below or in the middle of the glyph to be accented. There are also cases where accents must be placed between letters, as in the German abbreviation for the ending *-burg*. Here are some examples:

É   f̃   Ẋ   Ō̄A   X̓   St. Petersb̆g

Even more impressive is the fact that a standard font change, such as switching to italics, preserves the position of the accents. Here is the example above after issuing \itshape:

É̃   f̃   Ẋ   Ō̄A   X̓   St. Petersb̆g

This means that the package is capable of doing arithmetic and calculating the proper position of the accents after a slope change.

Let us see how the package is used. The first command that one wants to know is

\upaccent{*accent* }{*character* }

This command places the *accent*'s box above the box of the *character*. Most of the time, we need to separate these two boxes by some vertical space. This is achieved with the command \aboxsplit{*accent* } (see in Figure 3.6 the letter T). On the other hand, the accent may already be raised in its box. This additional vertical space must be eliminated; this can be done with the command \abox-shift{*accent* } (see in Figure 3.6 the letter E). We may also have an accent that extends below the baseline (has depth), such as the comma character; these need additional shift. The command \aboxbase{*accent* } shifts the accent so that it is entirely above the baseline (see in Figure 3.6 the letter X).

Middle accents, as in the *-burg* abbreviation, are produced by inserting a zero-width rule of height equal to the height of the letter x and accenting this rule. The rule is produced by the command \markchar. Thus, the code for the *-burg* abbreviation is

```
St.~Petersb\upaccent{\aboxshift{\symbol{'10}}}{\markchar}g
```

```
       | \fontencoding{T1}\selectfont
  Ṫ    | \upaccent{.}{T}
  Ṭ    | \upaccent{\aboxsplit{.}}{T}
       |
  Ë    | \upaccent{\symbol{'001}}{\"E}
  É̈    | \upaccent{\aboxshift{\symbol{'001}}}{\"E}
  X̧    | \upaccent{\aboxsplit{,}}{X}
  X̧̓    | \upaccent{\aboxsplit{\aboxbase{,}}}{X}
```

Figure 3.6: Code of the examples.

Of course, one may define shorter forms of these long commands with the definition of a new command, for example

```
\newcommand{\burg}{%
   b\upaccent{\aboxshift{\symbol{'10}}}{\markchar}g}
```

and then type St.Peters\burg. If we want to use the same construction for capitals, then the invisible rule of height equal to the height of x must now have the height of X. The corresponding command is \MARKCHAR:

```
\newcommand{\BURG}{%
   \upaccent{\aboxshift{\symbol{'10}}}{\MARKCHAR}G}
```

It is useful to substitute the \aboxshift by its robust version \Aboxshift, especially when we change the case of the words with the commands \MakeUppercase and \MakeLowercase (for example, in running heads). Then, the commands

```
\MakeUppercase{st. peters\burg} and
\MakeLowercase{ST. PETERS\BURG}
```

will give the correct output.

Putting accents on letters may break other properties of a font. One of them is kerning information. The package also provides the means to correct this by the command \akern. The following example shows its use:

```
  ÅW̊  | \upaccent{\aboxsplit{\tiny w}}{A}%
       | \upaccent{\aboxsplit{\tiny a}}{W}
  ÅW̊  | \upaccent{\aboxsplit{\tiny w}}{A}\akern AW%
       | \upaccent{\aboxsplit{\tiny a}}{W}
```

Several commands are available with the package for fine-tuning accents. We refer the interested user to the package's documentation or to the article by the author of the package in [2] (available online from http://obelix.ee.duth.gr/eft).

➤ **Exercise 3.8** Write down the commands that created the phonetic transcriptions of Navaho words in Section 3.2.3. ☐

4

LISTS AND CATALOGS

In many cases, one may want to write a series of names, words, or other items one after the other. This is commonly known as a list. Moreover, in many cases, users want to be able to write an itemized display (e.g., titles, course offerings, or articles for exhibition or sale) usually including descriptive information or illustrations. This is commonly known as a catalog. Since lists and catalogs are very frequently used in printed text, LaTeX offers a variety of environments that can be used to produce any possible list or catalog. Furthermore, it provides environments suitable for the typesetting of poetry and quotations, among others. In this chapter, we present these environments and their uses. Although alignment is a subject that does not comfortably fit in this chapter, we include the relevant discussion here as we think this is the most appropriate place for this material.

4.1 Units of Measure

Donald Knuth designed TeX so that one can fully use the metric system of units. Since most English-speaking countries still use the imperial system of units (contrary to their legislation most of the time), TeX was designed to be able to work with this system, too. Moreover, since the English-speaking world uses the period to separate the integer part of a number from the decimal part while the rest of the world uses the comma, TeX recognizes both forms of number writing to facilitate its use all over the world. The various units of measure that TeX understands are presented in Table 4.1. Note that TeX cannot handle any length whose absolute value is greater than or equal to 2^{30} sp; that is approximately, 575.83 cm. Now that we know all of the units of length that TeX understands, it is important to say how we can write a particular length. Any valid TeX length consists of an optional sign, followed by an optional decimal or integer, followed by a unit of length. The unit of length can be separated from the number by one or more spaces. As we mentioned above, when we have a decimal number, the integer part can be separated from the decimal part by a comma or a period. Here are a few examples of valid TeX lengths:

Table 4.1: Units of measure and their relationships.

Symbol	Unit Name	Equivalence
pt	point	$1\,\text{pt} = 0.013837\,\text{in}$
pc	pica	$1\,\text{pc} = 12\,\text{pt}$
in	inch	$1\,\text{in} = 72.27\,\text{pt}$
bp	big point	$72\,\text{bp} = 1\,\text{in}$
cm	centimeter	$1\,\text{in} = 2.54\,\text{cm}$
mm	millimeter	$1\,\text{mm} = 0.1\,\text{cm}$
dd	didot point	$1157\,\text{dd} = 1238\,\text{pt}$
cc	cicero	$1\,\text{cc} = 12\,\text{dd}$
sp	scaled point	$65536\,\text{sp} = 1\,\text{pt}$

$$3\text{in} \quad -.5 \text{ cm} \quad +3,14 \text{ dd} \quad -,\text{pt}$$

Although the fourth length looks quite peculiar, it is actually equal to 0 pt!

Apart from the *absolute* units above, TeX supports two more *relative* units, which depend on the font that is currently in use. This means that their actual length depends on which font we use. These units are the em and the ex. One em is a distance equal to the type size. In 6 point type, an em is 6 pt; in 12 pt type, it is 12 points, and so on. Thus, a one-em space is proportionately the same in any size.

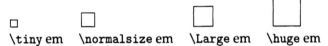

\tiny em   \normalsize em   \Large em   \huge em

One ex is a distance equal to the "x-height" of the current font. This particular unit of measure is not widely used outside of the TeX world. If we are using the Computer Modern fonts by Knuth, one ex is equal to the height of the small letter x.

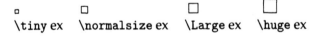

\tiny ex   \normalsize ex   \Large ex   \huge ex

In general, the em is used for horizontal spacing and the ex for vertical spacing.

▶ **Exercise 4.1** Put the following lengths in ascending order:

$$123456789\,\text{sp} \quad 101.7\,\text{dd} \quad 2.5\,\text{in}$$
$$33\,\text{pc} \quad\quad\quad 45\,\text{mm} \quad 0.89\,\text{cm}$$
$$1.9\,\text{cc} \quad\quad\quad 536\,\text{bp} \quad 674\,\text{pt}$$

▶ **Exercise 4.2** How many points are in 254 cm?

4.2   Typesetting Poetry

Although LaTeX is a typesetting system particularly well-suited for the preparation of manuscripts containing a lot of mathematics, it can be used to typeset poems by using the verse environment, which has been designed to facilitate the typesetting of poems. Here is a simple example that demonstrates the environment:

To a Young Poet	`\begin{verse}`
	`\emph{To a Young Poet}`
Time cannot break the bird's wing from the bird.	
Bird and wing together	`Time cannot break the bird's`
Go down, one feather.	`wing from the bird.\\`
	`Bird and wing together\\`
No thing that ever flew,	`Go down, one feather.`
Not the lark, not you,	
Can die as others do.	`No thing that ever flew,\\`
	`Not the lark, not you,\\`
	`Can die as others do.`
	`\end{verse}`

The poem above is by Edna St Vincent Millay, 1892–1950.

   While in the verse environment, verses are separated by the command \\ and stanzas by a blank line. Note that one must not put the command \\ at the end of a stanza. Moreover, the title of the poem is separated from the poem itself by a blank line. Certainly, one can use other environments to produce fancy results, but we have not introduced them yet, so we skip this issue. We have already used the \\ command, but here we present all three forms of this command as well as the functionality of each form:

\\   This command simply forces TeX to prematurely break a line at the point where the command is placed.

*   In case we just want to break a line and prevent a possible page break, we use this form of the command.

\\[length]   When TeX breaks a line, usually it does not add any additional vertical space between the two consecutive lines. However, if for some reason we want to put some additional vertical space, we use this form. The length is just a valid TeX length (e.g., 1 cm, 0.5 in, etc.).

Note that it is possible to use a combination of the second and third cases. However, there is a special case that we must be able to deal with. Suppose that we want to force LaTeX to break a line at some point and then we want to write something enclosed in square brackets, for example,

$$\ldots \text{ \textbackslash\textbackslash [yeah!] } \ldots$$

The code above will make LaTeX complain that a number is missing. It is important to note that the space between the \\ command and the character [is actually ignored by LaTeX so it assumes that the user wanted to specify a length. The solution to this unusual problem is to enclose the left square bracket in curly brackets, that is,

$$... \text{\textbackslash\textbackslash \{[\}yeah!]} ...$$

Admittedly, using LaTeX can be quite tricky sometimes!

▶ **Exercise 4.3** Typeset the following poem such that the line space between verses is exactly 3.5 pt.

Autumn Song

Soon we will plunge ourselves into cold shadows,
And all of summer's stunning afternoons will be gone.
I already hear the dead thuds of logs below
Falling on the cobblestones and the lawn.

All of winter will return to me:
derision, Hate, shuddering, horror, drudgery and vice,
And exiled, like the sun, to a polar prison,
My soul will harden into a block of red ice.

I shiver as I listen to each log crash and slam:
The echoes are as dull as executioners' drums.
My mind is like a tower that slowly succumbs
To the blows of a relentless battering ram.

It seems to me, swaying to these shocks, that someone
Is nailing down a coffin in a hurry somewhere.
For whom? –It was summer yesterday; now it's autumn.
Echoes of departure keep resounding in the air.

–Charles Baudelaire
(translated from French by Steven Monte)

The poem was originally published in the *Boston Reviews* and is reproduced with kind permission from the translator. ☐

4.3 Lists

LaTeX provides three different environments that can be used to typeset lists and catalogs. The environment `enumerate` can be used to typeset numbered lists such as those usually found in the exercise sections of a textbook. Here is a simple example:

1. Write the missing dates in words:
   (a) (17/5) I Norge firas...
   (b) (14/7) Fransmänne firas...
2. Answer the following questions using **dit/där/hit/här**:
   (a) Var är du? (*here*)
   (b) Var ligger staden? (*there*)
	`\begin{enumerate}`
	`\item Write the missing dates...:`
	`\begin{enumerate}`
	`\item (17/5) I Norge firas\ldots`
	`\item (14/7) Fransm\"anne...`
	`\end{enumerate}`
	`\item Answer the following...`
	`\textbf{dit/d\"ar/hit/h\"ar}:`
	`\begin{enumerate}`
	`\item Var \"ar du?(\textit{here})`
	`\item Var ligger staden?...`
	`\end{enumerate}`
	`\end{enumerate}`

As is evident, each list item starts with the command \item. Note that if the first thing that you type in the body of the environment is not a \item command, LATEX will not be able to process your file. This holds for all environments presented in this section. Moreover, it is possible to have *nested* environments, such as those in the example above. LATEX supports up to four levels of nesting. In order to avoid confusion, the numbering in each nested environment changes so that each item is uniquely numbered. However, there are cases where we are not interested in numbering the items of a list; that is, we want to create an unnumbered list. In this case, we have to use the itemize environment:

The i/o system consists of
 	
• A buffer-caching system
• A general device-driver interface
• Drivers for specific hardware devices
	`The \textsc{i/o} system consists of`
	`\begin{itemize}`
	`\item A buffer-caching system`
	`\item A general device-driver`
	`      interface`
	`\item Drivers for specific`
	`      hardware devices`
	`\end{itemize}`

The last supported list environment is the description environment, which can be used to create simple glossaries, among other things. Here is a simple example:

Ångstrom A unit of measure corresponding to one ten-billionth of a meter.
Amino acid Basic building blocks of proteins.
Articulation Movements of the vocal tract to produce speech sounds.
	`\begin{description}`
	`\item[\AA ngstrom] A unit of...`
	`\item[Amino acid] Basic building...`
	`\item[Articulation] Movements of...`
	`\end{description}`

The only thing that we must note is that each term must follow the command \item in square brackets.

➤ **Exercise 4.4** Explain how one might typeset a dictionary using LATEX. ☐

The enumerate package by David Carlisle redefines the enumerate environment and gives it an optional argument that determines the style in which the label is printed. An occurrence of one of the characters A, a, I, i, or 1 causes the environment to produce labels that are either uppercase or lowercase alphabetic or Roman numerals or numbers, respectively. These letters may be surrounded by any string involving other LATEX expressions; however, in this case, the letters A, a, I, i, and 1 must be enclosed in curly brackets if we do not want them to be taken as special. Figure 4.1 shows a usage example of the environment provided by the enumerate package.

EX i. one one one one
 	
EX ii. two two two two
 	
     example a) one of two
 	
     example b) two of two
	`\begin{enumerate}[EX i.]`
	  `\item one one one one`
	  `\item two two two two`
	    `\begin{enumerate}[{example} a)]`
	      `\item one of two`
	      `\item two of two`
	    `\end{enumerate}`
	`\end{enumerate}`

A-1. one
 	
A-2. two
	`\begin{enumerate}[{A}-1]`
	`\item one`
	`\item two`
	`\end{enumerate}`

Figure 4.1: A usage example of the enhanced enumerate environment.

4.3.1 Customizing the Standard Lists

It is relatively easy to customize the three standard LATEX list environments, and in this section we describe how we can modify these environments to suit our needs. The changes can be either local or global (i.e., we can modify the environments so that the changes are visible only at certain parts of our document or throughout the whole document).

Customizing the enumerate environment

The enumeration labels depend on the value of the counters enumi, enumii, enumiii, and enumiv. Counter enumN controls the numbering of the Nth-level enumeration. The label for each level of enumeration is generated by the corresponding \labelenumN command, which is usually defined by the document class in use. The output of a \ref command is a sequence of characters (also called a string) generated by the \p@enumN\theenumN commands. The command \p@enumN is provided by the document class in use. For instance, the following definitions appear in the article document class:

```
\renewcommand\theenumii{\alph{enumii}}
\newcommand\labelenumii{(\theenumii)}
\renewcommand\p@enumii{\theenumi}
```

Note that the command \p@enumN actually prints the prefix of the reference. For example, when referring to the second item of the third sublist, the actual label will be "3(b)," and the "3" is printed by the \p@enumN command. If we want to redefine any of the p@enumN commands in our document's preamble, we first have to type the command \makeatletter, then the redefinition, and then the command \makeatother. Since in any LaTeX document the symbol @ has a category code equal to 12, we use the first command to make @ a letter (category code 11) and the second to switch back to the original category code.

➤ **Exercise 4.5** Provide definitions for the command \makeatletter and \makeatother. □

We now give a complete example to demonstrate how one can modify enumerations. Suppose that we want the labels in the first enumeration level to appear as capital Roman numerals prefixed by the paragraph symbol (§) while we want to have a hyphen between the paragraph symbol and the Roman numeral in the labels generated by the \ref command. Here is how this can be done:

§I First item

§II Second item

§III Third item

label w1=§-I,
label w2=§-II

```
\makeatletter
\renewcommand{\theenumi}{\Roman{enumi}}
\renewcommand{\labelenumi}{\S\theenumi}
\renewcommand{\p@enumi}{\S-}
\makeatother
\begin{enumerate}
\item\label{w1} First item
\item\label{w2} Second item
\item\label{w3} Third item
\end{enumerate}
label w1=\ref{w1}, label w2=\ref{w2}
```

Customizing the *itemize environment*

The itemization within an itemize environment is controlled by four commands: \labelitemi, \labelitemii, \labelitemiii, and \labelitemiv. To make the second-level list use an em dash as its label, we use the following command:

```
\renewcommand{\labelitemii}{\normalfont ---}
```

If we want to use any of the dingbat symbols on page 49, we have to use the pifont package by Sebastian Rahtz. This package provides the command \Pisym-bol, which must be supplied with two arguments: the name of the font and a slot number (corresponding to the symbol that we want to access). Usually, the font is pzd (PostScript Zapf-Dingbats), and the slot number (in octal) can be determined from the table on page 49. Here is a simple example:

❍ First item	`\renewcommand{\labelitemi}{\Pisymbol{pzd}{'112}}`
❍ Second item	`\begin{itemize}`
	`\item First item`
❍ Third item	`. . . . . . . . . . . . . . .`
❍ Fourth item	`\item Fourth item`
	`\end{itemize}`

Customizing the `description` environment

The appearance of the labels used in a description environment is controlled by the `\descriptionlabel` command. The standard definition is as

```
\newcommand{\descriptionlabel}[1]{%
   \hspace{\labelsep}\normalfont\bfseries #1}
```

where `\labelsep` is the space between the end of the label box and the text of the item. In the following example, the description label is in boldface sans serif.

Perl A program-ming language	`\renewcommand{\descriptionlabel}[1]{%`
	`   \hspace{\labelsep}\textsf{\textbf{#1}}}`
Java Another pro-gramming lan-guage	`\begin{description}`
	`\item[Perl] A programming language`
	`. . . . . . . . . . . . . . . . . . . . . . . . . . . . . .`
TEX A typesetting system	`\end{description}`

➤ **Exercise 4.6** For each list level LATEX defines the length `\leftmarginN`, which controls the amount of white space that it leaves before it typesets the label of a particular item. Produce the enumeration of page 66 by changing the value of the length `\leftmarginii`. ☐

4.4 Quotations

A quotation is an explicit reference or allusion in an artistic work to a passage or element from another, usually well-known work. Since the use of quotations is very frequent in all kinds of scientific and scholarly writings, LATEX provides two environments for the correct typesetting of quotations. The environment quote is used mainly for short quotations. On the other hand, the environment quotation is used mainly for multi-paragraph quotations. Certainly, one can also use the quote environment for multi-paragraph quotations, but then LATEX does not do proper paragraph indentation (i.e., the first word of a new paragraph is not indented). This difference is evident in the following example:

Before the quote environment.	`Before the quote...`

This is a quote. This is a quote. This is a quote. This is a quote. This is a quote. This is a quote. This is a quote.	`\begin{quote}`
	`This is a quote.`
	`This is a quote...`

After the quote environment and before the quotation environment.	`This is a quote...`
	`\end{quote}`
	`After the quote...`

This is a quotation. This is a quotation. This is a quotation. This is a quotation.	`\begin{quotation}`
	`This is a quotation.`
	`This is a quotation...`

This is a quotation. This is a quotation.	`This is a quotation...` `\end{quotation}`
After the quotation environment.	`After the quotation...`

As is obvious from this simple, yet complete example, both environments leave some extra vertical space before and after the quotations. Moreover, special care is taken so that the quotation text appears centered on the page.

4.5 Footnotes

A footnote is a note placed at the bottom of a page of a book or manuscript that comments on or cites a reference for a designated part of the text. However, it is assumed to be a bad practice to use footnotes in a text, mainly because they do not allow the reader to focus on the main text. Despite this, LATEX does provide the command \footnote to facilitate the creation of footnotes in a text:

Text[1] with footnotes[5]. –––––––––– [1] A footnote [5] Another footnote	`Text\footnote{A footnote} with` `footnotes\footnote[5]{Another footnote}.`

From the example above, one can easily deduce that all one has to do in order to create a footnote is to write the command \footnote next to the word where the footnote mark is to appear. The text of the footnote follows the command and must be enclosed in curly brackets. A \footnote command can have an optional argument, enclosed in square brackets, which, when present, will be used to typeset the footnote mark.

Suppose that in one document we have a complicated enumerated list with very many footnotes. In addition, suppose that we want to have the text of all footnotes placed at the end of this enumerated list. In order to provide a solution to problems such as these, LATEX provides the following commands:

```
\footnotemark[number]
\footnotetext[number]{text}
```

The \footnotemark command produces just the footnote mark in the text, but not the footnote text. The optional argument is present and it is used to create the footnote mark. Otherwise, LATEX increments the footnote counter before generating the actual text. The \footnotetext command produces the actual footnote text. If the optional argument is present, it is used to produce the footnote mark inside the footnote.

text text text[66] text text text	`text text text\footnotemark[66]`
	`text text text`
_____	`\footnotetext[66]{footnote...}`
[66]footnote footnote footnote	

In many instances, people want to be able to use footnotes even in section headers, as in the following example:

```
\documentclass{article}
\begin{document}
\section{Text\footnote{footnote}}
text text text.
\end{document}
```

If we feed this file to LATEX, we get the following error message:

```
! Argument of \@sect has an extra }.
<inserted text>
                \par
l.3 \section{Text\footnote{footnote}}

?
```

Admittedly, this error message is not helpful at all. But, LATEX complains with this weird error message simply because we are not allowed to use the \footnote command as an argument of the \section command. (If you have actually tried this example, press x after the ? to force LATEX to abandon the processing of your file.) So, is it impossible to have a footnote in a sectioning command or is there something else we can do? Actually, we can have a footnote in a sectioning command, but we must *protect* it since the \footnote command is a *fragile* command. Any command that accepts an optional argument (i.e., an argument in square brackets) is a fragile command. Non-fragile commands are called *robust*. So, how do we protect a fragile command when it happens to be the argument of another command? We put the \protect command just before the command that we want to protect. Let us now correct the file above:

```
\section{Text\protect\footnote{footnote}}
```

Suppose now that we want to generate the table of contents for this document. If we inspect the resulting DVI file, we will notice that the footnote mark is included in the table of contents. The solution to this problem is as follows:

$$\texttt{\textbackslash section[Text]\{Text\textbackslash protect\textbackslash footnote\{footnote\}\}}$$

That is, we simply use the optional argument of the sectioning command.

4.5.1 Customizing Footnotes

Until now, we have seen how to use the various footnote-related commands, but it is possible to customize the appearance of footnotes (i.e., we can change the appearance of the footnote mark, the footnote text, and the footnote rule). By default, the \footnote command uses the \footnotesize font size to typeset the body of a footnote. However, one can easily change this behavior. Let us suppose that we want the footnote text to be typeset in \normalfont size. Before we go on with the solution to this problem, we must explain the functionality of the \let command. If \a and \b are two commands, the command \let\a\b makes command \a behave exactly like command \b. Now, we give the command that achieves the desired functionality:

$$\texttt{\textbackslash let\textbackslash footnotesize\textbackslash normalsize}$$

➤ **Exercise 4.7** The solution above permanently alters the meaning of the command \footnote so one cannot typeset text in \footnotesize. Can you devise a solution that remedies this drawback? ☐

$$\texttt{\textbackslash let\textbackslash myfootnotesize\textbackslash footnotesize}$$
$$\texttt{\textbackslash let\textbackslash footnotesize\textbackslash normalfont}$$

Now, a footnote can appear in the original size if we type it as follows:

$$\texttt{\textbackslash footnote\{\textbackslash myfootnotesize footnote\}}$$

The space between the main text and the footnotes is equal to the length stored in the length variable \skip\footins. Moreover, the space between footnotes is equal to the length stored in the length variable \footnotesep. The \footnotesep is actually a strut that is placed at the beginning of each footnote. The height of the footnote rule is roughly equal to

$$\texttt{\textbackslash skip\textbackslash footins} \qquad \texttt{\textbackslash footnotesep} \cong 3\,\text{pt}$$

The values of these lengths can be altered with the commands \setlength and \addtolength.

If we want to change the appearance of the footnote rule we must redefine the \footnoterule command. The following definition is equivalent to the definition provided by the LaTeX kernel:

$$\texttt{\textbackslash newcommand\{\textbackslash footnoterule\}\{\textbackslash vspace*\{-3pt\}}$$
$$\texttt{\textbackslash noindent\textbackslash rule\{2in\}\{0.4pt\}\textbackslash vspace*\{2.6pt\}\}}$$

The interesting thing with the command above is that it makes TEX believe that the footnote rule has height equal to zero! So, suppose now that we want a footnote rule with width equal to 5 in and height equal to 1.5 pt; then, if we put the following redefinition in the preamble of our LATEX document

```
\renewcommand{\footnoterule}{\vspace*{-3pt}
    \noindent\rule{5in}{1.5pt}\vspace*{1.5pt}}
```

we will get the desired effect. Note that if we want to have a footnote rule with a height greater than 3 pt, we have to modify the first vertical spacing command accordingly and the values of the variables \skip\footins and \footnotesep.

▶ **Exercise 4.8** How can we instruct LATEX to draw a footnote rule with length equal to the text line length and height equal to 4 pt?   □

The appearance of the footnote mark is controlled by the command \@make-fnmark. The following definition is equivalent to the standard one:

```
\newcommand{\@makefnmark}{%
    \mbox{\textsuperscript{\normalfont\@thefnamrk}}}
```

The counter \@thefnmark is used to number footnotes. If we want to redefine the command above in our document's preamble, we first have to type the command \makeatletter, then the redefinition, and then the command \makeatother. The following code sets the footnote number in boldface, surrounded by parentheses:

```
\renewcommand{\@makefnmark}{\mbox{%
    \textsuperscript{\normalfont({\bfseries\@thefnmark})}}}
```

Note that the character % is used to avoid unwanted space.

The appearance of the text of a footnote is controlled by the command \@make-fntext, whose definition, in the case of the article document class, is equivalent to the following:

```
\newcommand\@makefntext[1]{%
    \setlength\parindent{1em}%
    \noindent
    \makebox[1.8em][r]{\@makefnmark}#1}
```

The control sequence \parindent is a predefined length whose value is used for paragraph indentation (i.e., the amount of space that TEX leaves at the beginning of a new paragraph). So, if we want the text of footnotes to appear in italics, and the footnote mark to be just the footnote number followed by a period, we can use the following redefinition:

```
\renewcommand\@makefntext[1]{%
    \setlength\parindent{1em}%
    \noindent
    \makebox[1.8em][r]{\@thefnmark.\ }{\itshape#1}}
```

4.5.2 Endnotes

In scholarly work, it is common practice to have all footnotes at the end of a chapter or section, or even at the end of a document. Since such a facility is not provided by LATEX, one has to use the endnotes package, originally developed by John Lavagnino. The package has been further modified by Jörg Knappen and Dominik Wujastyk. The package provides the following commands:

\endnote{*Note*} This is the user command to produce an endnote, and *Note* is the text of the endnote.

\endnote[*Num*]{*Note*} A command to produce an endnote numbered *Num*.

\endnotemark[*Num*] A command to produce just the endnote mark in the text but not the endnote. With no argument, it steps (increases by one) the endnote counter before generating the mark.

\endnotetext[*Num*]{*Text*} A command to produce the endnote but no mark. The command \endnote is equivalent to

$$\text{\endnotemark \endnotetext}$$

\addtoendnotes{*Text*} A command to add text or commands to the current endnotes file and used for inserting headings, page breaks, and so on The Text must be \protected.

From the preceding description of the user commands, one can conclude that the package writes all endnotes in a separate file that is included in the document after all of the input files have been processed.

Endnotes can be customized exactly like footnotes. Here, we present only the commands that affect the appearance of endnotes. Readers interested in experimenting with these commands should read the corresponding discussion for footnotes in Section 4.5.1.

\enotesize With this command, we can change the font size used for endnotes.

\theendnote This command is used to produce the endnote number.

\@theenmark This holds the current endnote's mark.

\@makeenmark A macro to generate the endnote marker from \@theenmark.

\@makeentext{Note} Produces the actual endnote using \@theenmark as the mark of the endnote and Note as the text.

4.6 Simulating Typed Text

If you browse a computer program manual, you will notice that it contains simulated typed text, which shows what the user types to achieve a particular effect. For example, a Unix manual may contain simulated typed text to demonstrate the use of the various commands. The \ttfamily declaration produces a typewriter text style, but it does not stop TEX from breaking the text into lines as it sees fit. For this reason, LATEX provides two environments and two commands that perform no formatting upon their arguments (i.e., they allow you to type the text exactly the way you want it to appear in the document). The commands are used for short pieces of text that can fit on one line, whereas the environments are used for longer pieces of text. The two environments are called verbatim and verbatim*. Their only difference is that the second makes spaces visible by substituting each space character with the symbol ␣. Below, we give a simple example to demonstrate the use of these two environments:

<table>
<tr><td>

A little Perl program:

print "Hello World!\n" if 1;

Another little Perl program:

print␣"Hello␣World!\n"␣while␣1;

</td><td>

```
A little Perl program:
\begin{verbatim}
print "Hello World!\n" if 1;
\end{verbatim}
Another little Perl program:
\begin{verbatim*}
print "Hello World!" while 1;
\end{verbatim*}
```

</td></tr>
</table>

The two commands are \verb and \verb*. They do exactly what the corresponding environments do. Their argument is delimited by a single character. However, one must be careful to choose as a delimiter a character that does not occur in the text:

<table>
<tr><td>

Special characters: $%^&_
Special␣characters:␣$%^&_

</td><td>

```
\verb|Special characters: $%^&_|
\verb*+Special characters: $%^&_+
```

</td></tr>
</table>

From the example above it is evident that the special characters described in Section 2.1 are not "special" when used as arguments of either the commands or the environments. When using these commands or the environments, one must:

- Avoid using them as arguments of other commands and
- Keep in mind that there should be no space(s) between either the command \verb and the delimiting character or the tokens \end and {verbatim}.

There are two things that one can do to customize either the environments or the commands presented so far in this section. First, we can tell LATEX to use a different fixed-width font and second to prepend white space to each line. The LATEX kernel defines the command \verbatim@font that defines the font to be used to typeset the text of the verbatim commands and environments. The standard definition of this command is

```
\newcommand{\verbatim@font}{\normalfont\ttfamily}
```

For instance, if you want to use the Courier font instead of the default Computer Modern typewriter font as the verbatim font, then the following command will do the job:

```
\renewcommand{\verbatim@font}{\usefont{OT1}{pcr}{m}{n}}
```

Prepending white space to each line of a verbatim environment means that we first define a new length and then change the definition of the command \verbatim accordingly. In the example that follows, we tell LATEX to prepend each line with 1 cm of white space:

```
\newlength\verbatimindent
\setlength{\verbatimindent}{1cm}
\renewcommand{\verbatim}{%
    \addtolength{\@totalleftmargin}{\verbatimindent}
    \@verbatim \frenchspacing\@vobeyspaces \@xverbatim}
```

It is important to stress that one must copy verbatim (!) the code fragment above into a LATEX file and change only the length of the white space. Anything else may have unpredictable consequences.

4.6.1 Advanced Typed Text Simulation

Although the various commands and environments presented above are adequate for most cases, still there are many situations where a user might want additional functionality. For example, in a book on computer programming, it is convenient to prepend a line number to each line of a program listing. Here, we briefly present three packages that provide new environments and additional functionality to the commands presented above.

A new implementation of the verbatim environments

Rainer Schöpf, Bernd Raichle, and Chris Rowley have designed the verbatim package that reimplements the verbatim| and the verbatim* environments. This reimplementation solves a few problems of the original implementation. In particular, they can better handle very long texts that are supposed to be output verbatim. In addition, the package provides a comment environment that skips any commands or text between \begin{comment} and \end{comment}. This command is useful when one wants to comment out certain parts of a LATEX file and, naturally, has no effect if used inside a verbatim environment. The package also defines the command \verbatiminput, which can be used to input a whole file verbatim. The command has one argument, which is the name of a file that inputs verbatim; that is, the command \verbatiminput{xx.yy} has the same effect as

```
\begin{verbatim}
Contents of file xx.yy
\end{verbatim}
```

The *moreverb* package

The moreverb package designed by Robin Fairbairns, Angus Duggan, Rainer Schöpf, and Victor Eijkhout provides things in three broad areas:

- Tab expansion and related stuff.
- Line numbering.
- Miscellaneous: writing verbatim to a file and "boxed" verbatim.

When using a verbatim environment, LATEX treats tabs as single-space characters. However, there are many instances where we want each tab to expand to a specific number of spaces. This functionality is provided by the verbatimtab environment. By default, the *tab width* is equal to eight space characters. Certainly, one can change this behavior easily by simply specifying the tab width as an optional argument. Here is a simple example:

```
text      text      text           \begin{verbatimtab}
          text      text      text  text▷      text▷      text
          text      text             ▷                    text▷      text▷      text
text text text   text                ▷▷                              text▷      text
                                     text text text▷ text
                                     \end{verbatimtab}
```

In the previous example, the symbol ▷ denotes the tab key. When using this environment, one must have in mind that the size of the tabs persists between uses of the environment; that is, an optional argument to one of them applies to all subsequent ones.

The listing environment numbers the lines of its body. The user must specify the *start line* and can also specify the number of lines between numbered lines. If the *start line* is a number other than one, then the environment makes the assumption that the first line of its body has a line number equal to a *start line*.

```
4 line one                \begin{listing}[2]{4}
  line two                line one
6 line three              line two
  line four               line three
8 line five               line four
                          line five
                          \end{listing}
```

The environment listingcont continues from the place where the last listing left off. Both environments also expand tabs. Starred versions of both listing environments are provided; these print visual spaces (i.e., ␣) instead of ordinary spaces, and they do not expand tabs. The command \listinginput is a file input version of listing, and there is no starred version. So, if the text in the previous example is stored in a file, say text.txt, then the following command achieves the same visual effect with the command of this example:

$$\listinginput[2]{4}{text.txt}$$

The environment verbatimwrite takes one argument, the name of a file, and writes all text in its body to this file. The boxedverbatim puts its body in a framing box, but it makes no sense to use this environment in a naïve way (i.e., outside any environment that reduces the line width).

The *alltt* package

The package alltt was designed by Leslie Lamport and has been adapted to LATEX 2ε by Johannes Braams. The package provides the alltt environment, which is like the verbatim environment except that \, {, and } have their usual meanings. Thus, other commands and environments can appear in the body of an alltt environment. So, one can change fonts (e.g., by typing {\emph text}), include files, and insert a mathematical formula (see the next chapter). In particular, math mode can be started with either \(or \[. Naturally, one ends math mode with a \) or \] correspondingly. Moreover, superscripts and subscripts can be produced with the commands \sp and \sb, respectively, so the simple formula x_j^i will be produced by the code

$$\(x\sp{i}\sb{j}\)$$

4.7 Centering and Flushing Text

When we say that a piece of text, a paragraph, and so on is *flushed* either left or right, this means that the text is aligned either to the left or the right edge of the page. Moreover, when we say that a piece of text is horizontally centered, this means that each line of the text is centered on the page. Naturally, the term *page* does not include the margins of the paper but only the area that is reserved by the typesetting system, LATEX in our case, for the text body. Since these features are extremely useful, LATEX provides three environments that allow its users to produced flushed or centered text. In Figure 4.2, we give examples of the environments flushleft, flushright, and center that are used to typeset text flushed left, flushed right, and centered, respectively. Note that LATEX inserts some space before and after each environment. It is possible to produce either a flushed left or flushed right document by using the declarations \raggedright or \ragged-left, respectively. Moreover, a completely centered document can be produced with the declaration \centering. If we want to affect the typesetting part of a document, we have to use these commands in a local scope. If we want to produce only one line

Lars Valerian Ahlfors was born in Helsingfors, Finland, on April 18, 1907. He studied mathematics with Ernst Lindelöf at Helsingfors University and earned his doctorate in 1928.

```
\begin{flushleft}
Lars Valerian Ahlfors was born
in Helsingfors, Finland, on
April 18, 1907. He studied
...........................
\end{flushleft}
```

Lars Valerian Ahlfors was born in Helsingfors, Finland, on April 18, 1907. He studied mathematics with Ernst Lindelöf at Helsingfors University and earned his doctorate in 1928.

```
\begin{flushright}
Lars Valerian Ahlfors was born
in Helsingfors, Finland, on
April 18, 1907. He studied
...........................
\end{flushright}
```

Lars Valerian Ahlfors was born in Helsingfors, Finland, on April 18, 1907. He studied mathematics with Ernst Lindelöf at Helsingfors University and earned his doctorate in 1928.

```
\begin{center}
Lars Valerian Ahlfors was born
in Helsingfors, Finland, on
April 18, 1907. He studied
...........................
\end{center}
```

Figure 4.2: Examples of the flushleft, flushright, and center environments.

that is either centered or flushed left or right, we can use the corresponding commands \centerline, \leftline, or \rightline, as the following example demonstrates:

A centered line.	\centerline{A centered line.}
A flushed left line.	\leftline{A flushed left line.}
A flushed right line.	\rightline{A flushed right line.}

Note that these commands should not be used in the middle of a paragraph, as they generate a horizontal box with width equal to \linewidth (for more information on boxes, see Section 6.10).

4.8 Alignment

When we speak about alignment, we speak about the arrangement or position of words, numbers, and so on, in a straight line or in parallel lines. LaTeX provides two basic environments that can perform alignment: the tabular and the tabbing environments. In this section, we present the basic functionality of these two environments.

4.8.1 The tabbing Environment

The tabbing environment can be used to align text in columns by setting tab stops and tabbing to them, just like people do when using an ordinary typewriter. Let us see a simple example:

If today is Sunday	`\begin{tabbing}`
then I will go to	`If today \= is Sunday \\`
the beach,	`    \> then I \= will go to\\`
else I must go to	`    \>         \> the beach,\\`
the office	`    \> else I must go to\\`
Boy, I love Sundays!	`    \>         \> the office\\`
	`Boy, I love Sundays!`
	`\end{tabbing}`

The command \= defines a tab position. Tab positions are usually defined on the first line of a table, but as the example demonstrates, it is possible to define new tab positions in subsequent lines. The command \\ is used to change lines. However, if one uses the tabbing environment within another environment and would like to break a line only within the tabbing environment, one can use the command \tabularnewline. The command \> is used to jump to the next tab stop. Suppose that we want to define tab stops on a line that will only serve as the tab stops definition line (i.e., we do not want this line to appear in the final manuscript). Then we can prevent LATEX from including this line in the DVI file by appending the \kill command to this particular line, as the following example demonstrates:

	`\begin{tabbing}`
	`1234\=1234\=1234\=\kill`
a b	`a \> b \\`
c d	`\> c \> d \\`
e	`\>\> e`
	`\end{tabbing}`

The tabbing environment provides a few more commands that are used less commonly. The command \+ makes all following lines start from the second tab stop so that one has to specify one less \>. On the other hand, the command \- cancels the effect of the previous \+ for the following lines, while the command \< cancels the effect of the previous \+ command only for the current line. Note that this command can be used only at the beginning of a line. The command \' causes the text preceding the command to be aligned to the right (the space between columns can be defined by the \tabbingsep, which is a predefined length variable). The corresponding command \` moves the rest of the current line to the right. The line must be ended with \\. One can create a local scope for defining new tab stops with the command \pushtabs. The command \poptabs closes the local scope opened with the last \pushtabs command.

Since the tabbing environment redefines the meaning of the commands \=, \', and \',
it provides the commands \a=, \a', and \a', respectively, whose functionality is the
same as that of the original commands. We conclude the section with an example that
demonstrates the use of the commands just described:

```
\begin{tabbing}
12 \= 123 \= 1234 \= \kill
1 \> 2 \> 3 \> 4 \\
a \> b \+ \\
b\> c\> \- \\
\a'{o} \> \a'{e} \> \a={a}\\[12pt]
\pushtabs
1234\=123\=12\=\kill
1 \>  2 \> 3 \> 4\\
\poptabs
1 \> 2 \\
a \' b \\
b \> c \' a
\end{tabbing}
```

```
1 2  3    4
a  b
     b  c
ó  è    ā

1    2 3 4
1 2
a  b
     b  c
                    a
```

4.8.2 The tabular Environment

The tabular environment can be used to typeset tables with optional horizontal and
vertical lines that separate rows and columns, respectively. Since LATEX is a markup
language, it automatically determines the width of the columns. As usual, we start
with a simple example:

7d2	Hexadecimal
3722	Octal
11111010010	Binary
2002	Decimal

```
\begin{tabular}{|r|l|}
\hline
7d2          & Hexadecimal\\
3722         & Octal\\
11111010010 & Binary\\
\hline \hline
2002         & Decimal \\
\hline
\end{tabular}
```

The *table spec* of the command

$$\backslash\text{begin\{tabular\}\{}\textit{table spec}\text{\}}$$

defines the format of the table. We use the letter l for a column of left-aligned text,
the letter r for a column of right-aligned text, and the letter c for a column of centered
text. The | symbol denotes that a vertical line will be drawn on the left-hand and/or
the right-hand side of a column. In the body of the tabular environment, the character

& is used to separate the elements of each row, the command **** starts a new line, and the command **\hline** draws a horizontal line between rows of the table. Note that the **\hline** command must be specified immediately after the **** command. The *table spec* allows the use of the p{*width*} construct for a column containing justified text with line breaks; the text will be typeset on lines with length equal to *width*. For example,

```
\begin{tabular}{|p{100pt}|}
\hline
Here is how one can
create a boxed
paragraph.\\
\hline
\end{tabular}
```

Here is how one can create a boxed paragraph.

With the @{...} construct, it is possible to specify a column separator. This column specifier ignores the intercolumn space and replaces it with whatever is specified inside the curly brackets. If we want to suppress leading space in a table, we can use the @{} column specifier:

```
\begin{tabular}{@{}c@{}}
\hline
no leading space\\
\hline
\end{tabular}
```

no leading space

```
\begin{tabular}{c}
\hline
leading and trailing space\\
\hline
\end{tabular}
```

leading and trailing space

We will come back to this column specifier later in this section. The column specifier *{*num*}{*cols*} is equivalent to *num* copies of *cols*, where *num* is any positive integer and *cols* is any list of column specifiers, including the * column specifier. In the following example, we show how one can draw a poor man's commutative diagram[1] using the column specifier that we just described. The example uses the command \vline to draw the vertical lines. This command, when used within an r, l, or c item, produces a vertical line extending the full height (and depth) of its row. Note also that we use negative vertical spacing to bring the arrowhead closer to the arrow body.

---

1. Of course, we can produce much nicer diagrams using special LATEX packages (see Section 5.4.11).

```
A  - - - ->  B          \begin{tabular}
^            ^          {c*{4}{@{---}}@{\texttt{>}}c}
|            |          A & B \\ \begin{tabular}[b]{c}
C  - - - ->  D          {\Large\textasciicircum}\\[-11pt]
                        \vline\\[10pt]  C \end{tabular} &
                        \begin{tabular}[b]{c}
                        {\Large\textasciicircum}\\[-11pt]
                        \vline\\[10pt] D \end{tabular}
                        \end{tabular}
```

The \multicolumn{num}{col}{item} makes *item* the text of a single column that temporally replaces *num* columns and is positioned as specified by col. In case *num* is equal to one, the command serves simply to override the item positioning specified by the environment argument. The col argument must contain exactly one of r, l, or c and may contain one or more @{...} column specifiers. The \cline{colA-colB} command draws a horizontal line across columns colA through colB. Two or more \cline commands draw their lines in the same vertical position. The following example demonstrates the use of these commands:

SUN Microsystems Stock	
	Price
Month	low|high
Jan 2001	25.438–34.875
Dec 2000	26.938–45.875
Nov 2000	39.875–56.532

```
\begin{tabular}{|r||r@{--}l|}
\hline
\multicolumn{3}{|c|}
{SUN Microsystems Stock}
\\ \hline\hline
& \multicolumn{2}{c|}{Price} \\
\cline{2-3}
\multicolumn{1}{|c||}{Month} &
\multicolumn{1}{r@{\vline}}{low}
& high\\ \hline
Jan 2001 & 25.438 & 34.875\\
\hline
Dec 2000 & 26.938 & 45.875\\
\hline
Nov 2000 & 39.875 & 56.532\\
\hline
\end{tabular}
```

(The data have been collected from http://www.nasdaq.com.)

The following exercise assumes familiarity with TEX math mode. Do it after you have read enough on the subject.

➤ **Exercise 4.9** Draw the following *proof-net* [7]:

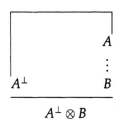

$$A^\perp \otimes B$$

[Hint: Use the \phantom command to fool LaTeX (see page 186).] ☐

The array environment is the math mode equivalent of the tabular environment. As such, it can be used only in math mode. Both environments can take an optional argument that specifies the vertical positioning: the default is alignment on the center of the environment. Letters t and b (i.e., the possible optional arguments) denote alignment on the top and bottom rows, respectively. The following example clearly shows the difference:

1			\begin{tabular}[b]{c}
1			1 \\ 1 \\ 1 \\ 1 \\ 1
1	2		\end{tabular}
1	2		\begin{tabular}{c}
1	2	3	2 \\ 2 \\ 2 \\ 2 \\ 2
	2	3	\end{tabular}
	2	3	\begin{tabular}[t]{c}
		3	3 \\ 3 \\ 3 \\ 3 \\ 3
		3	\end{tabular}

Note that each tabular environment is treated by TeX as a box (see Section 6.10). There is also a starred version of the tabular environment, which has the following general form:

\begin{tabular*}{*width*}[*pos*]{*table spec*}   rows \end{tabular*}

where *width* is the width of the table.

The following parameters can be changed anywhere outside an array or tabular environment. They can be changed locally within an item. but the changes must be delimited by braces or an environment. By changing these parameters, one affects how the array and tabular environments create tables.

\arraycolsep One-half of the width separating columns in an array environment. This parameter can be set with

\setlength\arraycolsep{\textit{len}}

The same applies to all other lengths presented in this paragraph.

\tabcolsep One-half of the width separating columns in a tabular environment.

\arrayrulewidth The width of rules.

\doublerulesep The space between adjacent rules in array or tabular environments.

\arraystretch Line spacing in array and tabular is done by placing the
strut

$$\text{\arraystretch} \times \text{\strut}$$

in every row. The default definition of this parameter is as follows:

$$\text{\newcommand\{\arraystretch\}\{1\}}$$

Consequently, its value can be changed with a redefinition.

\extrasep{*width*} This parameter is for use inside an @{...} column specifier. It causes *width* space to be added between columns for the rest of
the columns.

4.9   More on Alignment

The array package by Frank Mittelbach and David Carlisle provides an extended
reimplementation of the LaTeX array and tabular environments. Based on the
extended functionality provided by this package, David Carlisle has created the
packages dcolumn, delarray, hhline, and tabularx. In this section, we will describe all
of these packages and two other packages that are equally important: the package
supertabular by Theo Jurriens and Johannes Braams, and the package longtable by
David Carlisle.

The dcolumn package

This package provides a mechanism for defining column entries in an array or
tabular environment that are to be aligned on a "decimal point." The package
defines D to be a column specifier with three arguments:

$$\text{D\{}sep\text{-}tex\text{\}\{}sep\text{-}dvi\text{\}\{}decimal\ places\text{\}}$$

sep-tex will usually be '.' or ',', but, in general, it can be any single character. The
argument *sep-dvi* is used as the separator in the output file. It should be noted that
the package always uses math mode for the digits and the separator. The maximum
number of decimal places in the column is specified by *decimal places*. One may
not want to use all three arguments in the *table spec* of an array or a tabular
environment. In this case, one can create a column specifier using the command
\newcolumntype:

```
\newcolumntype{d}[1]{D{.}{\cdot}{#1}}
```

Here is a simple demonstration:

Pi expression	Value
π	3.1416
π^{π}	36.46
$(\pi^{\pi})^{\pi}$	80662.7

```
\begin{tabular}{cD{.}{.}{4}}
Pi expression & \mbox{Value}\\
$\pi$         & 3.1416\\
$\pi^{\pi}$   & 36.46\\
$(\pi^{\pi})^{\pi}$ & 80662.7
\end{tabular}
```

The *delarray* package

This package allows the construction of arrays surrounded by delimiters without the need to explicitly use the commands \left and \right. One has to put the left and the right delimiters before and after the column specifier, respectively:

$$\left\lfloor \begin{array}{cc} a & b \\ c & d \end{array} \right\rfloor$$

```
\begin{array}\lfloor{cc}\rfloor
a & b\\
c & d
\end{array}
```

As one would expect, a null delimiter is denoted by ".". In case one wants to provide a function definition, it is possible to write a construct such as the following one:

$$\chi_A(x) = \begin{cases} 1 & \text{if } x \in A \\ 0 & \text{otherwise} \end{cases}$$

```
\newcolumntype{L}{>{$}l<{$}}
\chi_{A}(x)=
\begin{array}\{{lL}.
1 & if $x\in A$\\
0 & otherwise
\end{array}
```

(See the discussion on the array package for an explanation of the >{$}l<{$} column specifier.) In the case where we want to use the optional b and t placement specifiers, this package takes special care that the brackets have the expected height and depth.

The *hhline* package

This package provides the command \hhline, which produces a line like the command \hline, but it takes special care of intersections with vertical lines. The arguments to \hhline are similar to the column spec of an array or a tabular environment. It consists of a list of symbols with the following meanings:

= A double \hline the width of a column.

- A single \hline the width of a column.

~ A column with no \hline.

| A \vline that "cuts" through a double (or single) \hline.

: A \vline that is broken by a double \hline.

\# A double \hline segment between two \vlines.

t The top half of a double hline segment.

b The bottom half of a double hline segment.

* *{3}{==\#} expands to ==\#==\#==\#, as in the *-form for the column specification.

If a double vertical line is specified, then the horizontal lines produced by \hhline are broken. To obtain the effect of a horizontal line "cutting through" the double line, one has to use a \# or omit the vertical line specifiers. An example using most of the features of the \hhline command is

a	b	c	d
1	2	3	4
i	j	k	l
w	x	y	z

```
\begin{tabular}{||cc||c|c||}
\hhline{|t:==:t:==:t|}
a&b&c&d\\
\hhline{|:==:|~|~||}
1&2&3&4\\
\hhline{#==#~|=#}
i&j&k&l\\
\hhline{||--||--||}
w&x&y&z\\
\hhline{|b:==:b:==:b|}
\end{tabular}
```

The *array* package

This package provides a new implementation of the tabular and array environments. It introduces a number of new column specifiers, which are shown in Table 4.2. The new parameter \extrarowheight can be used to enlarge the normal height of every row of a table. This parameter is a length variable so it can be set with \setlength. We will now briefly discuss a few examples that can be done using the new column specifiers.

- If one wants to use a special font in a left-justified column, one can use the >{font}l column specifiers, where font is any font selection command.

- One can change the paragraph indentation in any column generated with the p, m, or b column specifiers with the command \setlength. Remember that the paragraph indentation is stored in parameter \parindent.

- If one uses the idiom >{$}c<{$} as a column specifier in an array environment, then a centered column in text mode is produced.

In the case where one wants to create a shorthand for a long column specifier, one can use a \newcolumntype. For example, the command

$$\newcolumntype{L}{>{\$}l<\{\$\}}$$

is used to define a new column specifier, namely L, that can be used in an array environment to get left-justified columns in text mode. A list of all current active \newcolumntype definitions is sent to the terminal and the log file if the \showcols

Table 4.2: New column specifiers introduced by the array package.

New Column Specifiers	
m{width}	Defines a column of width width. Every entry will be centered in proportion to the rest of the line. It is somewhat like \parbox{width}.
b{width}	Coincides with \parbox[b]{width}.
>{decl.}	Can be used before an l, r, c, p, m, or a b option. It inserts decl. directly in front of the entry of the column.
<{decl.}	Can be used after an l, r, c, p{..}, m{..}, or a b{..} option. It inserts decl. right after the entry of the column.
|	Inserts a vertical line. The distance between two columns will be enlarged by the width of the line, in contrast to the original definition of LATEX.
!{decl.}	Can be used anywhere and corresponds to the | option. The difference is that decl. is inserted instead of a vertical line, so this option does not suppress the space normally inserted between columns, in contrast to @{...}.

command is given. If a tabular environment occurs inside another and starts with a \hline and ends with the same command, then it is a good practice to replace the first \hline with the command \firsthline and the last one with the command \lasthline.

The tabularx package

This package defines a new environment, tabularx, which takes the same arguments as the tabular* environment. The column specifiers of tabularx are essentially the same as those of the standard tabular* environment. Their real difference lies in the fact that tabular* adds space between columns to achieve the desired width and tabularx adjusts the width of some of the columns, which are specified with the column specifier X. If one tabularx environment occurs inside another, the inner one must be enclosed in curly brackets. Since the body of the environment is actually the argument of a command, one must avoid certain constructs that cannot be command arguments (e.g., \verb).

By default, the X column specifier is turned into p{...}. Since such columns often require a special format, this may be achieved using the >{...} column specifier provided by the array package (e.g., >{\small}X). A format that is useful in cases such as this is ragged right. But, the declaration \raggedright redefines \\ in a way that conflicts with its use in a tabular environment. For this reason, the package introduces the command \arraybackslash, which may be used after a \raggedright. If you want to see all of the computations performed by the package, try the declaration \tracingtabularx.

The *supertabular* package

This package defines the `supertabular` environment. The environment is an extension of the `tabular` environment that solves one major drawback of this environment: a `tabular` environment must fit on one page. If the environment becomes too large, the text overwrites the page's bottom margin and one gets an Overfull vbox message. The new environment accepts the same column specifiers as the `tabular` environment, but one must be careful to end each line with a \\.

The package also defines the following environments: `supertabular*`, `mpsupertabular` and `mpsupertabular*`. The environment `supertabular*` works much like the `tabular*` environment, whereas the two other environments are for use within a `minipage` environment.

The `supertabular` package allows the use of three options that control the amount of information written in the log file:

`errorshow` It is the default option that does not actually write any extra information.

`pageshow` When the environment decides to break a page if this option is active, then the environment explains where and why it actually breaks a page.

`debugshow` This option also shows information regarding each line that is added to the tabular arrangement.

The package introduces a few new commands, which are explained below:

`\tablefirsthead` This command takes a single argument and defines the contents of the first occurrence of the tabular head.

`\tablehead` This command takes one argument and defines the contents of all subsequent occurrences of the tabular head.

`\tabletail` It takes one argument and defines something that will be appended to tabular just before a page break occurs.

`\tablelasttail` This command takes a single argument and defines something that will be appended at the end of the table.

The package provides three commands to introduce captions in a table: `\topcaption`, `\bottomcaption`, and `\tablecaption`. The command `\shrinkheight` can be used to adjust the allowed maximum and minimum height of a part of the `supertabular` on a page. It has one argument, the length that will shrink (positive value) or grow (negative value) the allowed height. Here is a rather complete usage example:

```
\tablefirsthead{%
\hline
\multicolumn{4}{c}{Perl's Operator}\\
\hline
\multicolumn{1}{c|}{Associativity}&
\multicolumn{1}{l|}{Arity} &
\multicolumn{1}{l|}{Precedence Class}&
```

```
\multicolumn{1}{1}{Precedence}\\
\hline}
%
\tablehead{%
\hline
\multicolumn{4}{c}{\small\slshape continued from previous page}\\
\hline
\multicolumn{1}{c|}{Associativity}&
\multicolumn{1}{l|}{Arity} &
\multicolumn{1}{l|}{Precedence Class} &
\multicolumn{1}{l}{Precedence}\\ \hline}
%
\tabletail{%
\hline
\multicolumn{4}{c}{\small\slshape continued on next page}\\
\hline}
%
\tablelasttail{\hline}
%
\bottomcaption{Perl's Operator}
%
\begin{supertabular}{c|l|l|l}
None & 0      & terms and list oper. (leftward) & 0\\
Left & 2      & \verb|->| & 1 \\
..................................................
\end{supertabular}
```

The output of this example is shown rotated in Figure 4.3 (page 91).

The *longtable* package

This package defines the environment longtable. The functionality of this environment is similar to that of the supertabular environment. It is possible to have a table caption generated by a \caption command that can appear in the list of tables (generated by a \listoftables command). The value of the counter \LT-chunksize is used to break the table into chunks that contain a number of rows equal to this value. This way, TeX is not forced to keep the whole table in its memory and therefore makes it possible to handle large tables in TeX installations with limited memory. The default value of this counter is 20. At the start of the table, one may specify lines that are to appear at the top or bottom of every page. The lines are entered as normal, but the last \\ is replaced by the appropriate command: \endhead (for each table head), \endfirsthead (for the first table head), \endfoot (for each table foot), and \endlastfoot (for the last table foot). It is possible to have an effect similar to that achieved by the \kill command of the tabbing environment by using a reimplementation of the \kill command, provided by the longtable package.

Here is the example table of the previous section coded using longtable:

```
\begin{longtable}{c|l|l|l}
```

```
\hline
\multicolumn{4}{c}{Perl's Operator}\\
\hline
Associativity & Arity & Precedence Class & Precedence\\
\hline
\endfirsthead
\hline
\multicolumn{4}{c}{\small\slshape continued from previous page}\\
\hline
Associativity & Arity & Precedence Class & Precedence\\
\hline
\endhead
\hline
\multicolumn{4}{c}{\small\slshape continued on next page}\\
\hline
\endfoot
\hline
\endlastfoot
None & 0     & terms and list oper. (leftward) & 0\\
Left & 2     & \verb|->| & 1 \\
...................................................
```

The output of this example is shown rotated in Figure 4.3 (page 91).

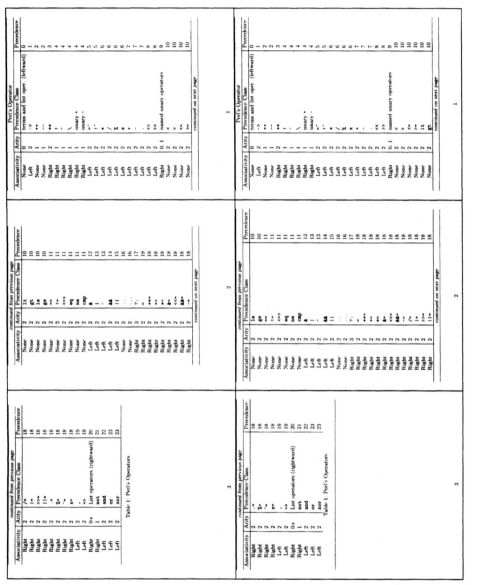

Figure 4.3: The same table typeset with supertabular (left column) and longtable (right column).

5

Typesetting Mathematics

One of the most demanding jobs in the typesetting business is the typesetting of mathematical text. Here, TeX really excels. Its output is incomparable to the output of any other document preparation system. LaTeX, as usual, adds an easier interface to the TeX typesetting engine, making the writing of even the most demanding mathematical text straightforward.

The American Mathematical Society has made an enormous effort to create LaTeX document classes and packages that deal even better with several issues compared to the usual `article` and `book` document classes. They have created the `amsart` and `amsbook` classes, and packages that provide the basic functionality of these classes and work with any document class. The advantages are really important for mathematicians or others who use mathematics in their work, so we thoroughly discuss these classes and packages.

5.1   The Mathematics Mode

When we want to write mathematical text, we must inform LaTeX of our intentions. This is done by surrounding the mathematical text with the dollar character: $. LaTeX takes special care about spacing in math mode even if we write something simple, leading to professional mathematical typesetting. Take a look at the following:

$$10/5=2 \text{ and } 7\text{-}3\text{=}4$$
$$10/5 = 2 \text{ and } 7 \quad 3 = 4.$$

The second line looks much better, and it was produced by the input

```
$10/5=2$ and $7-3=4$.
```

When we talk about the "mathematics mode," we must bear in mind that TeX typesets in six different modes. A "mode" is the way TeX reads its input text. There are actually two mathematical modes. We have already shown the first of these—that is, text surrounded by single dollar characters. The other mode is when we want to typeset mathematics in a "display" (that is, outside the text paragraph) surrounded by (vertical) white space, like this:

$$1 + 2 + 3 + \cdots + n = n(n+1)/2.$$

This is achieved by using double dollar characters instead of single ones. The display above was produced by

$$1+2+3+\cdots+n=n(n+1)/2.$$

The commands \(and \) can be used to enter and leave the math mode. In addition, these commands check that we are not actually in math mode, so they prevent possible errors. Of course, if we are already in math mode, they cause LATEX to stop and print an error message (see Table 11.1). If you do not feel comfortable with these commands, then you can use the math environment. Similarly, the commands \[and \] enter and leave the display math mode and do similar checks. The corresponding environment is the displaymath environment.

5.2 Font Selection in Mathematics Mode

Selecting shape in math mode is achieved by the commands \mathrm, \mathit, and \mathcal, and it affects only text (not symbols). These commands work like \textrm or \textit but in math mode. In addition, spaces are not observed, and no hyphenation is applied. Thus, $\mathrm{offer it}$ will produce offerit and $\mathcal{OFFER IT}$ will give $\mathcal{OFFERIT}$.

The selection of series such as bold is not always supported. The default (Computer Modern) fonts do provide bold series for many mathematical symbols, but this may not be available for other fonts. The way to write mathematics in bold is to switch to bold before entering the math mode with the command \mathversion{bold}. This changes to bold for all mathematics (including symbols) until it is disabled with \math-version{normal}. If inside normal math we would like to enter text in bold, then the easiest way is to switch to text mode locally using an escape or to use the \mathbf command. Here is an example that uses all of the above:

$\forall x \in \mathcal{A}$ we have $x^2 = 1$	`$$\forall \mathbf{x}\in\mathcal{A}` `\textrm{ we have } \mathbf{x}^2 =1$$` `\mathversion{bold}`
$\forall x \in \mathcal{A}$ **we have** $x^2 = 1$	`$$\forall x\in\mathcal{A}` `\textbf{ we have } x^2 =1$$` `\mathversion{normal}`
$\forall x \in \mathcal{A}$ we have $x^2 = 1$	`$$\forall \mathbf{x}\in` `\mathcal{A} \textrm{ we have }\` `\mathbf{x}^2=1$$`

Note that the commands \boldmath and \unboldmath are shorthands for the bold and the normal \mathversions.

The selection of sans serif and typewriter fonts in math mode is done with the commands \mathsf and \mathtt, respectively. Thus,

$\mathsf{A}^i_{\mathsf{j}} =\mathtt{W}(\mathtt{\alpha})$

will be set as $A^i_j = W(\alpha)$. The commands \textrm, \textit, and so forth work fine in math mode, and they observe spaces. They also observe language, so

$x^2=-1 \textbf{ and thus } x=\pm i$
$x^2=-1 \textrm{ \textgreek{>'ara} } x=\pm i$
$f_n \stackrel{\textrm{\textgreek{oμ}}}{\longrightarrow} f$.

produces

$$x^2 = \quad 1 \text{ and thus } x = \pm i$$
$$x^2 = \quad 1 \text{ ἄρα } x = \pm i$$
$$f_n \xrightarrow{\text{ομ}} f.$$

We should also add here that if the family has been chosen, the commands \mathrm, \textrm, and so on, will observe this. Therefore,

\sffamily We get $x^2=-1 \textbf{ and thus } x=\pm i$

will be set as

$$\text{We get } x^2 = \quad 1 \text{ and thus } x = \pm i$$

that is, the "and thus" phrase was set in sans serif family and bold series.

5.3 Symbols for the Mathematics Mode

One of the most difficult parts of typesetting mathematics is the wealth of mathematical symbols needed even for simple texts. By "symbols" we mean glyphs such as ∀, ∃, →, and so on. In addition to symbols, mathematicians need different alphabets, as it is customary to use different fonts for certain tasks such as script capitals $\mathcal{A}, \mathcal{B}, \mathcal{C}, \mathcal{D}$ or Greek letters $\alpha, \delta, \epsilon, \varepsilon$.

5.3.1 Special Latin Alphabets

In Section 5.2, we saw a calligraphic alphabet provided by the \mathcal command. Sometimes, though, one needs an even more scripted font. Such a font is provided by the mathrsfs package (by Jörg Knappen). The package provides the command \mathscr, which gives access to script capitals. For example, $\mathscr{MATHEMATICS}$ produces $\mathscr{MATHEMATICS}$. A similar option is to use the eucal package of the $\mathcal{AMS}$. If the package is used, the option `mathscr` provides both the standard `\mathcal` and the `\mathscr` commands, producing a different script font: $\mathscr{MATHEMATICS}$.

Another commonly used font is the blackboard bold font:

$$\mathbb{ABCDEFGHIJKLMNOPQRSTUVWXYZ}$$

which is available with the amsfonts or amssymb package and can be accessed with the `\mathbb` command. Notice that this command will not work for Greek or lowercase letters. If one has such a need, one can use the package mathbbol by Jörg Knappen. The package provides lowercase and Greek blackboard bold. One should use the package as follows:

`\usepackage[bbgreekl]{mathbbol}`

We should also note that this package uses a different font for the `\mathbb` command. Here is an example:

ABCDEFGHIJKLM	`\begin{center}`
NOPQRSTUVWXYZ	`$\mathbb{ABCDEFGHIJKLM}$\\`
abcdefghijklm	`$\mathbb{NOPQRSTUVWXYZ}$\\`
nopqrstuvwxyz	`$\mathbb{abcdefghijklm}$\\`
0123456789	`$\mathbb{nopqrstuvwxyz}$\\`
ΔΘΛΞΠΣΦΨΩ	`$$\mathbb{0123456789}$\\`
αβγδε(ηθικλμ	`$\mathbb{\Gamma\Delta\Theta`
νξοπρστυφχψω	`\Lambda\Xi\Pi\Sigma\Phi\Psi`
	`\Omega}$\\`
	`$\bbalpha\bbbeta`
	`\bbgamma\bbdelta\bbespilon`
	`\bbzeta\bbeta\bbtheta\bbiota`
	`\bbkappa\bblambda\bbmu$\\`
	`$\bbnu\bbxi\mathbb{o}\bbpi`
	`\bbrho\bbsigma\bbtau`
	`\bbupsilon\bbphi\bbchi`
	`\bbpsi\bbomega$`
	`\end{center}`

Note that the command `\bbespilon` is misspelled in the current version of the mathbbol package (`\bbespilon` instead of `\bbepsilon`).

5.3.2 The Greek Letters

The Greek letters are shown in Table 5.1. The rest of the capital Greek letters are in common (in design) with the Latin letters. For example, it would make no sense to provide a command such as `\Alpha`, as in Greek this letter is identical to the Latin A.

Table 5.1: The Greek letters in math mode.

α	\alpha	β	\beta	γ	\gamma	δ	\delta
ϵ	\epsilon	ε	\varepsilon	ζ	\zeta	η	\eta
θ	\theta	ϑ	\vartheta	ι	\iota	κ	\kappa
λ	\lambda	μ	\mu	ν	\nu	ξ	\xi
o	o	π	\pi	ϖ	\varpi	ρ	\rho
ϱ	\varrho	σ	\sigma	ς	\varsigma	τ	\tau
υ	\upsilon	ϕ	\phi	φ	\varphi	χ	\chi
ψ	\psi	ω	\omega				
Γ	\Gamma	Δ	\Delta	Θ	\Theta	Λ	\Lambda
Ξ	\Xi	Π	\Pi	Σ	\Sigma	Υ	\Upsilon
Φ	\Phi	Ψ	\Psi	Ω	\Omega		

Table 5.2: $\mathcal{AMS}$ Greek and Hebrew (amssymb package).

F	\digamma	$\varkappa$	\varkappa	$\beth$	\beth	$\daleth$	\daleth	$\gimel$	\gimel

The amssymb package provides two additional Greek letters and three Hebrew letters (in math mode). These are shown in Table 5.2. One more Hebrew letter is available in the standard LATEX fonts. This is ℵ, which is accessed with the command \aleph.

5.3.3 Accents in Math Mode

When we want to put an accent above a letter in math mode, we use special commands such as \vec{a} in order to get $\vec{a}$. This practice is also used for "usual" accents such as the acute. We write \acute{a} to get $\acute{a}$. One may think that these commands are not needed for standard accents if one has a keyboard that can access these glyphs. This is wrong, and the reason is that these commands work even if you want to put an accent above a glyph that the fonts do not provide accented. For example, we can write $\acute{\gamma}$ or $a\dot{+}b$. Moreover, the commands can be combined to give multiaccented letters. Here is an example:

Han Thế Than | Han Th$\acute{\hat{\textrm{e}}}$ Than

Table 5.3 gives all of the relevant commands.

A better ring accent is provided by the yhmath package (by Yannis Haralambous). The command is \ring, and the output of, say, \ring{B}, is $\mathring{B}$. This package provides additional functionality that we will discuss later on.

Some accents have a wide form that is useful for more than one letter. These are \widehat and \widetilde, giving $\widehat{xy}$ and $\widetilde{xy}$, respectively.

Table 5.3: Accents in math mode.

â	\hat{a}	á	\acute{a}	ā	\bar{a}	ȧ	\dot{a}
ă	\breve{a}	ǎ	\check{a}	à	\grave{a}	ā	\vec{a}
ä	\ddot{a}	ã	\tilde{a}	å	\mathring{a}		

5.3.4 Binary Operators

Binary operators are symbols such as + or   that are used between two *operands*, which, in turn, are numbers, letters, or formulas. The predefined binary operators that LaTeX provides are shown in Table 5.4. Note that the commands marked with (*) are only available if we use the latexsym package. In particular, one can get the symbols •, · and ∗ in ordinary text mode by using the commands \textbullet, \textperiodcentered, and \textasteriskcentered, respectively.

Additional binary operators are provided by the packages amssymb and amsfonts. These are given in Table 5.5.

Table 5.4: Binary operators.

±	\pm	∩	\cap	◇	\diamond	⊕	\oplus
∓	\mp	∪	\cup	△	\bigtriangleup	⊖	\ominus
×	\times	⊎	\uplus	▽	\bigtriangledown	⊗	\otimes
÷	\div	⊓	\sqcap	◁	\triangleleft	⊘	\oslash
∗	\ast	⊔	\sqcup	▷	\triangleright	⊙	\odot
⋆	\star	∨	\vee	◁	\lhd(*)	○	\bigcirc
∘	\circ	∧	\wedge	▷	\rhd(*)	†	\dagger
•	\bullet	\	\setminus	⊴	\unlhd(*)	‡	\ddagger
·	\cdot	≀	\wr	⊵	\unrhd(*)	II	\amalg

Table 5.5: AMS binary operators (amssymb package).

∔	\dotplus	╲	\smallsetminus	⋒	\Cap
⋓	\Cup	⊼	\barwedge	⩒	\veebar
⩟	\doublebarwedge	⊟	\boxminus	⊠	\boxtimes
	\boxdot	⊞	\boxplus	⁂	\divideontimes
⋉	\ltimes	⋊	\rtimes	⋋	\leftthreetimes
⋌	\rightthreetimes	⋏	\curlywedge	⋎	\curlyvee
⊖	\circleddash	⊛	\circledast	⊚	\circledcirc
.	\centerdot	⊺	\intercal		

5.3.5 Variable-Size Operators

Variable-size operators are operators whose size changes according to the math mode in which they are used. Both

$$\sum \text{ and } \sum$$

are produced by the \sum command, but the first one is in text math mode and the second one is in display math mode. The available symbols that behave this way are shown in Table 5.6

Table 5.6: Symbols of variable size.

$\sum$	\sum	$\prod$	\prod	$\coprod$	\coprod
$\bigcap$	\bigcap	$\bigcup$	\bigcup	$\bigsqcup$	\bigsqcup
$\bigodot$	\bigodot	$\bigotimes$	\bigotimes	$\bigoplus$	\bigoplus
$\int$	\int	$\bigvee$	\bigvee	$\biguplus$	\biguplus
$\oint$	\oint	$\bigwedge$	\bigwedge		

5.3.6 Delimiters

Delimiters are symbols that are used mainly for grouping. These are shown in Tables 5.7, 5.8, and 5.9.

Table 5.7: Delimiters.

↑	\uparrow	⇑	\Uparrow	↓	\downarrow	⇓	\Downarrow
{	\{	}	\}	↕	\updownarrow	⇕	\Updownarrow
⌊	\lfloor	⌋	\rfloor	⌈	\lceil	⌉	\rceil
⟨	\langle	⟩	\rangle	/	/	\	\backslash
|	|	‖	|				

Table 5.8: $\mathcal{AMS}$ delimiters (amssymb package).

⌜	\ulcorner	⌝	\urcorner	⌞	\llcorner	⌟	\lrcorner

5.3.7 Arrows

The available arrow symbols are shown in Tables 5.10, 5.11, and 5.12.

Table 5.9: Large delimiters.

⎱	\rmoustache	⎰	\lmoustache	⎱	\rgroup	⎰	\lgroup
|	\arrowvert	‖	\Arrowvert	⎮	\bracevert		

Table 5.10: Arrow symbols (starred symbols are available with the latexsym package).

←	\leftarrow	⟵	\longleftarrow	↑	\uparrow
⇐	\Leftarrow	⟸	\Longleftarrow	⇑	\Uparrow
→	\rightarrow or \to	⟶	\longrightarrow	↓	\downarrow
⇒	\Rightarrow	⟹	\Longrightarrow	⇓	\Downarrow
↔	\leftrightarrow	⟷	\longleftrightarrow	↕	\updownarrow
⇔	\Leftrightarrow	⟺	\Longleftrightarrow	⇕	\Updownarrow
↦	\mapsto	⟼	\longmapsto	↗	\nearrow
↩	\hookleftarrow	↪	\hookrightarrow	↘	\searrow
↼	\leftharpoonup	⇀	\rightharpoonup	↙	\swarrow
↽	\leftharpoondown	⇁	\rightharpoondown	↖	\nwarrow
⇝ (*)	\leadsto				

Table 5.11: $\mathcal{AMS}$ arrows (amssymb package).

⇢	\dashrightarrow	⇠	\dashleftarrow	⇇	\leftleftarrows
⇆	\leftrightarrows	⇚	\Lleftarrow	↞	\twoheadleftarrow
↢	\leftarrowtail	↫	\looparrowleft	⇋	\leftrightharpoons
↶	\curvearrowleft	↺	\circlearrowleft	↰	\Lsh
⇈	\upuparrows	↼	\upharpoonleft	↓	\downharpoonleft
⊸	\multimap	↭	\leftrightsquigarrow	↬	\looparrowright
⇄	\rightleftarrows	⇉	\rightrightarrows	↻	\circlearrowright
↠	\twoheadrightarrow	↷	\curvearrowright	↾	\upharpoonright
⇌	\rightleftharpoons	⇊	\downdownarrows	⇛	\Rrightarrow
↱	\Rsh	⇝	\rightsquigarrow		
↓	\downharpoonright	↣	\rightarrowtail		

Table 5.12: $\mathcal{AMS}$ negated arrows (amssymb package).

↚	\nleftarrow	↛	\nrightarrow	⇍	\nLeftarrow
⇏	\nRightarrow	↮	\nleftrightarrow	⇎	\nLeftrightarrow

5.3.8 Relational Operators

Again, there are two sets of such operators: the standard ones (Table 5.13) and the ones provided by the amssymb package (Table 5.14).

Table 5.13: Relation operators (starred symbols are available with the latexsym package).

≤	`\leq or \le`	≥	`\geq`	≡	`\equiv`	⊨	`\models`
≺	`\prec`	≻	`\succ`	~	`\sim`	⊥	`\perp`
⪯	`\preceq`	⪰	`\succeq`	≃	`\simeq`	|	`\mid`
≪	`\ll`	≫	`\gg`	≍	`\asymp`	‖	`\parallel`
⊂	`\subset`	⊃	`\supset`	≈	`\approx`	⋈	`\bowtie`
⊆	`\subseteq`	⊇	`\supseteq`	≅	`\cong`	⋈(*)	`\Join`
⊏	`\sqsubset`	⊐	`\sqsupset`	≠	`\neq`	⌣	`\smile`
⊑	`\sqsubseteq`	⊒	`\sqsupseteq`	≐	`\doteq`	⌢	`\frown`
∈	`\in`	∋	`\ni`	∝	`\propto`	=	`=`
⊢	`\vdash`	⊣	`\dashv`	<	`<`	>	`>`

Table 5.14: 𝒜ℳ𝒮 relational operators (amssymb package).

≦	`\leqq`	⩽	`\leqslant`	⋖	`\eqslantless`
≲	`\lesssim`	⪅	`\lessapprox`	≊	`\approxeq`
⋖	`\lessdot`	⋘	`\lll`	≶	`\lessgtr`
⋚	`\lesseqgtr`	⪋	`\lesseqqgtr`	≑	`\doteqdot`
≓	`\risingdotseq`	≒	`\fallingdotseq`	∽	`\backsim`
⋍	`\backsimeq`	⫅	`\subseteqq`	⋐	`\Subset`
⊏	`\sqsubset`	≼	`\preccurlyeq`	⋞	`\curlyeqprec`
≾	`\precsim`	⪷	`\precapprox`	◁	`\vartriangleleft`
⊴	`\trianglelefteq`	⊨	`\vDash`	⊪	`\Vvdash`
⌣	`\smallsmile`	⌢	`\smallfrown`	≏	`\bumpeq`
≎	`\Bumpeq`	≧	`\geqq`	⩾	`\geqslant`
⋝	`\eqslantgtr`	≳	`\gtrsim`	⪆	`\gtrapprox`
⋗	`\gtrdot`	⋙	`\ggg`	≷	`\gtrless`
⋛	`\gtreqless`	⪌	`\gtreqqless`	≖	`\eqcirc`
≗	`\circeq`	≜	`\triangleq`	∼	`\thicksim`
≈	`\thickapprox`	⫆	`\supseteqq`	⋑	`\Supset`
⊐	`\sqsupset`	≽	`\succcurlyeq`	≿	`\curlyeqsucc`
≿	`\succsim`	⪸	`\succapprox`	▷	`\vartriangleright`
⊵	`\trianglerighteq`	⊩	`\Vdash`	∣	`\shortmid`
∥	`\shortparallel`	≬	`\between`	⋔	`\pitchfork`
∝	`\varpropto`	◀	`\blacktriangleleft`	∴	`\therefore`
϶	`\backepsilon`	▶	`\blacktriangleright`	∵	`\because`

The negated form of a relational operator (i.e., $\nleq$) is obtained by prepending the command \not to the relational operator. For instance, the code x\not\leq y is being typeset as $x \nleq y$. However, the $\mathcal{AMS}$ provides special fonts and commands to access negated relational operators, which are shown in Table 5.15.

Table 5.15: $\mathcal{AMS}$ negated relational operators (amssymb package).

$\nless$	\nless	$\nleq$	\nleq	$\nleqslant$	\nleqslant
$\nleqq$	\nleqq	$\lneq$	\lneq	$\lneqq$	\lneqq
	\lvertneqq	$\lnsim$	\lnsim	$\lnapprox$	\lnapprox
$\nprec$	\nprec	$\npreceq$	\npreceq	$\precnsim$	\precnsim
$\precnapprox$	\precnapprox	$\nsim$	\nsim	$\nshortmid$	\nshortmid
$\nmid$	\nmid	$\nvdash$	\nvdash	$\nvDash$	\nvDash
$\ntriangleleft$	\ntriangleleft	$\ntrianglelefteq$	\ntrianglelefteq	$\nsubseteq$	\nsubseteq
$\subsetneq$	\subsetneq	$\varsubsetneq$	\varsubsetneq	$\subsetneqq$	\subsetneqq
$\varsubsetneqq$	\varsubsetneqq	$\ngtr$	\ngtr	$\ngeq$	\ngeq
$\ngeqslant$	\ngeqslant	$\ngeqq$	\ngeqq	$\gneq$	\gneq
$\gneqq$	\gneqq	$\gvertneqq$	\gvertneqq	$\gnsim$	\gnsim
$\gnapprox$	\gnapprox	$\nsucc$	\nsucc	$\nsucceq$	\nsucceq
$\succnsim$	\succnsim	$\succnapprox$	\succnapprox	$\ncong$	\ncong
$\nshortparallel$	\nshortparallel	$\nparallel$	\nparallel	$\nvDash$	\nvDash
$\nVDash$	\nVDash	$\ntriangleright$	\ntriangleright	$\ntrianglerighteq$	\ntrianglerighteq
$\nsupseteq$	\nsupseteq	$\nsupseteqq$	\nsupseteqq	$\supsetneq$	\supsetneq
$\varsupsetneq$	\varsupsetneq	$\supsetneqq$	\supsetneqq	$\varsupsetneqq$	\varsupsetneqq

5.3.9 Miscellaneous Symbols

There are also some symbols that do not fit into any of the categories above, so we collectively present them in Tables 5.16 and 5.17. It is rather important to stress that the symbol ℧ presented in Table 5.16 is the archaic term for the unit of electrical conductance. The modern name of this unit is siemens (symbolized S).

Table 5.16: Miscellaneous (starred symbols are available with the latexsym package; double-starred symbols are available with the yhmath package.

…	\ldots	⋯	\cdots	⋮	\vdots	⋰	\ddots
ℵ	\aleph	′	\prime	∀	\forall	∞	\infty
ℏ	\hbar	∅	\emptyset	∃	\exists	∇	\nabla
√	\surd	□	\Box$^{(*)}$	△	\triangle	◇	\Diamond$^{(*)}$
ı	\imath	ȷ	\jmath	ℓ	\ell	¬	\neg
⊤	\top	♭	\flat	♮	\natural	♯	\sharp
℘	\wp	⊥	\bot	♣	\clubsuit	◇	\diamondsuit
♡	\heartsuit	♠	\spadesuit	℧	\mho$^{(*)}$	ℜ	\Re
ℑ	\Im	∠	\angle	∂	\partial	⋰	\adots$^{(**)}$

Table 5.17: $\mathcal{AMS}$ miscellaneous symbols (amssymb package; starred symbols not defined in old releases of the package).

ℏ	\hslash	△	\vartriangle	▽	\triangledown
◊	\lozenge	Ⓢ	\circledS	∠	\angle
∄	\nexists	⅃	\Finv$^{(*)}$	Ⅎ	\Game$^{(*)}$
‵	\backprime	⌀	\varnothing	▲	\blacktriangle
■	\blacksquare	◆	\blacklozenge	★	\bigstar
∁	\complement	ð	\eth	╱	\diagup$^{(*)}$
□	\square	∡	\measuredangle	𝕜	\Bbbk$^{(*)}$
▼	\blacktriangledown	∢	\sphericalangle	╲	\diagdown$^{(*)}$

5.3.10 More Math Symbols

More symbols for mathematics exist in add-on packages. For example the txfonts package by Young Ryu provides additional symbols for the Times font family. One might keep an eye on CTAN for other math fonts, as the collection there gets richer every day. In tables 5.18–5.23 we show the math symbols provided by the txfonts package.

Table 5.18: Delimiters provided by the txfonts package.

⟦ \llbracket	⟧ \rrbracket	⟨ \lbag	⟩ \rbag

Table 5.19: Math alphabets provided by the txfonts package.

g \varg	v \varv	w \varw	y \vary

Table 5.20: Binary operators provided by the txfonts package.

○ \medcirc	● \medbullet	⅋ \invamp
⦸ \circledwedge	⦶ \circledvee	⊝ \circledbar
⦸ \circledbslash	⊕ \nplus	⊞ \boxast
⊠ \boxbslash	⊟ \boxbar	⧄ \boxslash
≀ \Wr	⊔ \sqcupplus	⊓ \sqcapplus
▷ \rhd	◁ \lhd	⊵ \unrhd
⊴ \unlhd		

Table 5.21: Ordinary symbols provided by the txfonts package.

α \alphaup	β \betaup	γ \gammaup
δ \deltaup	ε \epsilonup	ε \varepsilonup
ζ \zetaup	η \etaup	θ \thetaup
ϑ \varthetaup	ι \iotaup	κ \kappaup
λ \lambdaup	μ \muup	ν \nuup
ξ \xiup	π \piup	ϖ \varpiup
ρ \rhoup	ϱ \varrhoup	σ \sigmaup
ς \varsigmaup	τ \tauup	υ \upsilonup
φ \phiup	φ \varphiup	χ \chiup
ψ \psiup	ω \omegaup	◇ \Diamond
◈ \Diamonddot	◆ \Diamondblack	ƛ \lambdaslash
ƛ \lambdabar	♧ \varclubsuit	♦ \vardiamondsuit
♥ \varheartsuit	♤ \varspadesuit	⊤ \Top
⊥ \Bot		

Table 5.22: Large operators provided by the txfonts package.

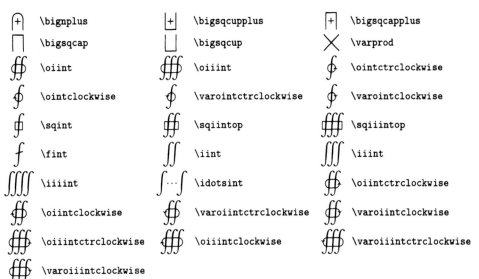

⊕	\bignplus	⊞	\bigsqcupplus	⊞	\bigsqcapplus
⊓	\bigsqcap	⊔	\bigsqcup	⨉	\varprod
∯	\oiint	∰	\oiiint	∮	\ointctrclockwise
∮	\ointclockwise	∮	\varointctrclockwise	∮	\varointclockwise
⨏	\sqint	∯	\sqiintop	∰	\sqiiintop
⨍	\fint	∬	\iint	∭	\iiint
⨌	\iiiint	∫⋯∫	\idotsint	∯	\oiintctrclockwise
∯	\oiintclockwise	∯	\varoiintctrclockwise	∯	\varoiintclockwise
∰	\oiiintctrclockwise	∰	\oiiintclockwise	∰	\varoiiintctrclockwise
∰	\varoiiintclockwise				

Table 5.23: Binary relations provided by the txfonts package (part 1).

⊄	\nsqsubset	⊅	\nsqsupset	⇠	\dashleftarrow
⇢	\dashrightarrow	⇠⇢	\dashleftrightarrow	↜	\leftsquigarrow
↠	\ntwoheadrightarrow	↞	\ntwoheadleftarrow	⇗	\Nearrow
↘	\Searrow	↖	\Nwarrow	⇙	\Swarrow
⊥⊥	\Perp	∼	\leadstoext	⤳	\leadsto
□→	\boxright	←□	\boxleft	⊡→	\boxdotright
←⊡	\boxdotleft	◇→	\Diamondright	←◇	\Diamondleft
◇→	\Diamonddotright	←◇	\Diamonddotleft	□⇒	\boxRight
⇐□	\boxLeft	⊡⇒	\boxdotRight	⇐⊡	\boxdotLeft
◇⇒	\DiamondRight	⇐◇	\DiamondLeft	◇⇒	\DiamonddotRight
⇐◇	\DiamonddotLeft	○→	\circleright	←○	\circleleft
⊙→	\circleddotright	←⊙	\circleddotleft	⦙	\multimapbothvert
⦙	\multimapdotbothvert	⦙	\multimapdotbothAvert	⦙	\multimapdotbothBvert

Table 5.23: Continued (part 2).

←	\mappedfrom	←	\longmappedfrom	⇒	\Mapsto
⟼	\Longmapsto	⇐	\Mappedfrom	⇐	\Longmappedfrom
↦	\mmapsto	⊪→	\longmmapsto	↤	\mmappedfrom
←⊪	\longmmappedfrom	⇒	\Mmapsto	⟼	\Longmmapsto
⇐	\Mmappedfrom	⇐	\Longmmappedfrom	//	\varparallel
\\	\varparallelinv	∦	\nvarparallel	∦	\nvarparallelinv
:≈	\colonapprox	:∼	\colonsim	::≈	\Colonapprox
::∼	\Colonsim	≐	\doteq	⊶	\multimapinv
⊶⊷	\multimapboth	⊸•	\multimapdot	•⊸	\multimapdotinv
•⊸•	\multimapdotboth	⊶•	\multimapdotbothA	•⊶	\multimapdotbothB
⊩	\VDash	⊪	\VvDash	≅	\cong
≦	\preceqq	≧	\succeqq	⋨	\nprecsim
⋩	\nsuccsim	⋦	\nlesssim	⋧	\ngtrsim
⋦	\nlessapprox	⋧	\ngtrapprox	⋨	\npreccurlyeq
⋩	\nsucccurlyeq	⋧	\ngtrless	≹	\nlessgtr
≠	\nbumpeq	≠	\nBumpeq	⋏	\nbacksim
⋏	\nbacksimeq	≠	\neq, \ne	≭	\nasymp
≢	\nequiv	≁	\nsim	≉	\napprox
⊄	\nsubset	⊅	\nsupset	≪̸	\nll
⋙̸	\ngg	≉	\nthickapprox	≆	\napproxeq
⋨	\nprecapprox	⋩	\nsuccapprox	⋨	\npreceqq
⋩	\nsucceqq	≄	\nsimeq	∉	\notin
∌	\notni, \notowns	⋢	\nSubset	⋣	\nSupset
⋢	\nsqsubseteq	⋣	\nsqsupseteq	:=	\coloneqq
=:	\eqqcolon	:−	\coloneq	−:	\eqcolon
::=	\Coloneqq	=::	\Eqqcolon	::−	\Coloneq
−::	\Eqcolon	⧙	\strictif	⧘	\strictfi
⧙⧘	\strictiff	⊘	\circledless	⊘	\circledgtr
⋉	\lJoin	⋊	\rJoin	⋈	\Join, \lrJoin
×	\openJoin	⋈	\lrtimes	×	\opentimes

The txfonts package provides variable blackboard bold and fraktur letters accessed by the commands \varmathbb and \mathfrak. Moreover, the \mathbb command provides a different shape according to the next table.

ABCD	\varmathbb{ABCD}	ABCD	\mathbb{ABCD}
k	\varBbbk	𝔄𝔅ℭ𝔇𝔞𝔟𝔠𝔡	\mathfrak{ABCDabcd}

5.3.11 Other Mathematics Font Families

Several other mathematics font families exist, some more complete than others. The free families, apart from the standard ones, are the families provided by the mathptm and txfonts packages for the Times family, the mathpazo package for the Palatino family, the cmbright package, which uses a sans serif math font family, and the euler package by Herman Zapf, which gives a handwritten accent to mathematics. Other free mathematics fonts exist in development, such as the kerkis fonts that we saw in Section 3.3, and thus one should check periodically in CTAN for updates.

There are also some commercial ones such as the lucida and mathtime font families by Y&Y Inc.

5.4 The Art of Typesetting Mathematical Text

In this section, we present how one can use all of these symbols to produce, in a simple yet efficient way, masterpieces of the art of typesetting mathematical text.

5.4.1 Exponents, Indices, Fractions, and Roots

Exponents, indices, fractions, and roots are probably the most common nontext objects that one wants to be able to typeset. LATEX provides an easy way to do it. Exponents are written using the ^ character. One writes x^y in order to get x^y. Similarly, one writes x_n in order to get x_n. In order to write something more complex such as x^{y+z}, a local scope must be used: x^{y+z}. Naturally, we can create both indices and exponents the expected way (i.e., x_{i+j}^{2+3} will be typeset as x_{i+j}^{2+3}). Furthermore, these operators can be nested: 2^{2^n} and $x^{y^z}_{k_n}$ will be typeset as

$$2^{2^n} \quad \text{and} \quad x^{y^z}_{k_n}.$$

There is also the possibility of creating an exponent outside math mode with the command \textsuperscript. The code math^{mode} will be typeset as math^{mode}.

Fractions are produced with the \frac command. This command has two arguments: the numerator and the denominator of the fraction. For example, the code

$\frac{x}{y}$ will be typeset as $\frac{x}{y}$. The reader may note that this use is not really proper. As is evident here, the fraction extends too much towards the lower line and consequently it looks poor. That is why it is better to use this command in display math mode, and in text mode it is better to write x/y (i.e., x/y).

One can simplify writing the commands if the numerator and/or the denominator is a single digit. For example, the code fragments $\frac23$, $\frac2{x}$, and $\frac{x}2$, will be typeset as

$$\frac{2}{3}, \quad \frac{2}{x}, \quad \text{and} \quad \frac{x}{2}$$

respectively.

\frac commands can be nested. Here is an example:

$$\frac{x + \frac{1}{x}}{y + \frac{1}{y}}$$
```
\begin{displaymath}
\frac{x+\frac1{x}}{y+\frac1{y}}
\end{displaymath}
```

In this example, one may want to increase the width of the main fraction line in order to make such a complex fraction more readable. Unfortunately, we cannot get the desired effect with the \frac command. For such a task, we have to use the \above command:

$$\frac{x + \frac{1}{x}}{y + \frac{1}{y}}$$
```
\begin{displaymath}
{x+\frac1{x} \above1pt y+\frac1{y}}
\end{displaymath}
```

Note that after \above, we specify the width of the fraction line and the denominator, whereas the numerator is specified before the command. Moreover, the code for the whole fraction must be enclosed in a local scope since \above is a primitive TeX infix operator.

In the example above one may note that the numerators and the denominators are set in small size when the \frac command is used. A simple way out is to use the \displaystyle which fools LATEX and sets the fraction parts as if the whole fraction was the only object of a math display.

$$x + \cfrac{a}{x + \cfrac{a}{x + \cfrac{a}{x + \cfrac{a}{\ddots}}}}$$
```
\begin{displaymath}
x+\frac{a}{\displaystyle x +
\frac{a}{\displaystyle x +
\frac{a}{\displaystyle x +
\frac{a}{\ddots}}}}
\end{displaymath}
```

TeX provides three more commands that can be used in situations such as this:

\textstyle This command sets math formulas in text math mode.
\scriptstyle This command sets math formulas in script style (i.e., $x+5$).
\scriptscriptstyle This command sets math formulas in script-script style (i.e., $x+5$).

Roots are produced with the command \sqrt. This command has one argument and possibly an optional argument to denote a root other than a square root.

$\sqrt[n]{x}$	`$\sqrt[n]{x}$`
$\sqrt{x}$	`$\sqrt{x}$`

Sometimes the roots will not look uniform in height. The \vphantom command helps by inserting a zero-width line with height equal to the height of the box of its argument. Thus, `$\sqrt{a}+\sqrt{b}$` will be typeset as $\sqrt{a} + \sqrt{b}$, but `$\sqrt{\vphantom{b}a}+\sqrt{b}$` will be typeset as $\sqrt{a} + \sqrt{b}$.

For roots and fractions, the $\mathcal{AMS}$ document classes and packages provide better control. We will present these in the relevant sections later.

5.4.2 Functions

The names of functions in math mode must appear in Roman face: compare $\log x$ to $log x$. That is why a command is provided for each of several function names. Table 5.24 shows these commands.

Table 5.24: Commands for accessing functions.

\arccos	\arcsin	\arctan	\arg	\cos	\cosh
\cot	\coth	\csc	\deg	\det	\dim
\exp	\gcd	\hom	\inf	\ker	\lg
\lim	\liminf	\limsup	\ln	\log	\max
\min	\Pr	\sec	\sin	\sinh	\sup
\tan	\tanh				

There are two more such commands: the \bmod and \pmod. The first is used as an infix operator, and the second has two arguments. Their use is made clear in the following example:

$\gcd(m, n) = a \bmod b$	`$\gcd(m,n)=a \bmod b$`
$x \equiv y \pmod{a + b}$	`$x\equiv y \pmod{a+b}$`

There are many cases where one wants to define another function name. For example, the inverse of the hyperbolic sine, arcsinh, is not listed in Table 5.24. The way to define such a new function name in a document's preamble is presented below:

`\newcommand{\arcsinh}{\mathop{\mathrm{arcsinh}}}`

After this definition, the code fragment `f(x)=\arcsinh x` will be typeset as $f(x) = arcsinh x$. The same trick can be used to define an operator using a math symbol and not necessarily text. For example, the definition

```
\newcommand{\doubleint}{\mathop{\int\!\!\int}}
```

can be used to create the formula $\displaystyle\iint_{\{x\in X:\,\|x\|\leq1\}}$.

➤ **Exercise 5.1** Write down the code that is necessary to typeset the formula above. ☐

Usually, function names are pretty much standardized. However, these names are not in common use all over the world. For example, in Greece, students learn to write $\eta\mu^2\,x + \sigma\upsilon\nu^2\,x = 1$ instead of $\sin^2 x + \cos^2 x = 1$. Here comes the slightly modified recipe—you declare the fonts for the operator names, writing in the preamble as follows:

```
\DeclareSymbolFont{groperators}{LGR}{cmr}{m}{n}
```

This way, we declare a new symbol font, `groperators`, with encoding LGR, name `cmr`, medium weight, and normal shape. Now, it is easy to define new function names that make use of the new symbol font:

```
\newcommand{\grsin}{\mathop{%
    \mathgroup\symgroperators hm}\nolimits}
```

The `\mathop` command has one argument and makes its argument a unary math operator, so TEX leaves the necessary space on both sides. Similar commands are:

`\mathord` This command makes its argument an ordinary mathematical object (e.g., a letter).

`\mathbin` Makes TEX treat its argument as a binary operator.

`\mathrel` Makes TEX treat its argument as a relational operator (e.g., the expression $x\ R\ y$ is produced by the code `$x\mathrel{R}y$`).

`\mathopen` With this command, we can create opening symbols such as left parentheses, and others. The various `\bigl` commands are defined as an application of this command.

`\mathclose` This command is used to create closing symbols such as right parentheses, and others. The various `\bigr` commands are defined as an application of this command.

`\mathpunct` Makes TEX treat its argument as a mathematics punctuation character.

The `\mathgroup` command is used to select a font family. The `\nolimits` command does not allow the setting of limits above and below variable-size operators and other operators. On the other hand, the `\limits` command has the opposite effect. We note that this solution also works with the `greek` option of the `babel` package, as it regards the name of the function. Actually, the first author has created the `grmath` package, which redefines most function name commands so that their names appear in Greek. With Λ, things are quite the same; you have to write the name of the function using a Latin transcription.

➤ **Exercise 5.2** Define a new function name for the Greek version of cos (use the "word" sun for the name of the function). ☐

5.4.3 One Above the Other

There are cases where we want to put one or more objects above one another. If there is a "main" object and a secondary one that needs to be put above the main object in order to characterize it, then we use \stackrel, which has two arguments:

$$f(x) \stackrel{\mathrm{def}}{=} \cos x$$

```
\begin{displaymath}
f(x)\stackrel{\mathrm{def}}{=}\cos x
\end{displaymath}
```

If we look carefully at the equation above, we will notice that the first argument of the \stackrel command is set using a smaller type size and the second argument is set at the baseline (or as it would be put if it was set alone).

There are cases where one wants to typeset objects one above the other, treating them in the same way; here is an example:

$$\sum_{\substack{I \subseteq \{1,2,\ldots,n\} \\ I' \geq k}} a_I$$

```
\begin{displaymath}
\sum_{{I\subseteq \{1,2,\ldots,n\} }
\atop {|I|\geq k}} a_I
\end{displaymath}
```

As is evident, we use \atop to achieve the desired effect. This command is a primitive TEX infix operator and must be used in a local scope.

➤ **Exercise 5.3** Write the code that sets the following display:

$$\sum_{\substack{0 \leq i \leq r \\ 0 \leq j \leq s \\ 0 \leq k \leq t}} C(i,j,k)$$

☐

Another need that often arises is to get the result of the \stackrel command but with the first argument *below* the second argument. For instance, one may want to write

$$a_n \xrightarrow[n \to \infty]{} 0.$$

In order to achieve this effect, we can define a new command:

```
\newcommand{\abottom}[2]{\mathrel{\mathop{#2}\limits_{#1}}}
```

Here is the code that has been used to typeset the display above:

```
\begin{displaymath}
a_n\abottom{n\to\infty}{\longrightarrow} 0
\end{displaymath}
```

Sometimes, there is a need to put one object above the other and to surround the whole construction with special delimiters. We can get this effect by using the primitive TEX command \atopwithdelims. The syntax is shown in the next example. Here are a couple of examples that demonstrate the use of this command:

$\left\langle\alpha\atop\beta\right\rangle$ | $\alpha \atopwithdelims<> \beta$
$\left\lfloor\alpha\atop\beta\right\rfloor$ | $\alpha \atopwithdelims\lfloor\rfloor \beta$

An alternative way to write $\alpha\atop b$ is $a\choose b$. The command \choose is provided by plain TEX and remains valid in LATEX.

Similar commands for setting fractions with delimiters are the \overwithdelims command, which has the same syntax as the \atopwithdelims, and the \abovewith-delims command, which has a third argument that specifies the width of the fraction line. Both are primitive TEX commands. Here is an example:

$\left(a\over b\right)$ | $a \overwithdelims() b$
$\left(a\over b\right)$ | $a \abovewithdelims() 1pt b$

Other stacking operations are those of underlining and overlining as well as those that put another symbol (e.g., a brace) above or below a collection of objects. In order to set such constructs, we have to use the commands \underline, \overline, and some others that are shown below:

$\overline{\overline{x}^2}$ | $\overline{\overline{x}^2}$
$\underline{\underline{x}+\underline{y}}$ | $\underline{\underline{x}+\underline{y}}$

$\widetilde{xyz}$ | $\widetilde{xyz}$
$\widehat{xyz}$ | $\widehat{xyz}$
$\overleftarrow{xyz}$ | $\overleftarrow{xyz}$
$\overrightarrow{xyz}$ | $\overrightarrow{xyz}$

$\overbrace{a,\ldots,z}$ | $\overbrace{a,\ldots,z}$
$\underbrace{\alpha,\ldots,\omega}$ | $\underbrace{\alpha,\ldots,\omega}$

In particular, for the commands \underbrace and \overbrace, we should add that they can have an index, as the following example shows:

$\overbrace{a,\ldots,z}^{26}$ | $\overbrace{a,\ldots,z}^{26}$
$\underbrace{\alpha,\ldots,\omega}_{24}$ | $\underbrace{\alpha,\ldots,\omega}_{24}$

One should avoid using the over- or under-bracing except in display mode, as this construction takes a lot of vertical space. Also, one should notice the difference between $\bar{x}_n$ and $\overline{x_n}$ (which were produced with `$\overline{x}_n$` and `$\overline{x_n}$`, respectively). Finally, notice that there is often a problem with the "overarrows." They will usually touch (or even cross) the letter z in the example above if special care is not taken. This problem is even worse with capital letters. It is a font design issue and is corrected for the default fonts if the package lamsarrow by Michael Spivak is used (as we did here).

5.4.4 Horizontal Space

When we write mathematical text, LATEX makes decisions about the spacing of the several mathematical symbols. However, there are cases where one wants to change the default behavior. The following display shows the common commands for changing the spacing in math manually.

x	x	`$x \qquad x$`
x	x	`$x \quad x$`
x	x	`$x \; x$`
x	x	`$x \: x$`
x	x	`$x \, x$`
xx		`$x x$`
xx		`$x \! x$`
x		`$x \!\! x$`

The command `\!` stands for a negative length and can be used repeatedly, as the last example demonstrates. Of course, we are allowed to use other LATEX space-changing commands (see Section 6.3.2).

5.4.5 Integrals and Series

Integrals are produced by the command `\int` and sums by the command `\sum`. The end points of integration or summation are declared by providing indices and exponents to these commands. Here are two examples:

$$\int_0^\infty e^{x^2}\, dx = \sqrt{2\pi} \qquad \text{\$\textbackslash int\_0\^{}\textbackslash infty e\^{}\{-x\^{}2\}\textbackslash,dx=\textbackslash sqrt\{2\textbackslash pi\}\$}$$

$$\sum_{n=1}^\infty \frac{1}{n^2} = \frac{\pi^2}{6} \qquad \text{\$\textbackslash sum\_\{n=1\}\^{}\textbackslash infty \textbackslash frac1\{n\^{}2\} =\textbackslash frac\{\textbackslash pi\^{}2\}6\$}$$

The position of the range of the sum index is not always as we described above. These are put above and below the sum only when we are in display mode. When we are in the first math mode, these are put as indices and exponents like this: $\sum_{n=1}^\infty 1/n^2 = \pi^2/6$. The purpose of this is to avoid forcing the spreading of the text in order to accommodate the summation indices and should not be bypassed. However, if we insist, we may do it using the `\limits` command. Once again, let us stress that this is poor practice, as it affects, in a bad way, the color balance of the page. Take a look at the following example:

This is an example of the sum given above, that is, $\sum\limits_{n=1}^{\infty} 1/n^2 = \pi^2/6$, where the default behavior of setting the sum indices is bypassed.

```
This is an example of the
sum given above, that is,
$\sum\limits_{n=1}^\infty
1/n^2 =\pi^2 /6$, where the
default behavior of setting
the sum indices is bypassed.
```

There are cases where one wants to reverse the behavior for the display math mode. For this, we must tell LATEX that we do not want limits by using the \nolimits command. The display

$$\sideset{}{'}\sum a_n$$

is produced by `$$\sum\nolimits^\prime a_n$$`.

As far as integrals are concerned, we note here that the integration range can be set under the integral using the \limits command. This is useful when we try to save space or improve the appearance when we integrate on a set whose description requires a lot of horizontal space. This behavior and the default one are compared in the next example:

$$\int_{\{x\in A:x\geq 0\}} f(x)\,dx \qquad \text{(the default)}$$

and

$$\int\limits_{\{x\in A:x\geq 0\}} f(x)\,dx.$$

The first one was produced by

$$\int_{ \{x\in A\,:\, x\geq 0\} } f(x)\,dx$$

and the second one by

$$\int\limits_{ \{x\in A\,:\, x\geq 0\} } f(x)\,dx.$$

We also note here that it is a good practice to add a little space before our first differential using the \, command. This is especially good if big parentheses are closing in front of the differential:

$$\int_a^b \left(\frac{1}{x}\right) dx$$

looks better balanced than

$$\int_a^b \left(\frac{1}{x}\right) dx.$$

The discussion above applies in the same way to all of the symbols of variable size, such as $\prod$, $\bigcup$, and so on.

However, there are cases where one wants to use another symbol that is not of variable size and pretend it is. For example, one may want to use a symbol such as

$$\ast_{i=1}^{n} f_i$$

for the convolution of functions. For such a task, we need to define a new operator (such as the \sum operator), but since the asterisk that we used above is not of variable size, we must decide what size we want. Therefore, the example above was typeset using

```
\newcommand{\astop}{\mathop{\LARGE\mathrm{\ast}}}
```
. . . .
```
$$\astop\limits_{i=1}^n f_i.$$
```

5.4.6 Matrices, Arrays, and Nonanalytically Defined Functions

The way to write a matrix in LaTeX is to use the array environment. This environment arranges the elements of the matrix in rows and columns without taking care of the delimiters. If we want to use, say, the symbols [ ] for the matrix delimiters, LaTeX must be informed that the size of these delimiters must be adjusted to fit nicely with the matrix they enclose. This is done by the \left and \right commands. These commands take care of the automatic size adjustment of the delimiters. Here is the syntax for a matrix:

$$\begin{bmatrix} x & \lambda & 1 & 0 \\ 0 & x & \lambda & 1 \\ 0 & 0 & x & \lambda \end{bmatrix}$$

```
$$\left[
  \begin{array}{lcr}
    x-\lambda & 1          & 0         \\
    0         & & x-\lambda & 1          \\
    0         & & 0         & & x-\lambda
  \end{array}
\right]$$
```

There are cases where we want only one of the two delimiters, and this is the case for the nonanalytically defined functions. If we just omit one of the two delimiters, LaTeX will complain about improper grouping. That is why the \left and \right commands accept the dot as a special delimiter that produces nothing in the output but satisfies LaTeX's fussiness about proper grouping. Here is an example:

$$\chi_A(x) = \begin{cases} 1 & \text{if } x \in A \\ 0 & \text{otherwise} \end{cases}$$

```
$$\chi_A (x) = \left\{
  \begin{array}{ll}
    1 & \mathrm{if}\ x\in A \\
    0 & \mathrm{otherwise}
  \end{array}
\right.$$
```

There are cases where this simple construction is not satisfactory. Here is an example:

$$a_n = \begin{cases} 1+2+3+\cdots+n, & \text{if } n \text{ is even} \\ 0, & \text{otherwise.} \end{cases}$$

It would look much better if most of the white space of the second line of the array was not there. For instance,

$$a_n = \begin{cases} 1+2+3+\cdots+n, & \text{if } n \text{ is even} \\ 0, \text{ otherwise} \end{cases}$$

was produced by putting a \quad after the comma and placing everything from the comma to the end of the line in a \mbox. In addition, the array now has only one left-justified column.

Sometimes, we need to write a matrix with rows and columns being labeled with the number of the row or column. This is done with the command \bordermatrix. Here is an example:

$$\begin{matrix} & 1 & 2 & 3 \\ 1 & a_{11} & a_{12} & a_{13} \\ 2 & a_{21} & a_{22} & a_{23} \\ 3 & a_{31} & a_{32} & a_{33} \end{matrix}$$

```
$$\bordermatrix{ & 1 & 2 & 3 \cr
1 & a_{11} & a_{12} & a_{13} \cr
2 & a_{21} & a_{22} & a_{23} \cr
3 & a_{31} & a_{32} & a_{33}}$$
```

The \bordermatrix command is not a LaTeX command, but it comes from plain TeX. That is why its syntax is not similar to the way that we created matrices and it uses the \cr command to denote the end of the line instead of the double backslash. Plain TeX provides also a \matrix command (with similar syntax) for matrices, but its use is obsolete in LaTeX. It can be used, but often it creates problems that cannot be overcome and that is why we do not describe it. Similarly, plain TeX provides the command \cases for writing nonanalytically defined functions and many mathematicians use it. Once again, the behavior of this command in LaTeX is unpredictable and very often does not work.

The array environment takes an optional argument that can be either t or b. These options align the top or the bottom of the array block to the baseline instead of centering it, which is the default. For example, the matrices

$$\begin{pmatrix} x & \lambda & 1 & 0 \\ 0 & x & \lambda & 1 \\ 0 & 0 & x & \lambda \end{pmatrix} \begin{pmatrix} x & \lambda & 1 & 0 \\ 0 & x & \lambda & 1 \\ 0 & 0 & x & \lambda \end{pmatrix}$$

were produced by using the lcr *table specs* and the t and b optional arguments, respectively. Notice that the parentheses are more cursive than the standard ones of

LaTeX because we use the yhmath package, which provides this feature. For better control of the optional arguments, we suggest that the reader check out the array package (see Section 4.9).

5.4.7 Theorems

LaTeX provides an easy interface to create the environments for a theorem, a proposition, a lemma, and so on. The preamble of a document must contain the command \newtheorem to define such an environment. The general form of this command is

$$\texttt{\textbackslash newtheorem\{}\textit{name}\texttt{\}\{}\textit{output name}\texttt{\}[}\textit{counter}\texttt{]}$$

or

$$\texttt{\textbackslash newtheorem\{}\textit{name}\texttt{\}[}\textit{counter}\texttt{]\{}\textit{output name}\texttt{\}}$$

where *name* is the name of the new environment, *output name* is the name that will be printed in the output, and *counter* is the counter with respect to which this environment will be numbered. Counters are special LaTeX variables that hold numerical values (see Section 6.4). Note that the two forms above are different.

For example, in a mathematics book or article, one can put (in the preamble) the following definitions:

```
\newtheorem{thm}{Theorem}[section]
\newtheorem{lemma}[thm]{Lemma}
\newtheorem{prop}[thm]{Proposition}
\newtheorem{cor}[thm]{Corollary}
\newtheorem{def}[thm]{Definition}
\newtheorem{remark}[thm]{Remark}
\newtheorem{axiom}[thm]{Axiom}
\newtheorem{exercise}[thm]{Exercise}
```

Now, one can write

```
\begin{thm}There are infinitely many primes.\end{thm}
We first prove the following:
\begin{lemma}
Every integer has at least one prime divisor.
\end{lemma}
```

in order to get

Theorem 5.4.1 *There are infinitely many primes.*

We first prove the following:

Lemma 5.4.2 *Every integer has at least one prime divisor.*

Note that the frame in all examples of this section is not produced by the code presented. We see that the defaults for the theorem environments are as follows. The *output name* is written in bold, then the number follows, which is the section's number, followed by the theorem's counter. The alternate position of the optional argument [*counter*] for the lemmata, corollaries, and so on, says that the counters for these are the same as the theorem's counter, which, in turn, is defined from the section's number *followed* by the theorem counter. If, for example, we had set \newtheorem{lemma}{Lemma}[thm], the number of our lemma above would not be 5.4.2 but 5.4.1.1. In other words, the theorem's number is *followed* by the theorem's counter. After that, the theorem's text is set in italics.

There is one optional argument for all of the theorem environments that is used to typeset the name of the people who proved the theorem or a theorem's name, if such a name exists. For example, we could write

```
\begin{thm}[Euclid]
There are infinitely many primes.
\end{thm}
```

to get

> **Theorem 5.4.3 (Euclid)** *There are infinitely many primes.*

or, if we are in a multilingual environment we can set the language with the relative command. For example, the code

```
\begin{thm}[\textgreek{Εὐκλείδης}]
There are infinitely many primes.
\end{thm}
```

can be used to get

> **Theorem 5.4.4 (Εὐκλείδης)** *There are infinitely many primes.*

Sometimes, it happens that immediately after the theorem header, we need a line break. This happens when the theorem's description is long and the first word of its statement does not fit on the line and cannot be hyphenated (it may be one syllable). This problem appears again when we typeset with a smaller text width, as in a multicolumn environment. The way to achieve this line break is to use an \hfill command after the \begin{theorem} and leave a blank line after that, like this:

<table>
<tr><td>

Theorem 5.4.5
There are infinitely many primes.

</td><td>

```
\begin{thm}\hfill
```

```
There are infinitely
many primes.\end{thm}
```

</td></tr>
</table>

Note that the \hfill command puts an infinite glue (see page 21).

Since the theorem environments use a counter, we can label them so that we can refer to them later on. For example, we write

```
\begin{lemma} \label{primeDivisors-lemma}
Every integer has at least one prime divisor.
\end{lemma}
Using Lemma \ref{primeDivisors-lemma} we can now prove
\begin{thm}[Euclid]
There are infinitely many primes.
\end{thm}
```

in order to get

> **Lemma 5.4.6** *Every integer has at least one prime divisor.*
>
> Using Lemma 5.4.6 we can now prove
>
> **Theorem 5.4.7 (Euclid)** *There are infinitely many primes.*

5.4.8 Customizing the theorem Environment

For a long time now, the theorem package by Frank Mittelbach has been the standard way for creating custom theorem environments. Recently, an extension of this package with the name ntheorem has appeared. It was written by Wolfgang May and Andreas Schlechte, and it incorporates many additional features while remaining compatible with the theorem package. Since it will take some time for this package to make it into the standard distributions of LATEX, we will mark with a [†] the commands that are in common with the standard theorem style. The package includes a standard file with definitions for theorems. This can be used as an example of how to set up your theorem environments.

The first customization point to note is that for every theorem defined with \newtheorem{name}{Name} a starred version is made available that implements the unnumbered version of the environment *name*. The package provides the following options:

standard It loads the standard definitions for theorem environments (contained in file `ntheorem.sdt`). It is good if you are using English or German, but for other languages you must define them yourself.

noconfig It does not load the `ntheorem.std` file but the `ntheorem.cfg` file that you must create with your preferences and language settings.

amsmath It provides a compatibility layer with the theorem commands of the amsthm package.

thmmarks Enables automatic placement of QED[1] symbols.

thref Extended theorem referencing.

hyperref Provides a compatibility layer with the `hyperref` package (see Section 6.2).

The ntheorem package uses the `\newtheorem` command as described in the previous section. However, the following customization commands are also defined:

`\theorempreskipamount`[†] This is the amount of vertical space before the theorem starts. It can be positive or negative.

`\theorempostskipamount`[†] This is the amount of vertical space after the theorem starts. It can be positive or negative.

`\theoremstyle{style}`[†] This defines a style for theorems. We will see how to define our own style but the package has several predefined styles. These are:

> plain This is like LaTeX's default but uses `\theorempreskipamount` and `\theorempostskipamount`.
>
> break This produces a line break after the theorem's header.
>
> change The number is set first, and then the theorem's name follows.
>
> changebreak The same as change but with a line break after the theorem's header.
>
> margin The number is set in the left margin.
>
> marginbreak The same as margin but with a line break after the theorem's header.
>
> nonumberplain Like plain without number (e.g., for proofs).
>
> nonumberbreak The same as nonumber but with a line break after the theorem's header.
>
> empty No number and no name; only the optional argument is printed.

`\theoremheaderfont{font-commands}`[†] Specifies the font to be used for the header.

`\theorembodyfont{font-commands}`[†] Specifies the font to be used for the body of the theorem.

`\theoremseparator{symbol}` The theorem separator can be " : ", " . ", or anything you like.

`\theoremindent{dimension}` Indentation for the theorem with respect to the surrounding text (no plus or minus here).

`\theoremnumbering{style}` Here, again, there are several styles: arabic (the default), alph, Alph, roman, Roman, greek, Greek, and fnsymbol.

---

1. QED is an acronym of the Latin phrase *Quod Erat Demonstrandum*, which means *which was to be demonstrated*. The Latin phrase is a translation of the Greek phrase ὅπερ ἔδει δεῖξαι.

\theoremsymbol{*symbol*} Theorem symbol to be set at the end of the statement only if the package is used with the thmmarks option.

\theoremclass{*type*} All parameters are set to values that were used when \new-theorem was issued. If type is set to LaTeX, the standard layout is chosen.

All of these must be set before we call \newtheorem. If none of the above are set, then the default scheme is the same as LaTeX's default. That is,

```
\theoremstyle{plain}
\theoremheaderfont{\normalfont\bfseries}
\theorembodyfont{\itshape}
\theoremseparator{}
\theoremindent{0cm}
\theoremnumbering{arabic}
\theoremsymbol{}
```

Another interesting feature of the package is its ability to create a list of theorems similar to the list of figures. This is done using the command \listoftheorems{*list*}, where *list* is a comma-separated list of the theorem environments to be listed. For example, \listoftheorems{theorem,conjecture} will produce a list of the theorems and conjectures. Notice that if a theorem is created with its starred version, it will *not* be listed. The appearance and selection of what theorems will appear in the list can be further customized with the \theoremlisttype{*type*} command. The *type* can be one of the following:

all List all theorems of the specified *list* by number, the optional name, if it exists, and the page number.

allname Like all but with the theorem name at the beginning.

opt Like all, but only the theorems that have an optional name will be listed.

optname Like opt with the theorem name at the beginning.

Similar to \addcontentsline for the contents file, the package provides a way to add information to the list of theorems by the command

\addtheoremline{*name*}{*text*}

where *name* is the name of a valid theorem and *text* is the text that should appear in the list. For example, the line

\addtheoremline{Conjecture}{The hyperplane conjecture}

has the same effect as if we had written in our document the environment

\begin{Conjecture}[The hyperplane conjecture] \end{Conjecture}

Note that the counter will not advance for such a command. The starred version \add-theoremline* is being set like the unstarred version but without the theorem number.

The \addtheoremfile[*name*]{*text*} command adds the *text* into the theorem file. The optional argument controls to which theorem lists the *text* will be added, and if it is omitted it is added to all of them.

The QED symbol is set automatically. If one would like to replace the standard symbol in an environment, then one can redefine the \qedsymbol by using \qedsymbol{*symbol*} and then call this symbol by the \qed command. The \qedsymbol can be reset anywhere in the document. This feature is useful for closing lemmata or corollaries that are easy and where no proof follows. On the other hand, in order to avoid setting the QED symbol, one can use the \NoEndMark command to turn off the automatic setting of the symbol. It can now be set manually with the command *name*Symbol like \theoremSymbol.

Extended reference features are activated when the package is used with the thmref option. The \label command is extended to use an optional argument \label{*label*}[*type*], which characterizes the *label* as belonging to a specific theorem environment. This additional information is used by the command \thref{*label*} as follows. If, for example, the *type* in the \label command is "Lemma" and the number of this environment is 4.3.7, then if we refer to that label with \thref{*label*}, it will produce "Lemma 4.3.7" instead of "4.3.7" as the \ref command produces. The optional argument is set automatically when the label is inside an environment. Thus, the labeling command \label{*name*}[*Lemma*] has the same effect as the label in

$$\text{\begin{Lemma}\label{\textit{name}}}$$

Notes on compatibility:

- The babel package must be loaded before the theorem package.
- If you want to use the amsmath package with the ntheorem package, use

  \usepackage{amsmath}
  \usepackage[amsmath,thmmarks]{ntheorem}

  in that order.

- If you want to use the features of the amsthm package, then instead of loading amsthm one should use the amsthm option of the ntheorem package. This option covers the theorem styles plain, definition, and remark as well as the proof environment. The \swapnumbers is not supported (see Section 5.5.21).

- When using the hyperref package, load the ntheorem package with the hyperref option to ensure compatibility.

  New theorem styles can be defined by the command \newtheoremstyle{*name*}{*head*}{*optional-head*}. The *head* must contain instructions on how to typeset the theorem's header and should use two parameters, the ##1 and ##2. The commands used to typeset ##1 are the commands that will be used to typeset the header of the theorem (theorem, lemma, and so on) and the commands used to typeset ##2 are the commands that will be used to typeset the number of the environment in the header. The *optional-head* has three arguments, and the commands for the third argument ##3 are used to typeset the optional theorem argument. Both header declarations must be of the form \item[...\theorem@headerfont ...] ..., where the user must insert the dotted parts. Implicit spaces must be taken care of in the case where there are statements producing output after the \item[...]. Here is an example of a new theorem style defined in the preamble of a document that would have been suitable for the days when people could be satisfied with just one text font:

```
\usepackage{letterspace,ntheorem}
\makeatletter %% needed since the special character @ is used below
\newtheoremstyle{oldstyle}%
  {\item[\hskip\labelsep%
     \upshape ##2\ \,\mbox{%
        \upshape\letterspace to 1.5\naturalwidth{##1:}}}%
     \theorem@separator]\upshape}%
  {\item[\hskip\labelsep%
     \upshape ##2\ \,\mbox{\upshape\letterspace to 1.5%
                          \naturalwidth{##3's\ ##1:}}}%
        \theorem@separator]\upshape}
\makeatother
\theoremstyle{oldstyle}
\newtheorem{thm}{Theorem}
```

Now, the code

```
\begin{thm}[Cauchy-Goursat]
If a function $f$ is analytic in the interior of a simple
curve $C$ as well as on the curve, then $\int_C f(z)\,dz=0$.
\end{thm}
```

will produce:

1 C a u c h y - G o u r s a t ' s T h e o r e m : If a function f is analytic in the interior of a simple curve C as well as on the curve, then $\int_C f(z)\, dz = 0$.

Similarly, one can define new theorem list environments with the command

\newtheoremlisttype{*name* }{*start* }{*line* }{*end*}

where *name* is the list's name, *start* refers to the commands to be executed at the beginning of the list, *end* refers to the commands to be executed at the end of the list and *line* is the part to be called for every list entry. The completed command must be a statement with four arguments: ##1 will be replaced by the theorem's name, ##2 with the number, ##3 with the theorem's optional name and ##4 with the page number. Defining new lists may break compatibility with the hyperref package. Finally, let us note that the theorem lists can be redefined with the command \renewtheoremlisttype with the same arguments as \newtheoremlisttype.

5.4.9 Equations

One of the most important advantages of LaTeX over plain TeX is the easy interface that it provides for typesetting equations plus the automatic numbering, labeling, and referencing for them. The easier way to produce a math display is with the double dollars, as in plain TeX. However, this does not provide automatic numbering, so LaTeX provides the environment equation. Here is an example:

Do you know this identity?	```Do you know this identity?```

$$a^2 = b^2 + c^2 \qquad (5.1)$$	```\begin{equation}```
	```a^2=b^2+c^2```
	```\end{equation}```

This is the Pythagorean theorem if a, b, c are the three sides of a right triangle.	```This is the Pythagorean```
	```theorem if a, b, c are```
	```the three sides of a right```
	```triangle.```

The equation above has (by default) the numbering at the right of the equation, and the equation is centered. The document class options leqno and fleqn change these two defaults. The first puts the equation number at the left of the equation, and the second one sets the equations flush left.

In some cases, we want to substitute the equation number with a word. For this, we use the \eqno command as in the following example:

$$\frac{\Gamma \vdash B}{\Gamma, !A \vdash B} \quad \text{(weakening)}$$	```$$\frac{\Gamma \vdash B}```
	```{\Gamma, !A \vdash B}```
	```\eqno\mathrm{(weakening)}$$```

Very often, we want to write several equations in one display aligned at some symbol. For this, LaTeX provides the eqnarray and the eqnarray* environments. The first numbers each equation, whereas the second does not put numbers. Here are some examples:

Some famous equations:
 	
$$E = \hbar \cdot \nu \qquad (5.2)$$
$$E = m \cdot c^2 \qquad (5.3)$$
$$\oint \vec{B} \cdot d\vec{S} = 0 \qquad (5.4)$$
$$\vec{S} = \frac{1}{\mu_0} \vec{E} \times \vec{B} \qquad (5.5)$$
 	
Do you recognize them?
	```Some famous equations:```
	```\begin{eqnarray}```
	```E &=& \hbar \cdot \nu \\```
	```E &=& m \cdot c^2 \\```
	```\oint \vec{B} \cdot```
	``` d\vec{S} &=& 0 \\```
	```\vec{S} &=& \frac1{\mu_0}```
	```\vec{E} \times \vec{B}```
	```\end{eqnarray}```
	```Do you recognize them?```

Notice that the double backslash denotes the end of a line, and the character & controls alignment. Moreover, notice that LaTeX puts some white space around the aligned symbol.

Naturally, we can customize the amount of white space that LATEX puts around the aligned symbol. To do this, we simply set the length variable \arraycolsep in a local scope that encloses the eqnarray environment, which is what we did in the next example. Of course, if we set this variable in our document's preamble, then the effect is global.

Sometimes, we want to write such an equation array but do not want to number all of the equations. This can be done with the \nonumber command:

$$\int_0^{\pi/2} \sin x \, dx = \left. \cos x \right|_0^{\pi/2} \qquad (5.6)$$

$$= \cos \frac{\pi}{2} \; (\quad \cos 0)$$

$$= 0 \; (\quad 1)$$

$$= 1$$

```
\begin{eqnarray}
\int_0^{\pi/2} \sin x \,dx
&=& -\cos x\biggm|_0^{\pi/2}\\
&=& -\cos\frac{\pi}2
     -(-\cos 0) \nonumber \\
&=& -0-(-1) \nonumber \\
&=& 1 \nonumber
\end{eqnarray}
```

However, we should use the eqnarray* environment if we want no numbers at all. One may note that the command \biggm above is not yet discussed. This is the subject of the next section.

Now, we will see how to typeset logic proofs since these require more attention. We need the proof package (by Makoto Tatsuta). The package provides the command \infer[*label*]{*lower*}{*upper*}, which draws an inference labeled with *label*. If we put an asterisk after the command (\infer*[...]), it draws a many-step deduction. If the star is changed to the = sign, it draws a double-ruled deduction. It also provides the command \deduce[*proof*]{*lower*}{*upper*}, which draws an inference without a rule with a *proof* name.

Thus, \infer{A}{B} and \deduce{A}{B} produce $\dfrac{B}{A}$ and $\overset{B}{A}$. To produce many steps, we just use the alignment character &, and the code for the first equation of the display

$$\pi = D \; \dfrac{\dfrac{F \, \& \, G \, \& \, H}{E}}{A \, \& \, B \, \& \, C} \begin{array}{l} (\to I) \\ (\&I) \end{array} \qquad \begin{array}{c} C \, \& \, D \\ \vdots \\ B \quad E \\ \hline A \end{array} \qquad (5.7)$$

is

```
\[  \pi = \vcenter{\infer[(\& I)]{A\,\&\,B\,\&\,C}{D &
        \infer=[(\to I)]{E}{F\,\&\,G\,\&\,H}\hspace{2em}}}\]
```

where \vcenter centers its argument vertically to the baseline and \hspace is used in order to make the second top argument bigger to force a bigger inference line. The code for the second equation is

```
\infer{A}{\infer*[(3)]{B}{C\,\&\,D} & E}
```

5.4.10 Size Selection in Math Modes

Size selection for delimiters has already been discussed through the \left and \right commands. These commands provide automatic adjustment of the size of the delimiters to the size of the expression that they enclose. However, there are cases where we want to force a specific size.

Suppose that we want to write $\big((n+1)/n\big)^2$. If we type $\left((n+1)/n \right)^2$, we will get $((n+1)/n)^2$, which is not as satisfactory. Here, we want to force LaTeX to make the size of the external parentheses bigger. The \left and \right commands do *not* fail, they do properly enclose the expression that they surround. The problem is with our aesthetics, which require bigger parentheses since the inner expression also uses parentheses and not because the inner part creates a larger box that the external parentheses failed to surround. The correct input for this task is $\bigl((n+1)/n \bigr)^2$. LaTeX provides the following commands for predetermining the size of a delimiter:

$$\bigg(\Big(\big(()\big)\Big)\bigg) \qquad \begin{array}{l} \texttt{\$\Biggl(\biggl(\Bigl(\bigl(} \\ \texttt{\textbackslash \bigr) \Bigr) \biggr) \Biggr)\$} \end{array}$$

The same commands work for all delimiters left or right. Another application of this facility is when writing big operators such as \sum. In the next display, the second expression (which used \biggl(\biggr) parentheses) looks better than the first (which used \left(\right):

$$\left(\sum_{i=1}^{\infty} a_n^2\right)^{1/2} \qquad \biggl(\sum_{i=1}^{\infty} a_n^2\biggr)^{1/2}. \tag{5.8}$$

There are some delimiters though that do not come in pairs, such as the character /. For these, the commands are

$$3\Bigg/4\bigg/5\Big/6\big/7/8 \qquad \texttt{\$3\textbackslash Biggm/ 4\textbackslash biggm/ 5\textbackslash Bigm/ 6\textbackslash bigm/ 7/8\$}$$

5.4.11 Commutative Diagrams

Commutative diagrams are often used in mathematics to depict a relation between mathematical entities. In this section, the term "diagram" will stand for commutative diagram. There are several ways that one can draw diagrams. One of them is presented as part of the $\mathcal{AMS}$ packages in Section 5.5.16. However, this is not a complete solution since it cannot produce diagonal arrows. A complete solution is provided by the kuvio package of Anders Svensson and the pb-diagram package of Paul Burchard.

Both packages work with a conceptual grid and attach nodes and arrows to it. The pb-diagram package provides the environment diagram. The two main commands are \arrow and \node. The syntax of the command \arrow is like this: \arrow{x,y}{z}. The {x,y} show the direction of the arrow. The parameter x can take all of the cardinal points of the compass; that is, it can be e (for east), w (for west), s (for south), and n (for north) and their intermediate positions: ne, nw, se, sw, nne, nnw, sse, ssw, ene, ese, wnw, and wsw. If pb-lams is used, we have additional directions available. These are nee, see, nww, sww, neee, nnne, nnnw, nwww, swww, ssse, seee, nnnee, nnnww sssww, sssee, nneee, nnwww, sswww, and sseee. The argument y sets the position of the label z with respect to the arrow. It can be t (for top), b (for bottom), l (for left), and r (for right). The \arrow command can be used with additional arguments:

$$\arrow[s]\{x,y1,y2\}\{z\} \text{ or } \arrow[s]\{x,y1,y2\}\{z1\}\{z2\}$$

In the second case, we have two labels: the z1 and z2 that are set above and below the arrow (y1=tb) or left and right (y1=lr). The y2 argument in both of the syntaxes above specifies the arrow shaft head and tail to be used. Most of the following options are available if either the pb-lams or the pb-xy package is used. Shafts can be .. for dotted lines, = for double lines, and ! for invisible lines. The head can be - for no arrowhead, <> for arrowheads at both sides, A for double arrowhead, ' for left half arrowhead, and ' for right half arrowhead. The tail can be V for single arrow tail, J for left hook arrow tail, and S for square arrow tail. The s in the optional argument sets the number of columns or rows that the arrow will span.

The position of the label on the arrow length can be set by dividing the arrow into a number of pieces, adding a y3 argument to hold the arrow shape information, and giving to the y2 option the number of the piece on which the label will be set. The division is done by setting the \dgARROWPARTS (by default, \dgARROWPARTS=4; thus, the y2 can be 1, 2, 3, or 4).

The \node command is used in the form \node{nols}{formula}, where ncols is a number that sets how many columns after the last node the formula will be set.

The typesetting of diagrams is written in rows, and we move to the next row by \\. Here is a simple example:

```
\begin{displaymath}
\begin{diagram}
\node{X}\arrow{s,l}{p}
\arrow{e,t,..}{g}
\node{Y}\arrow{s,r}{q}\\
\node{A}\arrow{e,b,..}{r}
\node{B}
\end{diagram}
\end{displaymath}
```

and here is a rather complex example (we used the lamsarrow, pb-lams, and pb-diagram packages):

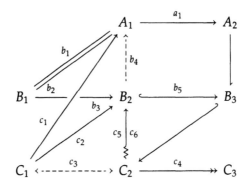

The code for this (and it will be good for the reader to study it) is

```
$$\begin{diagram}
\node[3]{A_1} \arrow[2]{e,t}{a_1} \node[2]{A_2}\arrow[2]{s,'}\\
\\
\node{B_1}\arrow[2]{ne,l,=}{b_1}\arrow{e,t,-}{b_2}
\node{}\arrow{e,b}{b_3}\node{B_2} \arrow[2]{n,r,..}{b_4}
\arrow[2]{e,t,J}{b_5}\node[2]{B_3}\arrow[2]{sw,L}\\
\\
\node{C_1} \arrow[2]{nne,l,1}{c_1} \arrow[2]{ne,r}{c_2}
\node[2]{C_2}\arrow[2]{w,t,<>,..}{c_3}\arrow[2]{e,t,A}{c_4}
\arrow[2]{n,lr,S}{c_5}{c_6} \node[2]{C_3}
\end{diagram}$$
```

If these capabilities of the pb-diagram package are not enough for the user (and this can happen), then, most probably, the capabilities of the kuvio package will cover the reader's needs. This is a big plain TeX macro package with a LaTeX wrapper package. The reader can check the documentation of this package, which provides more features than simply diagrams. For example, it provides some additional math symbols, including a circle symbol for composition of functions (the \circ command produces a circle that is too big for compositions). The package documentation can be found at ftp://ftp.math.ubc.ca/pub/svensson/kuvio/ with name tdwk-A4.ps.gz or tdwk.ps.gz. The package itself is available from the CTAN.

5.5 The $\mathcal{AMS}$ Classes and Packages

The $\mathcal{AMS}$ document classes amsart, amsbook, and amsproc for articles, books, and proceedings, respectively, provide a much better control for mathematical text than the standard document classes. This is why these are usually the preferred classes for texts containing a lot of mathematics. In a multilingual environment, the classes have a problem with accented letters in the running heads. The problem is solved if one uses the textcase package by David Carlisle. Actually, these classes provide the same functionality as the amsmath package plus the document design characteristics of the $\mathcal{AMS}$ publications. So, one may either directly use one of these classes or load the amsmath package with any other class. Note that this package defines the \AmS command, which prints the $\mathcal{AMS}$ logo.

In the following, we discuss the features provided by the amsmath package plus some additional packages provided by the $\mathcal{AMS}$, such as amsthm for customizing theorem environments and amscd for commutative diagrams. Since these are loaded by the $\mathcal{AMS}$ document classes, what we will say is valid for both strategies ($\mathcal{AMS}$ classes or any class plus amsmath and other $\mathcal{AMS}$ packages). We will refer to either of these strategies by saying "the $\mathcal{AMS}$ packages."

Information about these classes can be found in the $\mathcal{AMS}$ documentation of the $\mathcal{AMS}$-LaTeX distribution, as these packages and classes are usually called. In version 2.0 the file to look at is instr-l.dvi, which presents useful information about article preparation for journals and describes also the commands for titling (which are very similar to the standard article class of LaTeX). If your system has already installed these classes, this file is usually in texmf/doc/latex/amscls/.

5.5.1 Additional Symbols

The $\mathcal{AMS}$ packages provide bold symbols, Greek letters included, through \boldsymbol. Also, italic capital Greek letters are provided by using the letters "var" between the backslash and the name of the Greek letter. For example,

$$\boldsymbol{\pi}\boldsymbol{\in} \varPsi$$

gives $\pi \in \varPsi$.

They also provide what is known as "poor man's bold" for cases where the \boldsymbol command does not do anything (due to the lack of a bold glyph of the symbol needed). The poor man's bold command is \pmb and simulates bold by typing several copies of the symbol we want with slight offsets. Here are a few examples:

$$\sum\prod\bigvee$$

```
\begin{displaymath}
\pmb{\sum}\pmb{\prod}
\pmb{\bigvee}
```

Accents are also supported in bold. The command $\mathbf{\tilde{A}}$ will produce the right (bold) accent above the bold A in LaTeX: $\tilde{\mathbf{A}}$. But with the $\mathcal{AMS}$ packages we will get the right thing even for $\mathcal{\vec{A}}$ ($\vec{\mathcal{A}}$). Alternatives to \mathbb and \mathfrak are also provided, and they are \Bbb and \frak, respectively. They are useful when the standard LaTeX commands have been redefined (as in the case of the mathbbol package).

5.5.2 Accents in Math

The $\mathcal{AMS}$ packages provide better accents for math mode. The advantage is clear when one wants to use double accents. One should compare the following:

standard LATEX `$\hat{\hat A}$` : $\hat{\hat{A}}$,   AMS `$\hat{\hat A}$` : $\hat{\hat{A}}$.

The same holds true for all other accents (see Table 5.3). Double accents take a lot of processing time, and this is why, if we use them repeatedly, it is better to store the result of a double accent to a command using the \accentedsymbol available with the amsxtra package. This command introduces a shorthand and should be used only in the document's preamble. Here is an example:

$\hat{\hat{A}}$ and $\dot{\breve{Y}}$

```
\accentedsymbol{\Ahathat}{%
\hat{\hat{A}}}
\accentedsymbol{\Ybrevedot}{%
\dot{\breve{Y}}}
```

. . . .

`$\Ahathat$` and `$\Ybrevedot$`

The commands \dddot and \ddddot produce triple and quadruple dot accents in addition to the \dot and \ddot accents (which are already available with standard LATEX): `$\dddot{E}$` and `$\ddddot{T}$` give $\dddot{E}$ and $\ddddot{T}$, respectively.

Special symbols that are set as superscripts form another kind of accent. These are useful in math (for instance the Fourier transform uses a \hat as superscript unless the function is a single letter or a few letters). For example,

$$\bigl(\exp(-x^2)\bigr)\sphat$$

gives $\left(\exp(\ x^2)\right)\widehat{}$. Notice that we do not use the ^ character. The reader is recommended to try the commands \spcheck, \sptilde, \spdot, \spddot, \spdddot, and \spbreve. All of them are available with the amsxtra package.

5.5.3 Dots

The AMS packages provide five commands for accessing differently positioned ellipsis dots. \dotsc represents "dots with commas" like this $1, 2, \ldots, n$ (`$1,2,\dotsc,n$`). \dotsb stands for "dots with binary operators/relations" as in $1 + 2 + \cdots + n$ (`$1+2+\dotsb +n$`). \dotsm stands for "multiplication dots" as in $a_1 a_2 \cdots a_n$ (`$a_1 a_2 \dotsm a_n$`). \dotsi stands for "dots with integrals" as in $\int_A \int_B \cdots$ (`$\int_A\int_B \dotsi$`). Finally, \dotso covers "other dots," which are none of the above: $a \ldots b + \ldots + c$ (`$a\dotso b+\dotso +c$`).

5.5.4 Nonbreaking Dashes

There are cases (such as when we give the page range of a reference) when we do not want to allow a line break at the en dash point. This can be done with the command \nobreakdash. So, if you write "pages 321–345" as pages 321\nobreakdash--345, a line break will never occur between the dash and 345. The command can also be used for combinations such as `$p$-adic`. Naturally, one can define shorthands for commonly used constructs, but this is the subject of the next chapter.

5.5.5 Over and Under Arrows

Standard LaTeX, as we have shown in Section 5.4.3, provides the commands \over-rightarrow and \overleftarrow. Some additional commands including underarrows are now available. All of them are as follows:

```
\overleftarrow         \underleftarrow
\overrightarrow        \underrightarrow
\overleftrightarrow    \underleftrightarrow
```

For example, $\underleftarrow{xy}^{\overleftrightarrow{zw}}$ gives $\underleftarrow{xy}^{\overleftrightarrow{zw}}$. Note: This is *not* to be used for projective limits. See Table 5.5.12.

5.5.6 Multiple Integral Signs

The commands \iint, \iiint, and \iiiint give multiple integral signs with nice spacing between them in both text and display styles. The command \idotsint gives two integral signs with ellipsis dots between them. Also, the domain of integration is set nicely below these signs if the \limits command is written immediately following the integral command:

$$\iint_X f(x,y)\,dx\,dy \qquad \iiint_X f(x,y,z)\,dx\,dy\,dz$$

$$\iiiint_X f(x,y,z,w)\,dx\,dy\,dz\,dw \qquad \int_X \cdots \int f(x_1,\ldots,x_k)$$

The code that generates these formulas has the following general pattern:

$$\text{\(iii)int\limits\_X}$$

5.5.7 Radicals

A better control for the placement of the root index is provided through the commands \leftroot and \uproot. These commands shift the index of the root, giving a better appearance in certain circumstances. In the following example, we move the letter μ 3 units up and 1 to the right:

$\sqrt[\mu]{\nu}$	`$\sqrt[\mu]{\nu}$`
$\sqrt[\mu]{\nu}$	`$\sqrt[\leftroot{-1}\uproot{3}\mu]{\nu}$`

5.5.8 Extensible Arrows

The commands \xleftarrow and \xrightarrow provide extensible arrows in order to accommodate expressions above and below them:

$$0 \xleftarrow{x \to -\infty} f(x) \xrightarrow[x \to \infty]{x \notin \mathbb{Q}} 1 \qquad \Big| \qquad \texttt{\$0\textbackslash xleftarrow\{x\textbackslash to -\textbackslash infty\} f(x)}$$

```
$0\xleftarrow{x\to -\infty} f(x)
\xrightarrow[x\to\infty]{x\notin\mathbb{Q}} 1$
```

5.5.9 Affixing Symbols to Other Symbols

Standard LaTeX provides the \stackrel command for placing something above a binary relation. The $\mathcal{AMS}$ packages provide more general commands, \overset and \underset. These work with anything and not only with binary relations:

$$\overset{\circ}{X} \qquad \Big| \qquad$$

```
$\overset{\circ}{\textrm{X}}$
$\underset{\ast}{\textrm{X}}$
```

5.5.10 Fractions and Related Constructs

The command \genfrac provides an easy interface to define new fractions. Its syntax is as follows:

```
\genfrac{left-delim}{right-delim}{line-thickness}
        {dtyle}{numerator}{denominator}
```

The left and right delimiters are used, for example, for binomial expressions. The line thickness refers to the fraction line and is set to 0 pt for binomial expressions. To select the style, we use a number from 0 to 3. The number 0 is for display style, 1 for text style, 2 for script style, and 3 for script-script style.

By default, the following commands are defined:

Command	Expansion
\tfrac{x}{y}	\genfrac{}{}{}{1}{x}{y}
\dfrac{x}{y}	\genfrac{}{}{}{0}{x}{y}
\binom{x}{y}	\genfrac{(}{)}{0pt}{}{x}{y}
\dbinom{x}{y}	\genfrac{(}{)}{0pt}{0}{x}{y}
\tbinom{x}{y}	\genfrac{(}{)}{0pt}{1}{x}{y}

The commands \tfrac and \dfrac provide convenient abbreviations for {\textstyle\frac{..}{..}} and {\displaystyle\frac{..}{..}}, respectively. Here is an example:

$\frac{1}{x}\log x$	`$$\tfrac{1}{x}\log x$$`
$\dfrac{1}{x}\log x$	`$$\dfrac{1}{x}\log x$$`

Here is an example of \dbinom and \tbinom:

$\binom{n}{k} + \binom{n}{k}\sqrt{\frac{\binom{n}{k}}{k!}}$	`$$\dbinom{n}{k}+` `\frac{\tbinom{n}{k}}{k!}$$`

The special command \cfrac is for writing continued fractions:

$a + \cfrac{1}{a + \cfrac{1}{a + \ddots}}$	```` $$ \cfrac{1}{a+ \cfrac{1}{a+ \cfrac{1}{a+{ \above0pt \ddots} }}} $$ ```

You can request that the numerators to be set to the left or right of the fraction line. This is accomplished by using \cfrac[l] or \cfrac[r].

5.5.11 The \smash Command

The \smash command zeros the depth (option b) or height (option t) of characters and is useful for alignments. In the following example we present two different formulas typeset using the \smash command (odd rows) and without using the \smash command (even rows). The reader should have a close look at the result to see the difference.

$\sqrt{x} + \sqrt{y} + \sqrt{z}$	`$\sqrt{x}+\sqrt{\smash[b]{y}}+\sqrt{z}$`
$\sqrt{x} + \sqrt{y} + \sqrt{z}$	`$\sqrt{x}+\sqrt{y}+\sqrt{z}$`
$(1 \quad \sqrt{\lambda_j})X$	`$(1-\sqrt{\smash[b]{\lambda_j}})X$`
$(1 \quad \sqrt{\lambda_j})X$	`$(1-\sqrt{\lambda_j})X$`

5.5.12 Operator Names

We saw in Section 5.4.2 how to define new functions/operators with standard LaTeX. The $\mathcal{AMS}$ packages provide an easy interface for this. If you want to define the operator \random, all you have to say is

```
\DeclareMathOperator{\random}{random}
```

There is also a starred form:

$$\text{\textbackslash DeclareMathOperator*\{\textbackslash Lim\}\{Lim\}}$$

This means that the defined operator should have subscripts and superscripts placed in the "limits" positions (above and below like, say, the \max operator).

In addition to the ones already predefined by standard LaTeX (see Table 5.24), we also have the following available:

\injlim (inj lim)   \lg (lg)   \projlim (proj lim)   \varlimsup ($\overline{\lim}$)
\varliminf ($\underline{\lim}$)   \varinjlim ($\underrightarrow{\lim}$)   \varprojlim ($\underleftarrow{\lim}$)

There is also the command \operatorname such that \operatorname{xyz} can be used as a binary operator. You can use \operatorname* in order to get limits.

5.5.13 The \mod Command and its Relatives

The several space conventions for the mod notation are handled by the commands \mod, \bmod, \pmod, and \pod. The second and third commands are available in standard LaTeX as well. Here is an example:

$\gcd(m, n \bmod n)$	`$\gcd(m,n\bmod n)$`
$x \equiv y \pmod b$	`$x\equiv y \pmod b$`
$x \equiv y \mod c$	`$x\equiv y\mod c$`
$x \equiv y \pod d$	`$x\equiv y \pod d$`

5.5.14 The \text Command

The command \text is provided for writing text in math mode. If the text is to be written in sub/super-script position, the text size is adjusted automatically, and this is its main advantage over the previously described method using a \mbox:

$f(x) \stackrel{\text{def}}{=} x^2$	`$$f(x)\stackrel{\text{def}}{=} x^2$$`

In a multilingual environment, the command will use the current text language and will accept language-specific commands.

5.5.15 Integrals and Sums

We have seen how to deal with stacked expressions under a \sum symbol using the \atop command. The $\mathcal{AMS}$ packages provide the command \substack and the slightly more general environment subarray, which has a column specifier:

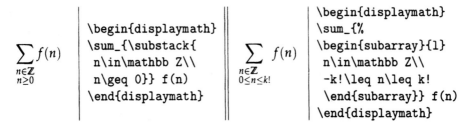

$$\sum_{\substack{n\in\mathbb Z\\ n\geq 0}} f(n)$$
```
\begin{displaymath}
\sum_{\substack{
   n\in\mathbb Z\\
   n\geq 0}} f(n)
\end{displaymath}
```

$$\sum_{\substack{n\in\mathbb Z\\ 0\leq n\leq k!}} f(n)$$
```
\begin{displaymath}
\sum_{%
\begin{subarray}{l}
   n\in\mathbb Z\\
   -k!\leq n\leq k!
\end{subarray}} f(n)
\end{displaymath}
```

If one wants to put accents and limits on a large operator, he or she can use the command \sideset. Here is an example that fully demonstrates the capabilities of this command:

$${}^2_1{\sum_a^b}{}^4_3 \qquad \texttt{\$\$\sideset\{\_1\^2\}\{\_3\^4\}\sum\_a\^b\$\$}$$

5.5.16 Commutative Diagrams

Commutative diagrams are supported with the amscd package. This is provided not as a complete solution but as a package for a quick diagram, drawn without diagonal arrows (for a complete solution, see Section 5.4.11). Consequently, we will not go to the trouble to describe the functionality of this package. Here is an example that demonstrates the use of the package:

$$\begin{CD} A @>a>b> B \\ @VcVV @AAdA \\ C @= D \end{CD}$$

```
\beg{displaymath}
\begin{CD}
A       @>a>b> B     \\
@VcVV          @AAdA\\
C       @=     D
\end{CD}
\end{displaymath}
```

5.5.17 Displayed Equations and Aligned Structures

Maybe the biggest advantage of using the *AMS* packages is the ability they give to better deal with displayed and aligned environments—much better than the already discussed eqnarray environment of LaTeX. These environments are:

equation	equation*	align	align*
gather	gather*	flalign	flalign*
multline	multline*	alignat	alignat*
split	gathered	aligned	alignedat

The eqnarray environment can still be used, but it is preferable to use the align environment or a combination of the equation environment plus the split environment.

The starred forms (except split) do not number the environment. You can also suppress the number of any line by putting \notag before the \\. The \nonumber command is still valid and can be used. If we want to give a specific tag to a line, we can use the \tag command:

$$x = y \qquad (x^y?)$$

```
\begin{equation}
x=y\tag{$x^y$?}
\end{equation}
```

Notice that \tag automatically puts the given text in parentheses. This can be avoided with its starred version: \tag*. The split environment can only be used inside some of the other environments (except multline).

The most important difference from the standard eqnarray environment is that here we have no extra white space around the aligned symbol, and the main syntactical difference is the use of a single & in front of the symbols to be aligned instead of surrounding the symbol by two &. Here are some examples:

$$\begin{aligned} a &= b+c+d \\ &+e+f \\ &= g+h \\ &= i \end{aligned} \qquad (5.9)$$

```
\begin{equation}
\begin{split}
a & = b+c+d \\
& \quad +e+f\\
& = g+h\\
& = i
\end{split}
\end{equation}
```

Notice here that the split environment is treated as one mathematical formula and therefore takes only one number. The same happens with multline:

$$a+b+c+d+e+f \\ +1+2+3+4+5 \qquad (5.10)$$

```
\begin{multline}
a+b+c+d+e+f\\
+1+2+3+4+5
\end{multline}
```

The multline environment is used for equations that do not fit on one line, and it always sets the first line of the equation flushed to the left and the last line flushed to the right (apart from the indent amount \multlinegap). If there are middle lines, they are centered independently within the display width. This can change with the commands \shoveleft and \shoveright. These commands take the entire equation line as an argument (except the final \\).

The gather environment allows us to write several displayed formulas without big vertical spaces separating them. Moreover, any of its equations can be a split environment:

$$a = b + c \qquad (5.11)$$
$$d + e = f + g$$
$$h = i \qquad (5.12)$$

```
\begin{gather}
a=b+c\\
  \begin{split}
    d+e & =f+g\\
    h &= i
  \end{split}
\end{gather}
```

The align environment is an enhanced version of the standard eqnarray. In addition to the better spacing around the aligned symbol, it also accepts many align points:

$$a = b + c \qquad h = i \qquad (5.13)$$
$$d + e = f + g \qquad j = k \qquad (5.14)$$

```
\begin{align}
a &= b+c & h &= i\\
d+e &= f+g & j &= k
\end{align}
```

The align environment can also be used for adding comments to equations:

$$a = b + c \quad \text{by axiom 5} \qquad (5.15)$$
$$d + e = f + g \quad \text{by the hypothesis} \qquad (5.16)$$

```
\begin{align}
a &= b+c & &
\text{by axiom 5}\\
d+e &= f+g & &
\text{by the hypothesis}
\end{align}
```

The variant alignat takes as an argument the number of columns that are to be aligned and leaves no space between them, which is useful for constructs such as

$$10x + 111y = 1 \qquad (5.17)$$
$$x + \quad y = 0 \qquad (5.18)$$

```
\begin{alignat}{2}
10&x+111&y=1\\
  &x+    &y=0
\end{alignat}
```

The flalign environment sets the equations flushed to the left and to the right of the display. That is why the starred version is to be preferred. If only one point of alignment is given, then it behaves like the align environment:

$$a = b + c \qquad\qquad h = i$$
$$d + e = f + g \qquad\qquad j = k$$

```
\begin{flalign*}
a &= b+c & h &= i\\
d+e &= f+g & j &= k
\end{flalign*}
```

The environments that we have seen thus far are designed to produce displays that occupy the full width of the display/page. The environments gathered, aligned, and alignedat occupy only the space needed by the equation. Thus, they can be used as building blocks for certain applications. For example:

$$
\left.\begin{array}{l} E = mc^2 \\ E = hv \end{array}\right\} \quad \text{well-known equations}
$$

```
\begin{equation*}
\left.\begin{aligned}
E &= mc^2\\
E &= h\nu
\end{aligned}\right\}
\qquad \text{%
well-known equations}
\end{equation*}
```

These -ed variants accept the optional arguments for vertical positioning [t] and [b] like the array environment.

Finally, the cases environment gives another way of typesetting nonanalytically defined functions with tighter spacing between the delimiter and the array:

$$
\chi_A = \begin{cases} 1 & \text{if } 1 \in A \\ 0 & \text{otherwise.} \end{cases}
$$

```
\begin{displaymath}
\chi_A$ =
\begin{cases}
1 & \text{if $1\in A$}\\
0 & \text{otherwise.}
\end{cases}
\end{displaymath}
```

Notice the absence of \{ in the input above!

LaTeX is not allowed to break the displays produced by the commands described thus far at the end of a page. To allow this in a particular line of an equation, you must use \displaybreak[n], where n can be either 0 or 1 or 2 or 3 or 4, immediately before the \\. If n is set to zero, it means that a break is permissible here, whereas if it is set to four, LaTeX is forced to break. If you want this policy to be used systematically for all equations, you can put the command \allowdisplaybreaks[n] in the preamble of your document, where n has the same meaning as the argument of the \displaybreak command. Recall here that * prohibits a page break.

Displays can also be interrupted for inserting text using the \intertext command. Align points are preserved:

$$
\begin{aligned}
x &= y & (5.19)\\
y &= z & (5.20)\\
\text{hence}&&\\
x &= z & (5.21)
\end{aligned}
$$

```
\begin{align}
x &= y\\
y &= z\\
\intertext{hence}
x &= z
\end{align}
```

5.5.18 Numbering Equations and Referencing

Equation numbers are usually set with respect to the section that the equation belongs to. Usually, the equation counter is used for this purpose; this counter is predefined for the book document class. For the article document class, we usually define

```
\renewcommand{\theequation}{\thesection.\arabic{equation}}
```

This works fine except that it must be reset at the beginning of each new section using the \setcounter command. The $\mathcal{AMS}$ packages make this easier by providing the \numberwithin command, so we can set

```
\numberwithin{equation}{section}
```

Of course, this command can be applied to any other counter. Adjusting the tag placement can be done using the \raisetag command. For example, \raisetag{6pt} will raise the tag by 6 pt. For cross referencing, we additionally have the \eqref command as well as the standard \ref command. The only difference is that \eqref also provides the parentheses around the equation number.

Finally, we can create subordinate equations with the subequations environment:

$$x = y \tag{5.22}$$

$$y = z \tag{5.23a}$$
$$z = w \tag{5.23b}$$

```
\begin{equation}
x=y
\end{equation}
\begin{subequations}
  \label{subeqnarray}
\begin{eqnarray}
y &=& z\\
z &=& w
\end{eqnarray}
\end{subequations}
```

Notice that we still need a math display environment inside the subequations environment. This inner environment can be any of the ones discussed in the previous section (here we used the standard eqnarray). In addition, you will see that the numbers start from the next number *after* the last equation. The \ref or \eqref command for these will not produce the subordinate numbering but the parent one instead if the label is immediately after the start of the subequations environment. Thus, if we refer to the last eqnarray, we get (5.23).

The counters involved in the subequations are parentequation and equation. So, if we want to change something, we can use standard LATEX commands. For example:

```
\begin{subequation}
\renewcommand{\theequation}{%
  \theparentequation \roman{equation}}
..............................
\end{subequation}
```

5.5.19 Matrices

For matrices, the environments differ from the standard `array` environment in that they have predefined delimiters and a more compact appearance. On the negative side is the fact that they do not allow alignment of the entries. For such a task, we must use the `array` environment. The available environments are the `pmatrix` (with parentheses as delimiters), `bmatrix` (with delimiters [ ]), `vmatrix` (with delimiters | |), and `Vmatrix` (with delimiters ‖ ‖). In addition to these, we have the `matrix` environment (for which delimiters must be provided) and the `smallmatrix` environment for matrices such as $\left[\begin{smallmatrix} a & b \\ c & d \end{smallmatrix}\right]$ that fit nicely inside text. Notice that for the `smallmatrix` environment the delimiters must also be provided. Thus, the small matrix above was

```
$\left[\begin{smallmatrix}
        a & b \\ c & d
\end{smallmatrix}\right]$
```

The command `\hdotsfor{number}` produces a row of dots that spans the given *number* of columns. Here is an example:

$$
\begin{matrix}
a & b & c & d \\
e & & \dots
\end{matrix}
$$

```
\begin{displaymath}
\begin{matrix}
a&b&c&d\\
e&\hdotsfor{3}
\end{matrix}$$
\end{displaymath}
```

The `\hdotsfor` command takes an optional argument that is used as a multiplicative factor for the distance between consecutive dots:

$$
\begin{bmatrix}
a_{11} & a_{12} & a_{13} & \dots & a_{1n} \\
a_{21} & a_{22} & a_{23} & \dots & a_{2n} \\
\hdotsfor{5} \\
\hdotsfor{5} \\
\hdotsfor{5} \\
a_{n1} & a_{n2} & a_{n3} & \dots & a_{nn}
\end{bmatrix}
$$

```
\begin{displaymath}
\begin{bmatrix}
a_{11} & a_{12} &
a_{13} & \dots  & a_{1n}\\
a_{21} & a_{22} &
a_{23} & \dots  & a_{2n}\\
\hdotsfor[.5]{5}\\
\hdotsfor{5}\\
\hdotsfor[3]{5}\\
a_{n1} & a_{n2} &
a_{n3} & \dots  & a_{nn}
\end{bmatrix}
\end{displaymath}
```

5.5.20 Boxed Formulas

The `\boxed` command puts a frame around its argument:

$$\boxed{a+b=c} \qquad (5.24)$$

```
\begin{equation}
\boxed{a+b=c}
\end{equation}
```

5.5.21 Customizing Theorems

Customization of the appearance of the theorem-related environments can be done by using the amsthm package. Like the theorem package discussed in Section 5.4.8, it recognizes the \theoremstyle specification and has a starred form for unnumbered environments. It also defines a proof environment that automatically enters the QED symbol at the end of the proof. The available theorem styles are plain, definition, and remark. The theorem styles are declared before the relative \newtheorem commands. For example, the preamble of a document may contain

```
\theoremstyle{plain}% default
\newtheorem{thm}{Theorem}[section]
\newtheorem{prop}[thm]{Proposition}
\newtheorem{lem}[thm]{Lemma}
\newtheorem{cor}[thm]{Corollary}

\theoremstyle{definition}
\newtheorem{defn}[thm]{Definition}
\newtheorem{exa}[thm]{Example}

\theoremstyle{remark}
\newtheorem{rem}[thm]{Remark}
\newtheorem*{note}[thm]{Note}
```

The \swapnumbers command is available in order to allow the theorem number to be printed before the theorem header:

5.1 Theorem. *Every bounded monotone sequence in* $\mathbb{R}$ *converges.*

Corollary 5.2. *A monotone sequence in* $\mathbb{R}$ *either converges or diverges to either* ∞ *or* ∞.

This was produced by issuing the \swapnumbers command before the definition of the theorem environment in the preamble and before the definition of the corollary environment.

The package also provides the command \newtheoremstyle, which enables us to define custom theorem styles. The syntax is

```
\newtheoremstyle{name of style}{space-above}
                {space-below}{body-font}
                {indent amount}{Thm head font}
                {punctuation after Thm head}
                {space after Thm head}{Thm head spec}
```

The *space-above* is the space between the theorem and the last line of the previous paragraph. The *space-below* is the space between the theorem and the next paragraph. If you leave these two empty, then the "usual" space will be used. The *body-font* declaration needs no explanation (you can use, for example, \itshape). The *indent amount* is the indentation space before the header begins. If you leave this empty, then no indentation will be used. You can also use an already defined length such as \parindent for paragraph-like indentation. For example, *Thm head font* can be set to \bfseries and the *punctuation after Thm head* is the punctuation that will be set after the theorem (for example, a dot: {.}). Finally, *space after Thm head* is self-explanatory (if set to \newline it will create a line break after the theorem head) and the *Thm head spec*, if set to [\thmnot#3], will produce the comment in the theorem's header, read from the square brackets of \begin{thm}[Thm Head spec].

The proof environment has an optional argument that customizes the head of the proof. For example, you may say

$$\text{\begin{proof}[Proof of the main theorem]}$$

The shape of the QED symbol is controlled by \qedsymbol. The default is □. It can also be accessed, if necessary, by the \qed command. It frequently happens in mathematics that a proof may end with a display equation or array. In these cases, the position of the QED symbol is problematic. The nice way is to use the command \qedhere at the end of the line where the display ends. For example,

Proof.

$$f(x) = x \quad x$$
$$= 0 \qquad\qquad\qquad\qquad\qquad\qquad □$$

was produced by

```
\begin{proof}
  \begin{align*}
    f(x) &=x-x\\
         &=0\qedhere
  \end{align*}
\end{proof}
```

It is important to note that the amsthm package must be loaded *before* the amsmath package. It is automatically loaded with the $\mathcal{A}_{\mathcal{M}}\mathcal{S}$ document classes. In case you get an error relating to the \qedhere command, try \mbox{\qedhere} instead.

5.5.22   Options of the **amsmath** Package

The following options are available for the amsmath style file:

centertags (default) place the equation number of a split environment vertically centered on the total height of the environment.

tbtags place the number of the split environment at the bottom of the environment if the tags are on the right and at the top when the tags are on the left.

sumlimits (default) place the sub/super-scripts of summation symbols above and below the symbol in displayed equations. This affects products, direct sums, and direct products as well, but not integrals (see two items below).

nosumlimits always place the sub/super-scripts to the side of a sum (or similar) symbol, even in displays.

intlimits The same as sumlimits but for integrals.

nointlimits (default) The opposite of intlimits.

namelimits (default) Like sum limits, but for certain operator names such as det, inf, lim, max, min, and so on.

nonamelimits The opposite of namelimits.

leqno Place equation numbers on the left.

reqno Place equation numbers on the right.

fleqn Position the equations at a fixed indentation from the left margin (not centered).

You can choose any of the options above with the optional argument of the \usepackage command; for example, \usepackage[nosumlimits]{amsmath}.

5.5.23 Converting from Standard LaTeX to the $\mathcal{AMS}$ Packages

If you have already written something but now you want to load the amsmath and other $\mathcal{AMS}$ packages, you usually have only to load them and your files should run successfully (provided that they are in LaTeX 2_ε). Some changes that you may want to do are to substitute all eqnarray environments with the align environment and to use the proof environment combined with the \qedhere command for your proofs that end with displays.

5.5.24 The amsart Top Matter Commands

The top matter of an $\mathcal{AMS}$ article document contains information about the author, the title subject classification, the key words, the abstract, and so on. In Table 5.25, we see the commands defined by the amsart class for such information. All arguments in square brackets are optional and are not always necessary. A short title should be provided for use in the running heads if the title is too long. In this case, we can force a line break in the title using \\ or let LaTeX take care of this. An author command is given for any author separately. The optional argument is for a shortened name, such as

\author[L. Euler]{Leonard Euler}

Table 5.25: amsart top matter commands

\title[*short-title*]{*title*}	\author[*short-name*]{*name*}
\address{*address*}	\curraddr{*current-address*}
\email{*email*}	\urladdr{*URL*}
\dedicatory{*dedication*}	\date{*date*}
\thanks{*thanks*}	\translator{*translator's name*}
\keywords{*comma separated key words*}	\subjclass[2000]{*Primary; Secondary*}
\begin{abstract}...\end{abstract}	\maketitle

 If there are many authors and their names do not fit into the running head, then we can replace the running head names with "FIRST AUTHOR ET AL." by using \markboth{FIRST AUTHOR ET AL.}{*short title in all caps*} (for more details on the \mathboth command, see the next chapter.) The \markboth command should come after the \maketitle command.

For each author, we must provide an address. Line breaks in the address are again by \\. The same applies for the \curraddr if an author is currently in another (temporary) address. The e-mail address is written as usual (user@server.domain) and for the URL address, if a ~ is needed, we use the command \textasciitilde. The class will automatically use labels like "*E-mail address:.*" In a multilingual environment, you should be careful, as these terms may not be translated in the main document language.

The \thanks field is provided for acknowledgments of grants and support, and it can appear more than once in the top matter.

The subject classification and keywords appear as footnotes but without footnote-mark. The 2000 in the subject classification's optional argument follows the 2000 Mathematics Subject Classification scheme (http://www.ams.org/msc). If the optional argument is omitted, the 1991 Subject Classification will be used.

Finally, for the abstract, let us note that it should be placed *before* the \maketitle command.

5.6 From Λ to MathML

This section shows how we can generate XML content, in general, and MathML content, in particular, from Λ input files. Therefore, the material presented here is not necessary for the understanding of the rest of the book.

XML, the eXtensible Markup Language, is a standard for document markup that is getting universal acceptance. Data can be marked up with simple, human-readable *tags*. In addition, XML is becoming the standard format for computer-related documents.

XML *elements* are delimited by start and end tags. Start tags begin with a <, and end tags begin with a </. Both of these are followed by the name of the element and are closed by a >. For example,

<title> My Article </title>

is a simple example of a *title* element. Note that this element has content. There are elements without content, which are called *empty* elements. Empty elements start with a <, the name of the tag, and close with a /> (e.g., <hr/>). XML itself does not specify any particular formatting; rather, it specifies the rules for tagging elements. These tags can then be interpreted to format elements in different ways. SGML, the Standard Generalized Markup Language, is a system for organizing and tagging elements of a document. SGML, like XML, is used to mark up documents, but, unlike XML, it is a very complex system.

MathML is an XML application that is primarily intended to facilitate the use and reuse of mathematical and scientific content on the Web. Of course, other applications, such as computer algebra systems and print typesetting, are possible as well. In general, MathML markup is embedded into HTML documents. But, currently, only Mozilla, Netscape's successor, can render MathML content.

Since many people would really love a tool that would allow them to write ordinary LATEX content and transform it very easily to MathML, the authors of Ω have extended this system so that it can directly produce MathML content. These new features are described in brief in [10]. Here, we will try to completely document these new features.

The command \MMLmode[2] tells Ω to enter MathML mode; that is, now mathematical formulas will be output as MathML instructions. The command \noMMLmode cancels the effect of \MMLmode so mathematical formulas are output as DVI instructions. Nevertheless, it is not enough to enter MathML mode in order to get MathML output: we must delimit each mathematical formula by the commands \MMLstarttext and \MMLendtext. Let us see a concrete example. Suppose that the code that is shown on the right of Table 5.26 is stored in file math.tex. Then, Λ will generate two output files—math.mml, which will contain the MathML content (or, in general, the XML content), and math.dvi, which is just a normal DVI file. The code at the left of Table 5.26 is the output generated by Λ: Now, the next step is to see how we can generate a complete HTML or XML file from a Λ input file. If we feed Λ with the following input file[3]

```
\documentclass{article}
\begin{document}
  \MMLmode
  \SGMLwrite{<!-- Generated by Omega version \OmegaVersion-->}
  \SGMLwriteln
  \SGMLstarttexttag{html}%
```

---

2. The discussion aplies to Ω version 1.15, new features introduced to versions 1.23 and later are discussed on Appendix E.

3. Whatever appears between <!-- and --> is considered a comment and it is ignored.

Table 5.26: A Λ input file and the generated MathML content.

```
<mtext>
  <inlinemath>
    <math>
      <mrow>                         \documentclass{article}
        <mi> E </mi>                 \begin{document}
        <mo> = </mo>                   \MMLmode
        <mi> m </mi>                   \MMLstarttext
        <msup>                           $E=mc^2$
          <mi> c </mi>                 \MMLendtext
          <mn> 2 </mn>                 \noMMLmode
        </msup>                          $E=mc^2$
      </mrow>                         \end{document}
    </math>
  </inlinemath>
</mtext>
```

```
  \SGMLstarttexttag{head}%
  \SGMLstarttexttag{title}%
  A Simple HTML document
  \SGMLendtexttag{title}%
  \SGMLendtexttag{head}%
  \SGMLstarttexttag{body}%
  \SGMLemptytag{hr}{}
  This is a simple HTML document
  \SGMLlonetag{hr width="50\SGMLpercent"}{}%
  \SGMLendtexttag{body}%
  \SGMLendtexttag{html}%
\end{document}
```

then we will get the following output:

```
<!-- Generated by Omega version 1.15-->
<html>
  <head>
    <title>
      A Simple HTML document
    </title>
  </head>
  <body>
    <hr/>
    This is a simple HTML document
    <hr width="50%">
```

```
    </body>
  </html>
```

The command \SGMLstarttexttag is used to specify start tags and the command \SGMLendtexttag is used to specify end tags. The commands \SGMLemptytag and \SGMLlonetag are used to specify empty tags and lonely tags, which are tags similar to the <p> and <br> tags of the HTML markup language. The command \SGMLwrite is used to output content to the .mml file, while the command \SGMLwriteln just changes line to the output file. The command \OmegaVersion prints the current version of the Ω typesetting engine. Since certain symbols have a predefined meaning but, at the same time, are frequently used in XML and SGML content, Ω provides the following commands, which generate the symbols on the left-hand side of each column.

\SGMLampersand	&	\SGMLbackslash	\
\SGMLcarret	^	\SGMLdollar	$
\SGMLleftbrace	{	\SGMLrightbrace	}
\SGMLhash	#	\SGMLpercent	%
\SGMLunderscore	_		

The commands \SGMLstartmathtag and \SGMLendmathtag are used to introduce MATHML elements. These commands are necessary when one wants to create macros and so forth.

Although the Ω MATHML engine is capable of handling a wide range of mathematical expressions, there are still cases where it fails. The main problem is that it cannot handle complex macros well. For example, the \sqrt command is such a case:

```
<mtext>
    <inlinemath>
        <math>
                    <mi> x </mi>
        </math>
    </inlinemath>

</mtext>
```

```
\documentclass{article}
\begin{document}
\MMLmode
\MMLstarttext
$\sqrt{x}$
\MMLendtext
\end{document}
```

To assist Ω to produce correct MATHML output, we need to redefine the \sqrt macro:

```
\renewcommand{\sqrt}{\@ifnextchar[\sqrttwo\sqrtone}
\newcommand{\sqrtone}[1]{%
    \SGMLstartmathtag{msqrt} #1 \SGMLendmathtag{msqrt}}
\def\sqrttwo[#1]{\sqrttwoend{#1}}
\newcommand{\sqrttwoend}[2]%
{\SGMLstartmathtag{mroot} {#2} {#1} \SGMLendmathtag{mroot}}
```

The command \@ifnextchar checks whether the next input character is the same with the character that follows the command and, if it is, it executes the first command that follows; otherwise, it executes the second command. Now, we try again the example above to see what we will get:

```
<mtext>
    <inlinemath>
        <math>
            <msqrt>
                <mi> x </mi>
            </msqrt>
        </math>
    </inlinemath>
</mtext>
```

```
\documentclass{article}
.....................
\begin{document}
\MMLmode
\MMLstarttext
$\sqrt{x}$
\MMLendtext
\end{document}
```

Of course, now we get correct output. The same technique can be used to create a new document class, which would be used to create output, for example, in the DocBook format.

The following exercise assumes familiarity with HTML, so readers not familiar with this markup can skip it.

➤ **Exercise 5.4** Design a Λ file that will generate a simple HTML file that will display the text *Planck's Equation: E = hv* centered. ☐

5.7 Generating OMDoc Files

MathML is not the only XML application related to mathematical markup. OMDoc is an XML application that allows us to represent the semantics and structure of various kinds of mathematical documents. On the other hand, OpenMath and MathML only partially fulfill the goal of establishing a basis for communication between mathematics and mathematical software systems (and humans) since they exclusively deal with the representation of the various mathematical objects in particular and not with mathematical documents in general. Roughly speaking, it is possible to transform any kind of XML content to LATEX by using XSLT. The latex2omdoc package (by Michael Kohlhase) is a tool that allows the generation of OMDoc content from LATEX markup. The following example shows a simple LATEX file that when processed will generate OMDoc content.

```
\documentclass[a4paper]{article}
\usepackage{latex2omdoc}
\begin{document}
  \begin{omdocout}
    \begin{ommetadata}
      \dctitle{Simple OMDoc Example}
      \dccreator[edt]{Jim Editor}
      \dccontributor{Steve Author}
```

```
    \dcdate{\today}
  \end{ommetadata}
  \omchapter{About programming}
  \omsection{About Perl}
  \ommigtheory{About Perl being a fine language}
  \begin{theorem}
    \begin{omverb}
      Perl is a fine programming language.
    \end{omverb}
  \end{theorem}
  \ommigtheory{About Perl being an effective language}
  \begin{lemma}
    \begin{omverb}
      Perl is also effective.
    \end{omverb}
  \end{lemma}
  \omref{http://www.mathweb.org/omdoc/}
  \end{omdocout}
\end{document}
```

Before we give you the OMDoc output, we should mention that LaTeX creates a file that has the same filename as the input file but with the .omdoc filename extension. The environment omdocout creates the output file and writes the file header. The ommetadata environment is used to add metadata information to the output file with the commands \dctitle, \dccreator, \dccontributor, \dcdate. The \dctitle command has an optional argument that should be used to specify the language of the title [default language is en(glish)]. Naturally, it is possible to include metadata information within the various environments that the package provides. The nonstandard environments theorem, lemma, corollary, conjecture, definition, and remark are used to generate the corresponding markup. The environment omverb should be used to enter content inside the tags. Of course, the package offers some other features and commands, but they are meaningless unless one is familiar with OMDoc.

Here is the OMDoc file that corresponds to the LaTeX file above:

```
<?xml version="1.0"?>
<!DOCTYPE omdoc SYSTEM
   "http://www.mathweb.org/omdoc/dtd/omdoc.dtd" []>
<!-- generated from t.tex, do not edit -->
<omdoc id="top">
<metadata>
 <dc:Title xml:lang="en"> Simple OMDoc Example
 </dc:Title>
 <dc:Creator role="edt">  Jim Editor
 </dc:Creator>
```

```
   <dc:Contributor role="aut"> Steve Author
   </dc:Contributor>
  <dc:Date>February 3, 2002</dc:Date>
  </metadata>
  </omgroup>
  <omgroup id="0{-{About programming}}" type="About programming">
  <metadata>
   <dc:Title xml:lang="en">
   chapter
   </dc:Title>
  </metadata>
  </omgroup>
  <omgroup id="0{-{About Perl}}" type="About Perl">
  <metadata>
   <dc:Title xml:lang="en">
   section
   </dc:Title>
  </metadata>
  <assertion id="1-theorem"
  theory="About Perl being a fine language" type="theorem">
  <CMP xml:lang="en" format="omtext">
  Perl is a fine programming language.
  </CMP>
  </assertion>

  <assertion id="2-lemma"
  theory="About Perl being an effective language" type="lemma">
  <CMP xml:lang="en" format="omtext">
  Perl is also effective.
  </CMP>
  </assertion>

  <ref xlink:href="http://www.mathweb.org/omdoc/"/>
  </omdoc>
```

6

MORE ON THE CORE

In this chapter, we describe additional topics not covered in the previous chapters: labels, horizontal and vertical spacing, page breaking, floating objects, marginal notes, page parameters and page setup, slide preparation, TeX boxes, lines, the definition of new commands, environments and catalogs, the file input family of commands, and the interactive use of LaTeX.

6.1 Labels and References

As we have already mentioned in the introduction, one of the benefits of using LaTeX instead of an ordinary WYSIWYG document preparation system is that one does not have to take care of the various labels (e.g., section labels, equation labels and others), as these are automatically generated by LaTeX. However, in many instances one wants to be able to refer to a particular label or page number. In this case, all we have to do is to put a \label command after the command that generates the label and to use a reference command at the place where we want to make the reference. The command \label has one argument, which is a symbolic name for a particular label. The symbolic name may consist of letters, digits, and punctuation symbols:

```
\section{On the Structure of DNA}\label{dna:struct}
```

To refer to this particular label anywhere in our document, we must use the command \ref:

```
As we describe on section~\ref{dna:struct}...
```

The character ~ creates a space that cannot be chosen by LaTeX to split a line. So, we make sure that the word before the reference and the reference itself will appear on the same output line. Of course, one may choose not to use this trick. But then the document may be highly unreadable. Note that in some languages this special character is used to accent letters. If the reader uses LaTeX to typeset a text in such a language, we suggest the use of the \nobreakspace command:

```
As we describe on section\nobreakspace\ref{dna:struct}...
```

If one wants to refer to the page number where a particular label occurs, the command \pageref needs to be used:

```
As we describe on page~\pageref{dna:struct}...
```

When processing a LATEX file that contains references to labels for the first time, LATEX is not able to correctly resolve the labels, so one has to run LATEX on the file a second time. If there is a reference to a nonexisting label, LATEX will complain that *there were undefined references*. Note that one will get this message anyway the very first time LATEX is run on a file.

Suppose that one wants to refer to both an equation and the page where this equation actually appears. A possible solution to this problem follows:

```
see~\ref{eqn} on page~\pageref{eqn}
```

However, this solution has the drawback that if the equation happens to be on the same page where we actually put the reference, the result looks odd and in that instance should be replaced by

```
see~\ref{eqn}
```

Since it is inconvenient to try to keep track of all references and labels, Frank Mittelbach has created the varioref package that takes care of such problems. When using this package, you will usually use the commands \vref and \vpageref instead of \ref and \pageref. The command \vref is equivalent to the \ref command when the reference and the \label are on the same page. In cases where the label and the reference are on pages that differ by one, the command will in addition produce a short phrase such as "on the following page." For example, in the excerpt

```
Equation \vpageref{eq:euler} is known as Euler's equation.
\newpage
\begin{equation}
e^{i\pi}+1=0\label{eq:euler}
\end{equation}
```

the first sentence will be typeset as

Equation 1 on the following page...

The command \newpage forces LATEX to start a new page. Finally, if the reference and the label are on pages that differ by more than one, it will produce both the \ref and the \pageref. The command \vpageref will produce the same phrase as \vref except that it does not start with the \ref. For example, in

```
The equation \vref{eq:euler} is known as Euler's equation.
\newpage
\begin{equation}
```

```
e^{i\pi}+1=0\label{eq:euler}
\end{equation}
```

the first sentence will be typeset as

> The equation on the following page...

Nevertheless, it is possible to control the output of this command by using the two optional arguments that this command can take. With the first argument, one can specify the text that should be used if the label and the reference fall on the same page. This feature is particularly useful when the label and the reference are near each other. When we know if the label is before or after the reference, we say something like

> `... see the example \vpageref[above]{ex:1} ...`

This will be typeset as "...see the example above..." if the label and the reference are on the same page, but as "...see the example on the page before..." if the reference comes after the label. The second optional argument can be used if we want finer control over the output phrase. For example, the input text

```
The \vpageref[following equation][equation ]{eq:euler}
is known as Euler's equation.
\begin{equation}
e^{i\pi}+1=0\label{eq:euler}
\end{equation}
```

will be typeset as

> The following equation is known...

On the other hand, the input text

```
The \vpageref[following equation][equation ]{eq:euler}
is known as Euler's equation.
\newpage
\begin{equation}
e^{i\pi}+1=0\label{eq:euler}
\end{equation}
```

will be typeset as

> The equation on the following page is known...

If we want to refer to a range of labels (i.e., labels that appear in consecutive order), we can use the \vrefrange command. So, if we want to collectively refer to labels eq:1, eq:2, and eq:3, we can use the following input text:

> `The equations~\vrefrange{eq:1}{eq:3} are called...`

The text above will be typeset as

The equations 3.1 to 3.3 are called. . .

The command above has an optional argument that can be used when all of the labels are placed on the same page. The command \vpagerefrange has functionality similar to the \vrefrange command. If both labels fall onto the same page, the command acts exactly like the \vrefrange. Otherwise, it produces something like "on pages 21–23." The command may take an optional argument that can be used in cases where all labels are placed on the current page.

The command \vrefpagenum is provided to allow users to write their own little commands that implement functions similar to those provided by the two previous commands. This command has two arguments: the first is the name of an arbitrary command that receives a page number related to the second argument, which is a label. So, if we have two (or more) labels, we can get their page numbers, compare them, and then decide what to output. Here is a slightly modified example from the documentation of the package:

```
\newcommand{\myvrefrange}[2]{%
    \vrefpagenum{\fstnum}{#1}%
    \vrefpagenum{\sndnum}{#2}%
    \ifthenelse{\equal{\fstnum}{\sndnum}}{%
        \ref{#1} and \ref{#2} \vpageref{#1}}{%
        \ref{#1} \vpageref{#1} and \ref{#2} \vpageref{#2}}%
}
```

The command makes use of the package ifthen. The first sentence in the following input text

```
The equations \myvrefrange{eq:euler}{eq:einstein} are Euler's
and Einstein's equations.
\newpage
\begin{equation}
e^{i\pi}+1=0\label{eq:euler}
\end{equation}
\newpage
\begin{equation}
E=mc^2\label{eq:einstein}
\end{equation}
```

will be typeset as

The equations 1 on the following page and 2 on page 3 are. . .

If we are not satisfied with the predefined phrases that the commands above produce, we can customize them by redefining the following commands:

\reftextbefore This command is used for backward references when the label is on the preceding page but is invisible.

\reftextfacebefore This command is used when the label is on the facing page (i.e., if the current page number is odd).

\reftextafter This command is used when the label comes on the next page but one has to turn the page.

\reftextfaceafter This command is used when the label is on the following facing page.

\reftextfaraway This command is used whenever the label and the reference are on pages that differ by more than one or when the page numbers are not Arabic numerals.

\vreftextvario This command has two arguments and outputs one of the two arguments depending on the number of \vref and/or \vpageref commands that are already seen. Here is an example:

```
\newcommand{\reftextafter}{%
  on the \reftextvario{following}{next} page}
```

\reftextpagerange This produces text that describes the page range of two labels denoting a "label" range.

\reftextlabelrange This command produces text that describes the range of equations, figures, and so forth.

The varioref package defines these commands so that they produce the correct phrases when used with the babel package.

6.2 Hyper-references

One of the main reasons that made the Portable Document Format (or PDF, for short) so popular is the fact that people can read and print the same document in a wide range of computing platforms. Another important reason is that PDF documents can contain hyperlinks (or hyper-references) to parts of the document or to external URLs. Additionally, recent versions of the PDF format allow the inclusion of *forms* (a template for data or text input) just like those found in many HTML documents. Since forms are useful only when they are interactive, PDF forms can be made interactive by using the JavaScript programming language or by invoking external cgi-scripts.

The hyperref package by Sebastian Rahtz, when used on LATEX sources that are processed by pdfLATEX, allows people to easily create PDF documents with hyperlinks and/or forms. However, there are many useful packages that cannot be used with pdfLATEX(e.g., the PSTricks packages; see section 9.9.2 on page 286), so the package allows users to generate PostScript files. When we transform these PostScript files to PDF files, the latter will actually have the "hyper" features that we described in our LATEX source file. The package can be used with more or less any normal LATEX document. However, the package provides a number of options. These options can be specified via the \usepackage mechanism using a single 'key=value' scheme, such as

\usepackage[linkcolor=blue,pdfpagemode=FullScreen]{hyperref}

Some of the options are characterized as Booleans. In this case, we can enable their use by simply specifying the corresponding key; otherwise, we can set them to either *true* or *false*. Alternatively, we specify the options with the command \hypersetup:

\hypersetup{backref,linkcolor=blue,pdfpagemode=FullScreen}

If we have a number of options that we always want to use, we can create a file called hyperref.cfg (which we store in a place accessible by pdfTEX) to which we will store only a \hypersetup command. If we want to override any of the options, we have to specify the changes in the options part of the \usepackage command.

We will now briefly describe most of the available options. The full list of options is available from http://www.tug.org/applications/hyperref/manual.html. Note that default options are overridden automatically when selecting some other option.

General options These options can be used to specify general behavior and page size.

draft (Boolean, default: false) All hypertext options are turned off.

debug (Boolean, default: false) Additional diagnostic messages are printed in the log file.

a4paper (Boolean, default: true) Sets paper size to A4.

a5paper (Boolean, default: false) Sets paper size to A5.

b5paper (Boolean, default: false) Sets paper size to B5.

letterpaper (Boolean, default: false) Sets paper size to letter paper size.

legalpaper (Boolean, default: false) Sets paper size to legal paper size.

executivepaper (Boolean, default: false) Sets paper size to executive paper size.

Configuration options

raiselinks (Boolean, default: true) The relevant commands that deal with hyperlinks reflect the real height of the links.

breaklinks (Boolean, default: false) Allows link text to break across lines.

pageanchor (Boolean, default: true) Determines whether every page is given an implicit anchor at the top left corner. If this is turned off, the table of contents will not contain hyperlinks.

plainpages (Boolean, default: true) Forces page anchors to be named by the Arabic form of the page rather than the formatted form.

nesting (Boolean, default: false) Allows links to be nested; this option is not actually supported.

Backend drivers

The package provides a number of drivers, but here we will use the pdftex driver and the dvips driver. The former has been set up for use with pdfTEX and the latter for use with DVIPS. In case we want to produce an HTML file from a LATEX document, we may opt to use the latex2html driver. In addition, the hypertex option is useful when previewing with xdvi or when generating a PostScript file with dvips with the -z option.

Extension options

extension (Text) Sets the file extension that will be appended to the links file created when one uses the hyper-xr package. This package is a modified version of the xr package by David Carlisle and Jean-Pier Drucbert. The xr package implements a system for eXternal References. If one file needs to refer to sections of another file, say aaa.tex, then this package may be loaded in the main file, and the command \externaldocument{aaa} must be given in the preamble. Then, one may use \ref and \pageref to refer to labels defined either in aaa.tex or in the main file.

hyperfigures (Boolean, default: false) Makes included figures hypertext links (see Chapter 9 for details on the inclusion of figures).

backref (Boolean, default: false) Adds "back-link" text to the end of each item in the bibliography as a list of section numbers. This works properly only if there is a blank line after each \bibitem.

pagebackref (Boolean, default: false) Adds "back-link" text to the end of each item in the bibliography as a list of page numbers.

hyperindex (Boolean, default: false) Makes the text of index entries into hyperlinks.

colorlinks (Boolean, default: false) Colors the text of links and anchors.

linkcolors (Color, default: red) The color for normal links.

anchorcolor (Color, default: black) The color for anchor text.

citecolor (Color, default: green) The color for bibliographic citations in text.

filecolor (Color, default: magenta) The color for URLs that open local files.

menucolor (Color, default: red) The color for Acrobat menu items.

pagecolor (Color, default: red) The color for links to other pages.

urlcolor (Color, default: cyan) The color for linked URLs.

PDF display and information options

baseurl (URL) Sets the base URL of the PDF document.

pdfpagemode (Text, default: empty) Determines how the file opens in Acrobat; the possible values are None, UseThumbs (show thumbnails), UseOutlines (show bookmarks), and FullScreen. If no mode is explicitly chosen but the bookmarks option is set, UseOutlines is used.

pdftitle (Text, default: empty) Sets the document information Title field.

pdfauthor (Text, default: empty) Sets the document information Author field.

pdfsubject (Text, default: empty) Sets the document information Subject field.

pdfcreator (Text, default: LaTeX with hyperref package) Sets the document information Creator field.

pdfproducer (Text, default: the current version of pdfTEX) Sets the document information Producer field. The default value depends on the version of pdfTEX.

pdfkeywords (Text, default: empty) Sets the document information Keywords field.

pdfview (Text, default: fitbh) Sets the default PDF "view" for each link. The possible values are shown in Table 6.1.

pdfstartpage (Integer, default: 1) Determines the page on which the PDF file is opened.

pdfstartview (Text, default: FitB) Sets the startup page view.

pdfpagescrop (Four numbers, no default value) Crops the page and displays/prints only that part. The four numbers correspond to the coordinates of the lower-left corner and the upper-right corner of the page.

Table 6.1: Outline and destination appearances.

Value	Description
Fit	fit the page in the window
FitH	fit the width of the page
FitV	fit the height of the page
FitB	fit the "bounding box" of the page
FitBH	fit the width of the "bounding box"
FitBV	fit the height of the "bounding box"
XYZ	keep the current zoom factor; can optionally be followed by a zoom factor that must be an integer

It is possible to activate the menu options of the Acrobat Reader by using the command \Acrobatmenu. This command has two arguments: the name of the Acrobat Reader menu option and the text that will be displayed on the screen. By clicking on a word, Acrobat Reader will activate the corresponding menu option. Some interesting menu options are: FirstPage, PrevPage, NextPage, LastPage, and Quit. The reader should consult the online manual for the full list of menu options. Here is a simple example:

```
\documentclass{article}
\usepackage[a4paper,pdfpagemode=FullScreen]{hyperref}
\begin{document}
\Acrobatmenu{Quit}{\LARGE QUIT}
\end{document}
```

If the PDF file corresponding to the code above is viewed with Acrobat Reader, it will be displayed in full screen. By clicking on the word QUIT, we terminate Acrobat Reader. We can create visually appealing PDF files if we use images instead of words for actions. Moreover, by using the \Acrobatmenu command, we can create slide shows.

If we need to make references to URLs or write explicit links, the following commands are provided:

\href{*URL*}{*text*} The *text* is made a hyperlink to the *URL*; the *URL* must be a full URL (e.g., http://www.tug.org). Special characters such as # and ˜ lose their special meaning and are treated as "letters." If we have defined a base URL with baseurl, then the *url* can be relative to it. Here is an example that starts the Web browser and points it to the TUG Web page:

\href{http://www.tug.org}{\LARGE TUG}

Suppose now that we have set

baseurl={http://www.tug.org}

Then, the following is a hyperlink to the applications section of the TUG Web pages:

\href{applications/index.html}{\LARGE Applications}

Note that the package makes use of the url package by Donald Arseneau to typeset URLs. This package provides the command \url, whose argument can be a URL, which will be typeset correctly (i.e., it will be split across lines if this is necessary).

\hyperbaseurl{*URL*} This is an alternative way to specify a base URL.

\hyperref{*URL*}{*category*}{*name*}{*test*} The *text* is made into a link to URL#category.name.

\hyperlink{*name*}{*text*} The *text* is made a link to the internal target *name*.

\hypertarget{*target*}{*text*} The *text* is made an internal link with symbolic name *target*. Here is a simple usage example:

```
\hyperlink{prop}{See proposition}
. . . . . . . . . . . . . . . . . . . . . .
\begin{proposition}
\hypertarget{prop}{Text of proposition}
\end{proposition}
```

However, in this particular case, it is easier to use the following to get the same effect:

```
See proposition~\ref{prop}
. . . . . . . . . . . . . . . . . . . . .
\begin{proposition}\label{prop}
Text of proposition
\end{proposition}
```

We will now present PDF forms and their usage. But what exactly is a form? According to the World Wide Web Consortium (W3C): A form is a section of a document containing normal content, markup, special elements called controls (checkboxes, radio buttons, menus, etc.), and labels on those controls. Users generally "complete" a form by modifying its controls (entering text, selecting menu items, etc.) before submitting the form to an agent for processing (e.g., to a Web server, to a mail server, etc.). So, one must know what controls are available and then how it is possible to process the modified controls. We start by briefly addressing the second issue.

There are two ways to process controls: either we send the data to some Web server by a specific transmission method (the available methods are Post and Get) or we process them locally with the JavaScript programming language. The second method is usually employed in simple cases. For example, it makes no sense to use the first method to get the square root of a number. The first method assumes that there is a so-called *cgi-script* (usually a program written in Perl) on the Web server that will actually process the data and send the processed data back to the client. Forms were initially introduced in HTML documents but can now be part of a PDF document. If we want to transmit data from a PDF form, we have to use the Acrobat Reader within a Web browser. This is necessary, as the Web browser will do all necessary actions to actually deliver the data.

We can create forms with the Form environment. The body of the environment contains several controls. Here is the list of the available controls:

\TextField[*parameters*]{*label*} A text field is a control where we can enter data from the keyboard.
\CheckBox[*parameters*]{*label*} Checkboxes are on/off switches that may be toggled by the user. A switch is "on" if the control element's checked attribute is set. When a form is submitted, only "on" checkbox controls can become successful.
\ChoiceMenu[*parameters*]{*label*} Menus offer users options from which to choose.
\PushButton[*parameters*]{*label*} Push buttons have no default behavior. Each push button may have client-side scripts associated with the element's event attributes. When an event occurs (e.g., the user presses the button, releases it, etc.), the associated script is triggered.
\Submit[*parameters*]{*label*} When a Submit button is activated, it submits a form. A form may contain more than one submit button.
\Reset[*parameters*]{*label*} When a Reset button is activated, it resets all controls to their initial values.

Before we proceed with a brief presentation of the most important parameters, we give a short example of a form that asks its user to enter two numbers in two text fields, submits the data, and returns their sum:

```
\documentclass{article}
\usepackage[bookmarks=false, pdfstartview={XYZ 4}]{hyperref}
\begin{document}
Enter two numbers in the following text fields and get their sum!
\begin{Form}[action={http://ocean1.ee.duth.gr/cgi-bin/sumnum},
            encoding=html, method=post]
\begin{center}
\TextField[width=1cm, height=4pt, name=num1, value={}]
   {First number: }\\
```

```
\TextField[width=1cm, height=4pt, name=num2, value={}]
    {Second number:  }\\[4pt]
\Submit{Send}\qquad \Reset{Clear}
\end{center}
\end{Form}
\end{document}
```

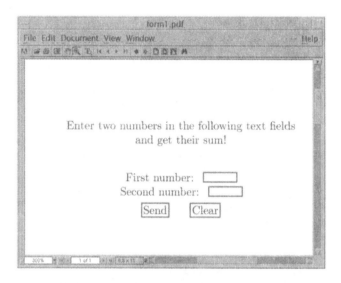

Figure 6.1: A simple PDF form.

The output of the code above is shown in Figure 6.1. Readers can download from the book's Web site the source code of a Perl script that can manipulate the data generated by the form above.

As is evident from the example above, it is important to give each control a name. The optional arguments of the Form environment are pretty standard, and we will make no attempt to explain them. Readers not familiar with them are advised to visit the URL http://www.w3.org/TR/html4/interact/forms.html. We now briefly present the most important parameters:

height **and** width The height and width of a text field.

multiline If set to true, the text field becomes a text area.

menulength The number of elements shown in a list.

charsize **and** color The font size and color of a text field. Note that the color must be expressed as R(red)G(reen)B(lue) triplets, in the range 0..1, that is,

```
color= 0 1 0
```

combo, popdown **and** radio The choice list is "combo", "pop-up" or "radio," respectively.
password Text field is "password" style.
value Initial value of a control.

If we want to process a form locally, we must use pieces of JavaScript code that
are activated when certain *events* happen (i.e., users push buttons, change the value of
text fields, etc). In the example that follows, the user enters the parameters of a simple
equation and by pressing "Solve Equation" gets the solution in a text field.

```
\begin{Form}
\begin{center}
\TextField[width=1cm,height=5pt,name=num1,value={}]{\quad}
$x$
\ChoiceMenu[popdown,default=+,name=sign]{}{+,-}
\TextField[width=1cm, height=5pt, name=num2,value={}]{}
$=$
\TextField[width=1cm, height=5pt, name=num3,value={}]{}
\\[0.5cm]
\PushButton[onclick={ var num1 = this.getField("num1");
var a=num1.value; var num2=this.getField("num2");
var b=num2.value; var sign=this.getField("sign");
var num3 = this.getField("num3"); var c=num3.value;
if (sign.value != "+") b=-b;
if (a == 0) app.alert("No solution!");
else this.getField("sol").value = (c-b)/a;}]{Solve Equation}
\TextField[width=2cm, height=5pt, name=sol,value={}]{Solution:}
\end{center}
\end{Form}
```

Figure 6.2 shows a captured screen of the form as it is displayed by Acrobat Reader.
As is evident from the example above, the JavaScript code is actually the value of a
parameter. Such parameters are called *event handlers*. Also note that the JavaScript code
must be surrounded by curly brackets. The available event handlers are:

onblur	onchange	onclick	ondblclick
onfocus	okeydown	okeypress	onkeyup
onmousedown	onmousemove	onmouseout	onmouseover
onmouseup	onselect		

At this point, it is really important to stress that the JavaScript language supported
by the PDF format is rather different from the language used in HTML documents.
The reader interested in learning more should read the *Acrobat Forms JavaScript Object
Specification*, Adobe Technical Note 5186, available from: http://partners.adobe.com/
asn/developer/technotes/acrobatpdf.html.

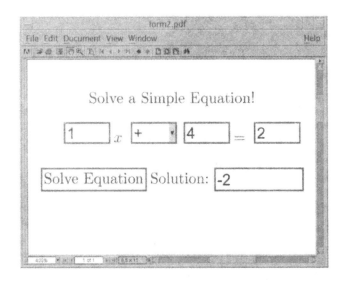

Figure 6.2: A simple JavaScript-enabled PDF form.

6.3 Horizontal and Vertical Space

In this section, we introduce length variables and their manipulation as well as the commands that allow users to put (additional) horizontal and vertical space in their document.

6.3.1 Length Variables

Length variables are either predefined or user-defined control sequences that can be used instead of a normal length. New length variables can be introduced with the command \newlength:

\newlength{\LenVar}

The name of a length variable consists of a leading backslash and one or more trailing ASCII letters. We can set a length variable with the command \setlength:

\setlength{\LenVar}{Expression}

Here, *Expression* is either an ordinary length (e.g., 5 cm or 35 pt) or the name of another length variable, possibly prefixed by a number or, more generally, the value of a counter. In the latter case, we set the length variable to the product of the number (or the value of the counter) times the second length. For example, after the execution of the commands

```
\newlength{\lenA}
\setlength{\lenA}{10pt}
\newlength{\lenB}
\setlength{\lenB}{5\lenA}
```

the value of \lenA will be 10 pt and the value of \lenB will be 50 pt. Moreover, we can increase or decrease the value of length variables by using the command \addtolength:

```
\addtolength{\lenB}{\lenA}
\addtolength{\lenB}{-2\lenA}
```

The first command increases the value of \lenB by the value of \lenA, so the value of \lenB becomes 60 pt. The second command decreases the value of \lenB by two times the value of \lenA, so the value of \lenB becomes 40 pt. Note that the leading minus sign indicates that LATEX must decrease the value of the length variable. Naturally, we are allowed to use ordinary lengths.

When TEX typesets a line, it puts letters and other objects, such as tables, on an imaginary line called the baseline. The height of a letter or a word refers to the vertical space the letter or the word occupies above the baseline. Similarly, the depth refers to the vertical space that the letter or the word occupies below the baseline. Now, it is even possible to set a variable to the width, height, or depth of a letter or a word. For example, after the execution of the following code, the three length variables will be set to the width, the height, and the depth of the tight frame that encloses the word Pipe, respectively.

```
\settowidth{\lenA}{Pipe}
\settoheight{\lenB}{Pipe}
\settodepth{\lenC}{Pipe}
```

In some cases it may be useful to be able to print the value of a length variable. The solution to this problem involves using a \the command. When we prefix a length variable or a counter with this command, we get its value. If we have a length variable, then we get its value in points. Here is a simple example:

The length of \lenA is 56.9055pt.	`\newlength{\lenA}`
	`\setlength{\lenA}{2cm}`
	`The length of \verb|\lenA|`
	`is \the\lenA.`

The interested reader may find more information about the \the command in [19].

6.3.2 Horizontal Space

Basically, there are two commands that allow one to put horizontal space in a document—\hspace and \hspace*. Both have one argument, which can be either a length or a length variable. The difference between the two commands is that the starred one can be used to put horizontal space at the beginning of a line; otherwise, this space is ignored:

Hello World	`Hello World\\`
Hello World!	`\hspace{1cm}Hello World!\\`
Hello World!	`\hspace*{1cm}Hello World!\\`
WHerllod!	`Hello\hspace{-1cm}World!`

The last example also shows what happens when we use a negative length. In addition to the two commands above, LATEX provides three more commands that produce a predefined horizontal space:

a b	`a b\\`
a b	`a\thinspace b\\`
ab	`a\negthinspace b\\`
a b	`a\enspace b`

In Section 2.3, we introduced the notion of glue lengths (i.e., lengths that can shrink and stretch). LATEX provides the glue `\hfill`, which flushes whatever follows to the right margin:

a b		`a b\\`
a	b	`a\hfill b\\`
a	b c	`a\hfill b\hfill c`

The commands `\dotfill` and `\hrulefill` have the same effect as `\hfill`, but they fill the space with dots and or a horizontal rule, respectively:

a b	`a b\\`
a....................b	`a\dotfill b\\`
a........b.........c	`a\dotfill b\dotfill c\\`
a_____b	`a\hrulefill b\\`
a_____b_____c	`a\hrulefill b\hrulefill c`

When we introduced the command `\hfill`, we told a little white lie: this command is not a glue but is equivalent to

$$\texttt{\textbackslash hspace\{\textbackslash fill\}}$$

Of course, `\fill` is the glue we were talking about. Suppose now that we want to put three letters on one line so that the distance between the first and the second is two times the distance between the second and the third, that is

a   b   c

This effect cannot be created with

$$\texttt{a\textbackslash hspace\{2\textbackslash fill\}b\textbackslash hfill c}$$

as the expression 2\fill will be transformed into an ordinary length equal to 0 pt! The correct code follows.

```
a\hspace{\fill}\hspace{\fill}b\hfill c
```

Since typing the command \hspace{\fill} two or more times is cumbersome, LATEX provides the command \stretch, which takes as argument an integer, say n, and produces a glue that is equal to n times \fill. So, the construct above can be written as

```
a\hspace{\stretch{2}}b\hfill c
```

▶ **Exercise 6.1** Typeset the following (without the underbraces!):

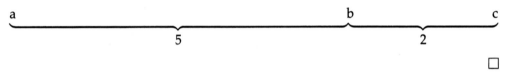

There are two more useful commands that produce horizontal space. The command \, produces a \thinspace and is used to correctly typeset quotes inside quotes:

" 'Hello', he said	`'' 'Hello', he said\\`
"'Hello', he said	`''\,'Hello', he said\\`
"'Hello', he said	`'''Hello', he said`

If the command \@ is placed before a '.', it makes it a sentence, ending with a period. This is necessary for American and British typography, as the space after a '.', that ends a period is longer than that after a '.', which ends an acronym. However, this is not a common practice in some European nations, where the space is the same in both cases. For this reason, LATEX provides the declaration \frenchspacing, which changes the default behavior described above.

6.3.3 Vertical Space

The space between paragraphs, sections, subsections, and so on, is determined automatically by LATEX. If necessary, more space *between two paragraphs* can be added with the command \vspace. If we need to add vertical space at the top or bottom of a page, we should use \vspace*. Both commands have one argument, which can be either a length or a length variable.

LATEX also supports three commands that insert a predefined amount of white vertical space:

Line 1	`Line 1 \\ \smallskip`
Line 2	`Line 2 \\ \medskip`
Line 3	`Line 3 \\ \bigskip`
Line 4	`Line 4`

The command \pagebreak encourages LATEX to break a page at a certain point. This command has an optional argument that is an integer from 0 to 4. The higher the number, the stronger the encouragement. The command \pagebreak is by definition equivalent to \pagebreak[4]. Likewise, the command \nopagebreak discourages LATEX from breaking a page at a certain point. The commands \linebreak and \nolinebreak are the corresponding line-breaking commands.

➤ **Exercise 6.2** Write a LATEX file that will create a one-page document with only three letters on three different lines that will occupy the whole page. The distance between the first and the second letter must be two times the distance between the second and the third. ☐

Extra vertical white space can be added with the command \addvspace. The sequence

$$\addvspace\{S_1\} \quad \addvspace\{S_2\}$$

is equivalent to

$$\addvspace\{\max(S_1, S_2)\}$$

The command \enlargethispage streches or shrinks a page by a specific length that is the argument of this command. Obviously, if the argument, which is a length or a length variable, is positive, then the page is stretched and vice versa. The command \flushbottom stretches the blank space on pages to make the last line of all pages appear at the same height. The opposite effect can be achieved with the command \raggedbottom.

If we find that the hyphenation algorithm fails to properly break lines, then we can assist it by manually hyphenating words. Suppose that TEX fails to correctly hyphenate the word *omnipresent*. Then, we can type it as follows:

om\-ni\-pres\-ent

that is, we put the symbol \- between the syllables of the word. Alternatively, if a word or a list of words appears frequently in our text, we can use the command

\hyphenation{om-ni-pres-ent Athenian}

Here, syllables are separated by - and words by space. Note that, in the example above, we tell TEX not to hyphenate the second word. If these "tricks" do not produce good results, we can manually break lines or force TEX to typeset *sloppy* paragraphs with the command \sloppy. This command forces TEX to increase the interword space if there is no other way to make the words fit on a particular line. In addition, the sloppypar environment should be used to typeset a \sloppy paragraph. The command \fussy forces TEX to switch back to its normal typesetting mode. These commands should be used with care and preferably in a local scope.

6.4 Counters

Counters are LATEX variables that can hold an integral value. We can create a new counter with the command

$$\newcounter\{newcnt\}[oldcnt]$$

If we use the optional argument, each time the counter oldcnt is increased by one, the value of newcnt becomes equal to zero. This feature is very useful and, for example, it is used to produce the section and subsection numbers. Initially, all counters are set to zero. We can set the value of a counter with the command \setcounter:

$$\setcounter\{newcnt\}\{7\}$$

When we want to set a counter to the value of another counter, we have to use the command \value:

$$\setcounter\{newcnt\}\{\value\{oldcnt\}\}$$

Another possibility is to advance the value of a counter by using the command \addtocounter:

```
\setcounter{newcnt}{5}                %% newcnt=5
\addtocounter{newcnt}{4}              %% newcnt=9
\addtocounter{newcnt}{-2}             %% newcnt=7
\addtocounter{newcnt}{\value{newcnt}} % newcnt=14
```

The command \stepcounter{newcnt} globally increments the counter newcnt by one and resets all subsidiary counters. Similarly, the command \refstepcounter{newcnt} globally increments the counter newcnt by one, resets all subsidiary counters, and also takes care so that a subsequent \label causes a \ref to generate the current value of the counter newcnt.

The following commands have one argument that must be a counter, and they print a visual representation of the value of the counter:

Command	Produces...
\arabic	an Srabic numeral.
\roman	a lowercase Roman numeral.
\Roman	an uppercase Roman numeral.
\alph	a lowercase letter that corresponds to the value of the counter.
\Alph	an uppercase letter that corresponds to the value of the counter.
\fnsymbol	the standard footnoting symbols: *, †, ‡, §, ¶, ‖, **, ††, and ‡‡.

Here are some examples:

XX
xx
T
 	
§
	`\newcounter{cnt}`
	`\setcounter{cnt}{20}`
	`\Roman{cnt}\\`
	`\roman{cnt}\\`
	`\Alph{cnt}\\`
	`\setcounter{cnt}{4}`
	`\fnsymbol{cnt}`

Usually, LATEX typesets page numbers as Arabic numerals, but in certain situations we want to change the page numbering. For example, in many instances, the first few pages of a book are numbered in Roman numerals. LATEX offers the command \pagenumbering, which has one argument that takes care of the visual representation of a counter, resets the page counter, and uses the argument to typeset the page numbers. For example, the command \pagenumbering{Roman} should be used to start typesetting the page numbers in capital Roman numerals.

When we define a new counter, say cnt, LATEX defines a new command that consists of the token \the and the name of the counter (e.g., \thecnt in our case). This command simply prints the value of the counter. However, one should not confuse the command \thecnt with the command \value{cnt}, as the latter provides access to the actual value of the counter. Usually, the command \thecnt prints the value of the counter in Arabic numerals; that is, the command \thecnt is actually equivalent to the following definition:

```
\newcommand{\thecnt}{\arabic{cnt}}
```

However, it is possible to redefine the command \thecnt to print whatever pleases us:

```
\renewcommand{\thecnt}{-\Roman{cnt}-}
```

This "trick" is employed to change how the predefined or user-defined counters will be printed. The list of predefined counters includes the following:

part	paragraph	figure	enumi
chapter	subparagraph	table	enumii
section	page	footnote	enumiii
subsection	equation	mpfootnote	enumiv
subsubsection			

The names of these counters make their usage evident. Below, as an example of a use of the predefined counters, we present how we can customize page numbers.

In many apparatus manuals, the page "numbers" consist of the chapter number, a dash, and the page number, while the page counter is reset each time we issue a \chapter command. To achieve this effect, we have to redefine the \pagenumbering command, whose standard definition follows:

```
\newcommand{\pagenumbering}[1]{%
\setcounter{page}{1}%
```

```
\newcommand{\thepage}{%
    \csname @#1\endcsname{\value{page}}}}}
```

Note that LaTeX uses internally the commands \@arabic, \@roman, etc., instead of \arabic, \roman, etc., respectively.

➤ **Exercise 6.3** Redefine the \pagenumbering command so that it prints page numbers for apparatus manuals. □

By using the calc package (see Section 7.1), one can better manipulate counters and length variables.

6.5 Floating Objects

Most modern publications contain a lot of figures and tables. There are instances where a table can be broken across pages, but this is unacceptable for figures. For this reason figures and short tables need special treatment. The naïve method of treating these objects is to start a new page every time a floating object is too large to fit on the present page. A more sophisticated method to tackle this problem is to "float" any object that does not fit on the current page to a later page while filling the current page with text. This is why these objects are called *floating objects*. LaTeX provides two environments that are treated as floating objects: the figure and the table environments. Both environments are written the same way; they differ only in the text that is prepended in the caption. Moreover, there are two environments that can be used in double column documents to generate floats that may occupy both columns: the figure* and the table* environments. Here is how we can begin a table or a figure:

$$\begin{aligned} &\texttt{\textbackslash begin\{table\}[\textit{placement specifier}]}\\ &\texttt{\textbackslash begin\{figure\}[\textit{placement specifier}]} \end{aligned}$$

The optional *placement specifier* is used to tell LaTeX where the float is allowed to be moved to. The *placement specifier* consists of a sequence of *float placing permissions*:

Placement Specification	Place the Float ...
h	*here* at the place where it occurred.
t	at the *top* of a page.
b	at the *bottom* of a page.
p	on a special page containing only floats.

Apart from the *float placing permissions* above there exists a fifth one, namely !, which forces LaTeX to actually ignore most of the internal parameters related to float placement. LaTeX also provides the command \suppressfloats, which prevents LaTeX from putting more floats on the current page. The command has an optional argument, which can be either b (for bottom) or t (for top). These arguments prevent LATEX from putting more floats on the bottom or the top of the current page.

Any float object can have a caption, which is generated by the \caption command:

\caption[*Short*]{*Long*}

The optional argument *Short* is useful when we want to have a list of figures or tables in our document and the caption text is quite long. The list of figures and tables can be generated with the commands \listoffigures and \listoftables, respectively. Naturally, we do not have to specify the optional argument to generate either a list of figures or a list of tables. If we want to have a reference to a floating object in our text, we have to put the \label *after* the caption.

Suppose that the placing mechanism has deferred the placement of some float objects. If we want to immediately place all floats deferred on a given page and start a new page, we have to use the command \clearpage. The command \cleardoublepage does the same job but starts a new odd-numbered page.

There are a number of parameters that control both the number of float objects that can appear on a page and the general appearance of floats. Here, we briefly present these parameters. The counters topnumber and bottomnumber can be used to specify the number of floats allowed at the top or bottom of a column, respectively. The counter totalnumber can be used to specify the total number of floats allowed in a single column. The commands \topfraction and \bottomfraction denote the fraction of the top or bottom of a column that can be devoted to floats. The command \dbltopfraction is used for double-column floats. The command \textfraction denotes the minimum fraction of column that must contain text. The commands \floatpagefraction and \dblfloatpagefraction specify the minimum fraction of a page that must be taken up by floats for single- and double-column documents, respectively. Here is an example that shows how one can set these counters and commands:

```
\setcounter{topnumber}{2}
\renewcommand\topfraction{.7}
```

The command \@makecaption is used to customize the appearance of the figure. The following definition is equivalent to the one provided by LATEX:

```
\newcommand{\@makecaption}[2]{%
  \vspace{\abovecaptionskip}
  \sbox{\@tempboxa}{#1: #2}%
  \ifthenelse{\wd\@tempboxa > \hsize}{%
    #1: #2\par}{%
    \setboolean{@minipage}{false}
    \centerline{\usebox{\@tempboxa}}
  }
  \vspace{\belowcaptionskip}
}
```

LaTeX leaves an amount of blank vertical space before and after the caption that is stored in the length variables \abovecaptionskip and \belowcaptionskip, respectively. The first argument of the command corresponds to the "number" of the floating object (e.g., "Figure 3.1"), and the second argument to the actual text of the caption. We store both arguments to the internal box variable \@tempboxa, and if its width is greater than the width of the line, we ask LaTeX to typeset them and then to change paragraphs. Otherwise, we ask LaTeX to typeset the contents of the box in a centered line. The internal Boolean variable \@minipage is set to false to ensure that the floating object is not part of a minipage.

▶ **Exercise 6.4** Modify the definition above so that the caption "number" does not appear in captions. □

There are a number of packages that can prove useful when dealing with floating objects. In what follows, we will briefly present these packages.

The *afterpage* package

This package by David Carlisle implements a command, \afterpage, that causes the commands specified in its argument to be "executed" after the current page is output. For example, the command

$$\text{\afterpage\{\clearpage\}}$$

can be used to fill the current page with text and force LaTeX to put all remaining floats onto the next page.

The *morefloats* package

This package by Don Hosek increases LaTeX's current limit of 18 unprocessed floats in memory at once to 36.

The *placeins* package

This package by Donald Arseneau keeps floats "in their place," preventing them from floating past a \FloatBarrier. A more convenient way to use this package is to redefine the sectioning commands and put the \FloatBarrier command somewhere. The package provides two options: below and section. The first option allows float objects to appear after the heading of a new section. The second "embeds" float barriers into \section commands.

The *endfloat* package

The purpose of the this package is to put all figures on pages by themselves at the end of an article in a section named Figures. The package has been designed by James Darrell

McCauley and Jeff Goldberg. The list of tables and figures can be suppressed by using the nofiglist and notablist options. Both can be suppressed with the nolist option. The default is list.

The package leaves notes on the text for each figure or table. The appearance of these notes is controlled by the commands \tableplace and \figureplace. Unfortunately, the current version of this package does not directly support the babel package. So, one has to manually redefine these commands. Here are the standard definitions:

```
\newcommand{\figureplace}{%
    \begin{center}
        [\figurename~\thepostfig\
            about here.]
    \end{center}
}
```
```
\newcommand{\tableplace}{%
    \begin{center}
        [\tablename~\theposttbl\
            about here.]
    \end{center}
}
```

Here is a redefinition that should be suitable for Italian-speaking people:

```
\renewcommand{\figureplace}{%
    \begin{center}
        [\figurename~\thepostfig\ circa qui.]
    \end{center}}
```

These and other redefinitions should be kept in the file endfloat.cfg.

The *float* package

This package provides an interface to define new float objects. Moreover, the package defines certain "float styles" that can be used to define new floating objects. The package was designed by Anselm Lingnau. New float objects can be defined with the command

\newfloat{*type*}{*placement*}{*ext*}[*within*]

Here *type* is the "type" of the new class of floats (e.g., program, diagram, etc.), *placement* gives the default placement specifier, and *ext* is the filename extension for the file that will keep the captions in cases where we want to have a list of programs, list of diagrams, or other lists. The optional argument *within* is used to number float objects within some sectioning unit (e.g., chapter, section). Here is a complete example:

\newfloat{program}{htb}{prg}[section]

Note that after each such definition, a new environment will be available. Naturally, its name depends on the "type" (e.g., the example code above will create the program environment). The "float style" can be specified with the \floatstyle command. The command has one argument, which is the name of a "float style":

plain This style is identical to one that LATEX applies to floats. The only difference is that the caption always goes *below* the body of the float, regardless of where it appears in the body of the new environment.

boxed The body of the float is printed inside a box. The caption goes below this box.

ruled The caption appears on the top of the float; ruled lines are put before and after the caption, and another rule is put after the float.

The command \listof is used to produce a list of all floats of a given class:

$$\text{\listof\{\textit{type}\}\{\textit{title}\}}$$

---

program 1.1	A trivial Java program.

```
public class test {
    public static void main(String[] args) {
        for(int i=0; i<args.length; i++) {
            System.out.println(args[i]);
        }
    }
}
```

Figure 6.3: Sample output of the float package.

So, if one wants to typeset the list of programs, the following command needs to be used:

$$\text{\listof\{program\}\{List of programs\}}$$

Figure 6.3 shows a sample output generated with the new program environment defined above and the ruled float style.

An important feature introduced by this package is the H placement specifier, which actually forces LATEX to put a float at the very position where it appears. H differs from h in that the latter just suggests that LATEX put the float *here*, whereas the former forces LATEX to put the float *exactly here*. Note that the H placement specifier cannot be used in the definition of a new class of float objects.

The *picinpar* package

By using the picinpar package (by Friedhelm Sowa), one can easily embed figures or tables inside paragraphs. The package provides the environment window plus the environments figwindow and tabwindow. The latter environments are just applications of the window environment. The window environment has four required arguments:

$$\text{\begin\{window\}[\textit{lines, placement, material, explanation}]}$$

Here, *lines* is the number of lines from the paragraph top, *placement* is either l, c, or r and is used to print the picture at the left-hand side, centered, or at the right-hand side of the paragraph, respectively, *material* is the figure to be displayed, and *explanation* is the caption for the figure. This environment is used mainly for graphics, and the environment tabwindow is for tables created with the tabular environment. Here is how one may create the dangerous paragraphs of this book:

```
\begin{window}[0,l,\includegraphics[scale=.2]{danger.eps},{}]
\small\noindent blah, blah, blah, blah...
\end{window}
```

The **wrapfig** package

Donald Arseneau has created the wrapfig package to allow people to place figures or tables at the side of a page and wrap text around them. The package provides the environments wrapfigure and wraptable. Both environments have two required and two optional arguments:

$$\text{\\begin\{wrapfigure\}[}\textit{nlines}\text{]\{}\textit{placement}\text{\}[}\textit{overhang}\text{]\{}\textit{width}\text{\}}$$

nlines is the number of narrow lines, and *placement* is one of r, l, i, o, R, L, I, or O for right, left, inside, and outside, respectively. The uppercase placement specifiers differ from their lowercase counterparts in that they force LATEX to put the float "here," whereas the lowercase placement specifiers just give a hint to LATEX to place them "here." The *width* argument is the width of the figure or table that appears in the body of the environment. Finally, *overhang* tells LATEX how much the figure should hang out into the margin of the page. Here is how one may create more dangerous paragraphs of this book:

```
\begin{wrapfigure}[4]{l}{1.5cm}
\includegraphics[scale=0.2]{danger.eps}
\end{wrapfigure}
{\small\noindent blah, blah, blah, blah...}
```

Note that the text can have more than one paragraph.

The **subfigure** package

This package provides support for the manipulation and reference of subfigures and subtables within a single figure or table environment. The package was designed by Steven Douglas Cochran. It provides a number of options, which are presented in Table 6.2.

In order to embed subfigures or subtables inside a figure or table, respectively, one has to use the following commands:

Table 6.2: Options of the subfigure package.

Option	Description
normal	Provides "normal" captions, this is the default.
hang	Causes the label to be a hanging indentation to the caption paragraph.
center	Causes each line of the paragraph to be separately centered.
centerlast	Causes the last line only to be centered.
nooneline	If a caption fits on one line, it will, by default, be centered. This option left-justifies the one-line caption.
scriptsize, ..., Large	Sets the font size of the captions.
up, it, sl, sc, md, bf, rm, sf or tt	Sets the font attributes of the caption labels.

$$\text{\subfigure[}caption\text{]\{}figure\text{\}}$$
$$\text{\subtable[}caption\text{]\{}table\text{\}}$$

If the *caption* is given (including the empty caption [ ]), the subfigure is labeled with a *counter* formatted by the command \thesubfigure. Similarly, the subtable is labeled with a counter formatted by the command \thesubtable. In Figure 6.4, we show the output generated by the following code fragment:

```
\begin{figure}%
  \centering
  \subfigure[First]{...}\hspace{\len}
  \subfigure[Second Figure]{...}\\
  \subfigure[Third]{\label{3figs-c}...}%
  \caption{Three subfigures.}
  \label{3figs}
\end{figure}
...
Figure~\ref{3figs} contains two top subfigures and
Figure~\ref{3figs-c}.
```

If one wants to have the captions of subfigures or subtables to be included in the list of figures or the list of the tables, respectively, the following commands must be put in the preamble of the input file:

```
\setcounter{lofdepth}{2}
\setcounter{lotdepth}{2}
```

The ccaption package

The ccaption package by Peter Wilson provides the necessary commands to restyle captions. In addition, it provides commands that produce "continuation" captions,

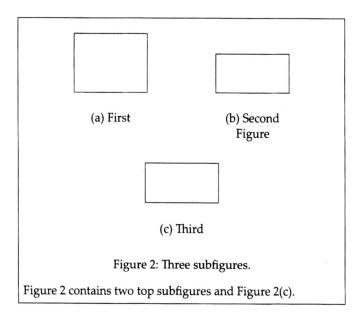

Figure 6.4: A typical figure with subfigures.

unnumbered captions, bilingual captions, or legends. The command \captiondelim should be used to change the symbol that is placed between the caption number and the caption text. To change the default symbol to an en dash, use the following command:

$$\texttt{\char92 captiondelim\char123--\char95\char125}$$

The \captionnamefont should be used to change the font that is used to typeset the caption title; that is, the float name, the number, and the delimiter. The argument of this command can be any font selection and/or size changing-commands, such as

$$\texttt{\char92 captionnamefont\char123\char92 Large\char92 sffamily\char125}$$

Similarly, the command \captiontitlefont should be used to change the font that is used to typeset the caption text. The \captionstyle command should be used to change the way the whole caption is being typeset. For example, if the argument of this command is the \raggedright declaration, then the whole caption will be typeset accordingly. Other possible arguments of this command are the declarations \ragged-left and \centering. In addition, the package defines the declaration \centerlast-line, which should be used to get a caption being typeset the usual way but with the last line centered.

The command \hangcaption will cause captions to be typeset with the second and later lines of a multiline caption text indented by the width of the caption title. The command \indentcaption takes a length or a length variable as argument and uses it to indent all caption text lines after the first line. To undo any effect produced by \hang-caption and \indentcaption, just use the \normalcaption command. Furthermore, it is possible to change the total width of a caption by using the \changecaption-width command. The total width must be set with the \captionwidth command. This command has one argument, which is just a length or a length variable. Each of the commands \precaption and \postcaption have one argument, which will be processed at the start and the end of a caption. For example, the commands

```
\precaption{\rule{\linewidth}{0.4pt}\par}
\postcaption{\rule{\linewidth}{0.4pt}}
```

will draw a horizontal line above and below the captions. The length of the line will be equal to the current line width. The ccaption package assumes that initially both commands have empty arguments.

The \contcaption command can be used to place a "continuation" caption into a float environment. It does not increment the float number nor make any entry into a float listing. The \legend command should be used to place a caption without a caption title in a float environment.

Bilingual captions can be produced with the following command:

$$\texttt{\textbackslash bitwonumcaption[}\textit{label}\texttt{]\{}\textit{shortA}\texttt{\}\{}\textit{longA}\texttt{\}\{}\textit{name}\texttt{\}\{}\textit{shortB}\texttt{\}\{}\textit{longB}\texttt{\}}$$

The *label* is an optional argument that is a normal label that can be used to refer to the caption. *shortA* and *longA* are the short and long forms of the caption in the primary language of the document, *shortB* and *longB* are the short and long forms of the caption in the second language, and *name* is the name of the caption title in the second language of the document. However, we have noticed that this command does not function when used with the babel package when the two languages use different scripts (e.g., the Latin and the Cyrillic). Fortunately, it does work with Λ! The package offers some other capabilities, which we feel are not of general interest so we do not describe them.

6.6 Marginal Notes

Some Marginal Notes

Marginal notes are used to label paragraphs . Usually, they consist of a few words (up to five or six) and give the reader an idea of what the paragraph deals with. Marginal notes can be created with the command

$$\texttt{\textbackslash marginpar[}\textit{left}\texttt{]\{}\textit{right}\texttt{\}}$$

The *left* text is optional, is used mainly for two-sided printing, and always appears on the left margin of a page. The *right* text always appears on the right margin of the page. One can easily reverse this functionality with the command \reversemarginpar. We can revert to normal marginal notes placement with the command \normalmarginpar.

There are three parameters that control the appearance of marginal notes:

\marginparwidth is the width of marginal notes.

\marginparsep is the distance between marginal notes and the text.

\marginparpush is the minimum vertical blank space between adjacent marginal notes.

Marginal notes are actually floating objects, and therefore one should be careful when using them.

6.7 Page Layout

As we have already explained, one can specify the paper size of a document in the \documentclass command. Then, LaTeX automatically calculates the right text margins. In case one is not satisfied with these values, one can either change some of these predefined values or use either the vpage package or the geometry package to define a new page size. Figure 6.5 on page 181 shows most of the parameters that can be changed. Variables \paperheight and \paperwidth are used to store the paper height and width, respectively. There are also some parameters (i.e., length variables) that are not shown in this figure:

\evensidemargin is used to store the width of the outer margin of even-numbered pages in two-sided documents.

\columnsep contains the width of space between columns of text in a multicolumn document.

\columnseprule refers to the width of the vertical line separating two adjacent columns in a multicolumn document.

\columnwidth holds the width of a single column in a multicolumn document.

\linewidth keeps the width of the current text line; usually it is equal to the current column width, but in many cases it is altered by environments, and so forth.

In addition, Ω provides the primitive length variables \pageheight and \pagewidth, which hold the height and the width of the page, respectively. The default values for these are for A4 pages. Furthermore, the Ω primitive commands \pagerightoffset and \pagebottomoffset should be used to move the page horizontally (from the left to the right) and vertically (upwards), respectively. Similarly, the command \hoffset and \voffset are use to move the page horizontally (from the left to the right) and vertically (downwards).

Figure 6.5 has been drawn with the layout package by Kent McPherson. In order to produce such a figure, simply create a LaTeX file that uses the layout package and contains only the \layout command in its body.

The package vmargin (by Volker Kuhlmann) provides an interface to change the layout of documents that are typeset using the metric paper sizes (i.e., A0, A1,..., B0, B1,..., C0, C1, and so on) and all of the standard American paper sizes. The package options include the paper size format (e.g., A6), the words portrait and landscape for a document that will be typeset either portrait or landscape, respectively, and the option nohf for a document without header and footer lines. When the last option is specified, the document will be produced without page numbers. If we want to use a custom paper size, we can define its dimensions with the command

\setpapersize{custom}{*width*}{*height*}

Once the paper size is selected, margins can be set by

\setmargins{*leftmargin*}{*topmargin*}{*textwidth*}{*textheight*}%
{*headheight*}{*headsep*}{*footheight*}{*footskip*}

or by

\setmarginsrb{*leftmargin*}{*topmargin*}{*rightmargin*}{*bottommargin*}%
{*headheight*}{*headsep*}{*footheight*}{*footskip*}

In the latter case, \textwidth and \textheight are calculated using the width and height of the selected paper. The commands \setmargnohf and \setmargnohfrb provide a page with no header and no footer. They work the same as \setmargins and \setmarginsrb except that they only need the first four parameters. The last four parameters are set to 0 pt. The commands \setmarg and \setmargrb are the same as \setmargnohf and \setmargnohfrb except that the last four parameters are kept unchanged instead of being set to 0 pt.

The package geometry, by Hideo Umeki, provides an easy and flexible user interface to customize the page layout. It implements auto-centering and auto-balancing mechanisms so that the users have only to give a few items of data for the page layout. Most of these data can be supplied as package options. For example, one has to give the command

\usepackage[body={8in,11in}]{geometry}

to obtain an 8 in × 11 in page. The \geometry command provided by the package can be used to achieve the same effect:

\usepackage{geometry}
\geometry{body={8in,11in}}

Here is another example borrowed from the package documentation. Suppose that we want a page layout described as follows:

The total allowable width of the text area is 6.5 inches wide by 8.75 inches high. The first line on each page should begin 1.2 inches from the top edge of the page. The left margin should be 0.4 inch from the left edge.

Then, we can specify the page layout above with geometry as follows:

\usepackage[body={6.5in,8.75in},
top=1.2in, left=0.4in, nohead]{geometry}.

The package offers many options, but here we only present the most significant ones:

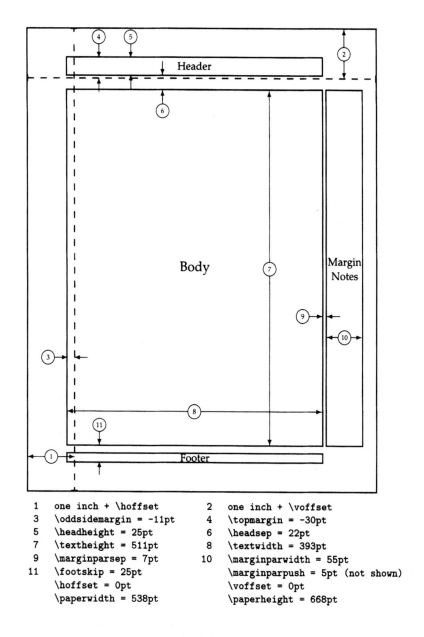

1	one inch + \hoffset	2	one inch + \voffset
3	\oddsidemargin = -11pt	4	\topmargin = -30pt
5	\headheight = 25pt	6	\headsep = 22pt
7	\textheight = 511pt	8	\textwidth = 393pt
9	\marginparsep = 7pt	10	\marginparwidth = 55pt
11	\footskip = 25pt		\marginparpush = 5pt (not shown)
	\hoffset = 0pt		\voffset = 0pt
	\paperwidth = 538pt		\paperheight = 668pt

Figure 6.5: Page layout parameters.

landscape switches the paper orientation to landscape mode.

portrait switches the paper orientation to portrait mode.

twoside switches on two-sided printing.

nohead sets the page head to 0 pt.

nofoot sets the page foot to 0 pt.

noheadfoot sets the page head and foot to 0 pt.

dvips writes the page layout information to the PostScript file generated by DVIPS.

pdftex sets the page layout when processing a LaTeX file with pdfLaTeX.

paper=*paper* specifies a *paper* name (i.e., a0paper,..., a6paper, b0paper,..., b6paper) and the American paper sizes.

paperwidth=*length* width of paper.

paperheight=*length* height of paper.

papersize={*width,height*} width and height of the paper.

total={*width,height*} width and height of the total body (i.e., head, body, foot and marginal notes).

body={*width,height*} text width and text height of the body of the page.

hmargin={*left,right*} left and right margins.

vmargin={*top,bottom*} top and bottom margins.

width=*length* text body width (including margins).

height=*length* text body height (including margins).

left=*length* left margin of text body.

right=*length* right margin of text body.

top=*length* top margin of text body.

bottom=*length* bottom margin of text body.

hscale=*scale* ratio of width of text body to paper width.

vscale=*scale* ratio of height of text body to paper height.

marginpar=*length* modifies \marginparwidth.

marginparsep=*length* modifies \marginparsep.

head=*length* modifies the \headheight.

headsep=*length* modifies the \headsep.

foot=*length* modifies the \footskip.

hoffset=*length* modifies the \hoffset.

voffset=*length* modifies the \voffset.

footnotesep=*length* modifies the \skip\footins.

6.8 Page Styles

Page styles define what goes in the header/footer of a page, provided, of course, that we have left enough room for headers and/or footers. LaTeX offers the following page styles:

empty both header and footer are empty.

plain header is empty and footer contains page number.

headings footer is empty and header contains the name of a chapter/section and the page number.

myheadings footer is empty and header contains the page number and user-supplied information.

A user can select a particular page style with the command \pagestyle. This command has one argument, which is the name of a page style. Moreover, it is possible to change the page style of an individual page with the command \thispagestyle, which also has one argument: the name of a page style. The myheadings page style makes it possible to define what will go in the header using the commands \markright and \markboth. The first command has one argument and is useful for one-sided printing, whereas the second has two arguments and is useful for two-sided printing. The first argument goes on even-numbered pages and the second on odd-numbered pages.

➤ **Exercise 6.5** Suppose that George Typesetter wants to prepare a manuscript and wants his name to appear on the header of even-numbered pages. Moreover, suppose that he wants the (short) title of his manuscript to appear on the header of odd-numbered pages. Write down the necessary commands that achieve the desired effect. ☐

Suppose that we prepare a book and want the word "chapter" to be in uppercase and the title of the chapter to be in lowercase. Then, we have to redefine the commands \chaptermark and \sectionmark as follows:

```
\renewcommand{\chaptermark}[1]{\markboth{%
  \MakeUppercase{\chaptername}\ #1}{}}
```

Note that the command \chaptername contains the word "chapter," or in the case where we are using the babel package, the corresponding word for "chapter" for the main language of the text. Moreover, the command \MakeUppercase forces its argument to be in uppercase. Now these definitions can be used, even with the fancyhdr package. Here is an example:

```
\fancyhead[LE]{%
  \scshape\thepage\ $\spadesuit$ \leftmark}
\fancyhead[RO]{%
  \scshape\rightmark $\spadesuit$ \thepage}
```

\leftmark contains the left argument of the last processed \markboth command; the \rightmark contains either the right argument of the last processed \markboth command or the only argument of the last processed \markright command. When an argument appears in uppercase by default, we can disable this feature by making any mark an argument of the command \nouppercase:

```
\lhead{\nouppercase{\rightmark}}
```

Although one can easily customize a page style, we recommend that people use the package fancyhdr by Piet van Oostrum, to create their own page style. The general page layout can be seen in Figure 6.6 (a). The LHdr and LFtr are left-justified; the CHrd and CFtr are centered; the RHdr and the RFtr are right-justified. We have to define each of the six elements and the two decorative lines separately.

Let us see how we can define the example page layout of Figure 6.6 (b):

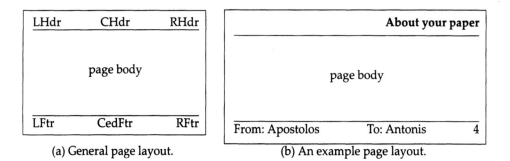

(a) General page layout.　　　　　(b) An example page layout.

Figure 6.6: Page layouts.

```
\pagestyle{fancy}
\lhead{} \chead{} \rhead{\bfseries About your paper}
\lfoot{From: Apostolos} \cfoot{To: Antonis} \rfoot{\thepage}
\renewcommand{\headrulewidth}{0.4pt}
\renewcommand{\footrulewidth}{0.4pt}
```

In the code above, we set the width of the two rules with a redefinition. For the moment, one does not need to know anything more to define simple page layouts. We first specify that we are going to use the fancy page style for the rest of the document, except of course for the first page. If we do want to have a fancy page, we have to select the fancy page style for the first page only. The commands \lhead, \lfoot, and so on, are used to set the six elements of the header and the footer.

In Figure 6.7 we show an example of a page layout suitable for two-sided printing.

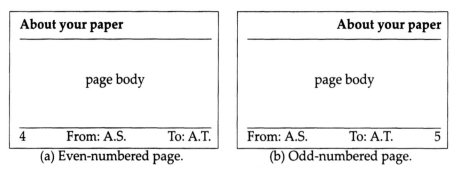

(a) Even-numbered page.　　　　　(b) Odd-numbered page.

Figure 6.7: An example of two-sided printing.

Here are the commands for this particular page layout:

```
\fancyhead{} % clear all fields
\fancyhead[RO,LE]{\bfseries About your paper}
```

```
\fancyfoot[LE,RO]{\thepage}
\fancyfoot[LO,CE]{From: A.S.}
\fancyfoot[CO,RE]{To: A.T.}
\renewcommand{\headrulewidth}{0.4pt}
\renewcommand{\footrulewidth}{0.4pt}
```

In the example above we use the commands \fancyhead and \fancyfoot to set the header and the footer. The optional arguments are called selectors, and the full list is: E(ven page), O(dd page), L(eft element), R(ight element), C(enter element), H(ead), and F(ooter). The last two elements can be supplied to the command \fancyhf, which is useful to combine the specifications for footers and headers.

Suppose that we want to get rid of the decorative lines on pages that have a float object on the top (or bottom) of the page. Then, the following command will achieve the desired effect:

```
\renewcommand{\headrulewidth}{\iffloatpage{0pt}{0.4pt}}
```

The first argument specifies the width of the decorative line if there is a float object on the top (bottom) of a page. The second one specifies the width of the decorative line in "normal" pages.

Another problem that can be tackled with this package is the redefinition of the standard page styles. Here is an example:

```
\fancypagestyle{plain}{%
    \fancyhead{} %
    \fancyfoot[C]{\bfseries \thepage}
    \renewcommand{\headrulewidth}{0pt}}
```

6.9 The Preparation of Slides

The production of simple transparencies ("slides") for use on overhead projectors can be done with the slides document class. However, one can create fancy transparencies by using special document classes that we will describe later.

An individual transparency is produced with a slide environment (see Figure 6.8). Any ordinary command can appear in the body of this environment. However, one should not use any sectioning commands, floating objects, or page breaks, as these commands make no sense in a slide environment. The overlay environment is used for an overlay (i.e., a slide that is meant to be put on top of another slide). This environment is the same as the slide environment except for numbering; suppose that the page number of a slide is 5. Then, the first overlay is numbered "5-a," the second one is numbered "5-b," and so on. Since overlays are put on top of a slide, the overlays must contain exactly the same text as the slide but, at the same time, we must make certain parts of the slide invisible. There are two approaches to this problem: (a) we use

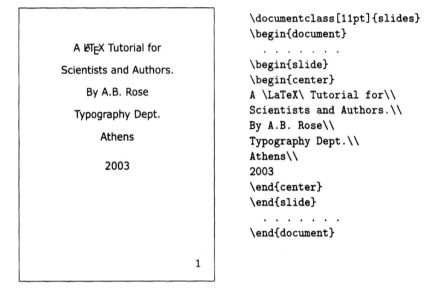

Figure 6.8: A slide and the code that generates it.

the colorackage and make the text invisible by coloring it, or (b) we use the \phantom command. This command has one argument, which is a piece of text that TEX typesets without printing any glyphs so the text is actually invisible. For example, the command

\underline{}

prints as _____. (Ω provides the primitive commands \leftghost and \rightghost, which do the same thing for the character that immediately follows the commands. The first command is useful for left-to-right typesetting and the second for right-to-left typesetting.) If one wants to write notes between slides, one can use the note environment. Notes that follow slide 3 are numbered "3-1," "3-2," and so on. The commands \onlyslides and \onlynotes can be used to print some of the slides and/or some of the notes. Both commands have as arguments a comma-separated list of slide/note numbers and possibly a range of slide/note numbers, respectively. For example, the commands

\onlyslides{1,4,6-8,15} \onlynotes{2,8-10}

will produce slides 1, 4, 6–8, and 15 and notes 2 and 8–10. These commands must be specified in the preamble of a LATEX file.

When giving a lecture, it is a good practice to have a strict time schedule, so one does not actually run out of time. To assist people, LATEX provides the command \addtime, which has one argument—the total time (in seconds) that the speaker should spend on a slide. This command can be placed before or after a slide. The total time that the speaker should have taken so far will be printed at the bottom of each note. One can reset the elapsed time with the \settime command. The command \settime{120} sets the total elapsed time to 120 seconds (2 minutes). This command must not appear in the body of any environment described in this section.

6.9.1 Advanced Slide Preparation

The seminar document class (by Timothy P. Van Zandt) provides an environment for advanced slide preparation. The document class provides two environments for slides: the slide environment for landscape slides and slide* for portrait slides. By default, the slides will be typeset in landscape mode, but if one specifies the portrait option, the slides will be typeset in portrait mode. Both environments have two optional arguments that let one change the width and the height of a slide. For example, the command

$$\begin{slide}[5cm, 7cm]$$

starts a slide in landscape mode that is 5 cm wide and 7 cm long.

The standard slide environment disables the use of commands that do page breaks in its body. Fortunately, this is not the case with the corresponding environments that the seminar document class provides. One can easily break a long slide into "sub"-slides with the command \newslide. However, this mechanism works only if we enlarge the standard slide height with the command

$$\extraslideheight{10cm}$$

Note that we just make slides long enough so that it makes sense to break them into "sub"-slides. This mechanism is particularly useful when we want to transform a manuscript into slides.

If one wants to customize a group of slides, then one has to change the value of the length variables shown in Figure 6.9.

Slides can be framed. The command \slideframe[command]{style} specifies the frame style to use. The predefined frame styles are none and plain. If one uses the fancybox, then the following frame styles are available: shadow, oval, Oval, and doublebox. It is even possible to create your own frame style, but we think that the six frame styles are more than enough.

The commands \onlyslides and \notslides can be used to include or exclude only those slides that are in the argument lists.

If we just want to preview our slides, we can force LATEX to put two slides per page with the command \twoup. This command should be placed in the preamble of a LATEX file. The command has an optional argument (an integer) that increases or decreases the two-up magnification. Additionally, if one wants four slides per page, the article document class option is used; then the \twoup command produces the desired effect. This option offers slide styles that can be used to specify where slide labels should go. The slide style can be selected with the command \slidestyle. There are three

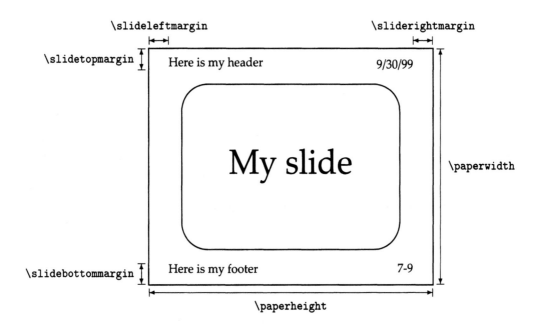

Figure 6.9: Slide margins.

predefined slide styles: empty, left, and bottom. Moreover, one can specify one's own slide style with the commands

\newpagestyle{*style*}{*header*}{*footer*}
\renewpagestyle{*style*}{*header*}{*footer*}

Here, *style* is the name of a (new or predefined) page style; the other arguments have the expected meaning. Naturally, if one chooses to use a predefined page style, then marks have to be set.

There is no special environment for notes. Consequently, one can write notes between slides. The document class has three options that handle the printing of notes: slidesonly (only the slides will be printed), notes (both notes and slides are printed), and notesonly (only the notes are printed).

Overlays are produced in a "natural way" (i.e., overlays are part of the slide environment, and the user just types the additional text in each overlay). Overlays are written in the body of the overlay environment. Each overlay has an argument that specifies the order in which overlays will be generated. This argument is a number from 0 to 9; number 0 denotes the slide. In order to be able to generate overlays, users have to specify the semlayer and the semcolor class options. Moreover, the frame style of an overlay can be specified with the command \overlayframe. Here is a small example:

```
\documentclass[semlayer,semcolor]{seminar}
  \usepackage{fancybox}
  \slideframe{Oval}
  \overlayframe{Oval}
\begin{document}
  \begin{slide}
    This a slide
    \begin{overlay}{1}
      This is the text of the first overlay.
    \end{overlay}
    \begin{overlay}{2}
      This is the text of the second overlay.
    \end{overlay}
  \end{slide}
\end{document}
```

Another document class that provides an environment for advanced slide preparation is the prosper document class (by Frédéric Goualard). When generating slides with prosper, you have to transform the resulting output file to PostScript. If you want to create a slide show, you have to transform the PostScript file to PDF using a program such as PS2PDF. To make fancy slide shows, prosper supports a number of *transition effects*, that is, visual effects employed when going from one slide to anothe, which are specified as optional arguments to slide enviroments (see Figure 6.10); the available transition effects are:

Split Two lines sweep across the screen revealing the next slide.

Box A (growing) box sweeps from the center, revealing the new slide.

Blinds A number of lines evenly placed on the screen appear and together sweep in the same direction to reveal the next slide.

Wipe A line sweeps across the screen from left to right revealing the new slide.

Dissolve The old slide *dissolves* to reveal the next slide.

Glitter Similar to Dissolve, except that the old slide *dissolves* from left to right as if a wave is moving in this direction.

Replace The effect of this option is just to replace the current slide with the next one.

The prosper document class accepts the following options:

draft Figures are not included in the resulting file and are replaced by frames that have the size of the figures. In addition, the caption at the bottom of every slide displays the processing date and time together with the filename.

final Figures are included in the resulting file; the text of captions appears as it was requested.

slideColor Slides will be colorful; this option should be used with caution when the slides are to be printed on a black and white printer.

slideBW Slides will be colored with a reduced number of colors and so they can be printed on a black and white printer without any problem.

total The caption at the bottom of every slide displays the number of the current slide along with the total number of slides.

nototal Only the number of the current slide appears in the caption.

nocolorBG The background of the slide is white whatever the style may be.

colorBG The background color depends on the current style.

ps Must be used when you want to print the slides on a PostScript printer.

pdf Must be used when you want to display the slides using Acrobat Reader.

accumulate The commands \onlySlide, \untilSlide, and \fromSlide operate in PostScript mode, also called ps mode.

noaccumulate The commands \onlySlide, \untilSlide, and \fromSlide do not operate in PostScript mode.

distiller This option should be used when the resulting PostScript file will be transformed to PDF by Adobe Distiller.

It is now time to present the commands and the environments that this document class provides. In Figure 6.10, one can see the general structure of a LATEX file that uses the prosper document class. The following commands should always appear in the preamble of the LATEX file:

\title The argument of the command becomes the title of the presentation.

\subtitle The argument of this command is the subtitle of the presentation. This command is optional.

\author The author(s) of the presentation.

\email The e-mail address(es) of the author(s) of the presentation.

\institution Author(s) affiliation (optional).

\slideCaption The argument of this command will be put at the bottom of every slide. If we do not use this optional command, the title of the presentation is used as each slide's caption.

\Logo The command has two forms: \Logo(x,y){ *logo* } and \Logo{ *logo* }. Here, *logo* is a set of commands that draws a figure or an image inclusion command (see Chapter 9 for more details). The optional arguments surrounded by parentheses specify the exact location where the *logo* will be placed. Point $(0,0)$ is located at the lower left-hand corner of the slide.

\displayVersion A draft caption is printed instead of the standard one.

\DefaultTransition The argument of this command is the name of a supported transition effect and will be the default transition effect employed.

\NoFrenchBabelItemize This command should be used only if the french option of the babel is used. This way, one can choose to have one's own itemization style.

The slide environment may have an optional argument that specifies the transition efect. The new Itemize environment corresponds to the standard itemize environment, and the itemize environment has been redefined so that the text is not justified.

```
\documentclass[Options]{prosper}

    \title{Title of presentation}
    \subtitle{Subtitle}
    \author{Names of the authors}
    \institution{Affiliation}
    \slideCaption{caption text}


    \begin{document}
    \maketitle

        \begin{slide}[transition]{slide title}
        Material for the slide
        \end{slide}

    ..........................

        \overlays{n}{
        \begin{slide}[transition]{slide title}
        Material for the overlay
        \end{slide}
        }

    \end{document}
```

Figure 6.10: Structure of a LaTeX file using the prosper document class.

The itemstep environment can be used in the slide of an overlay only. Each item of this environment is displayed incrementally in PDF mode. For example, the overlay

```
\overlays{3}{%
    \begin{slide}[Split]{Third Slide}
    Hello
    \begin{itemstep}
        \item One
        \item Two
        \item Three
    \end{itemstep}
    There
    \end{slide}
}
```

will initially display the first item. The second item will appear once we proceed to the next page by pressing Ctrl+PgDn and so on. The \overlays command should be used to create overlays. Note that here we use a command to create overlays instead of an environment. The \overlays command has two arguments: a number that indicates the number of overlays and a slide environment.

▶ **Exercise 6.6** Suggest a slide construction method that will allow readers to navigate in a slide show using only the mouse. □

The following commands may appear in the body of a slide:

\FontTitle This command should be used to change the font/size/color to be used for the slide titles. The command has two arguments: the first for color slides and the second for black and white slides. To change the color, one should not use the color package described in Section 9.11.1. Instead, one can use either the predefined colors \black, \darkgray, \gray, \lightgray, \white, \red, \green, \blue, \yellow, \cyan, or \magenta. If we want to use other colors, we can define them with the command \newrgbcolor, which is provided by the pstricks package. This command has two arguments: the name of a color and its RGB color specification (see Section 9.11.1). Here is a simple color definition:

$$\text{\textbackslash newrgbcolor\{gold\}\{0.804\ 0.498\ 0.196\}}$$

Note that the color is defined without a leading backslash, which, however, is necessary when using the color.

\FontText A command that can be used to change the font/color/size of the slide text. It has two arguments, exactly like \FontTitle.

\fontTitle Typesets its argument in the font/color/size specified with the \FontTitle command.

\fontText Typesets its argument in the font/color/size specified with the \FontText command.

\ColorFoot The footer of the slide is typeset with the color, which is the argument of this command.

\PDFtransition Can be used to alter the transition effect. The command has one argument, which is the name of a transition effect.

\myitem This command has two arguments and is used to redefine the itemization symbol. The first argument is the itemization level (possible values 1, 2, or 3), and the second argument is a command that defines the symbol. This command can be a graphics inclusion command or drawing.

The following commands may be used in the creation of a set of overlays:

\fromSlide Has two arguments and puts the second argument on all overlays starting from the one that the first argument (a number) specifies.

\onlySlide Puts the second argument on the overlay specified by the first argument.

\untilSlide Puts the second argument on all overlays starting from the first one and finishing with the one that is specified by the first argument.

`\FromSlide` All the material that follows the command will be placed on all overlays starting from the one that its only argument specifies.

`\OnlySlide` Adds the material that follows the command to the overlay that its only argument specifies.

`\UntilSlide` Similar to `\FromSlide` but adds the material to all overlays starting from the first one and finishing with the one that its only argument specifies.

As one might expect, all of these commands are meaningful in PDF mode. The commands `\onlySlide`, `\fromSlide` and `\untilSlide` have starred counterparts that typeset the material exactly in the same place. For example, the code

```
\overlays{3}{%
  \begin{slide}{Example}
    \onlySlide*{1}{\Huge\green A}%
    \onlySlide*{2}{\Huge\yellow A}%
    \onlySlide*{3}{\Huge\blue A}%
  \end{slide}
}
```

will create three overlays that will print the letter A at the same position in each overlay in three different colors. Note that the % is necessary to avoid unwanted white space that would distort the appearance of our slide show.

In certain cases, we want to place different material in an overlay or slide depending on whether it will be printed or displayed on a computer screen. The following commands have been designed to facilitate this procedure:

`\PDForPS` Has two arguments, and the first is used if we are preparing our slides in PDF mode. The second argument is used if we are preparing our slides in PostScript mode.

`\onlyInPS` The only argument of this command is used if we are preparing our slides in PostScript mode.

`\onlyInPDF` If we are preparing our slides in PDF mode, then LATEX will use the only argument of this command.

For example, one can add the following command in the overlay above in case it is necessary to print the overlays:

$$\text{\onlyInPS\{\Huge\gray A\}\%}$$

What makes the prosper document class exceptional is its ability to use a number of predefined presentation styles. The presentation styles currently available are presented on pages 194–195. Note that the document class declaration used to create these slides is

$$\text{\documentclass[slideBW,nocolorBG,ps,xxxxx]\{prosper\}}$$

where xxxxx is the name of a presentation style.

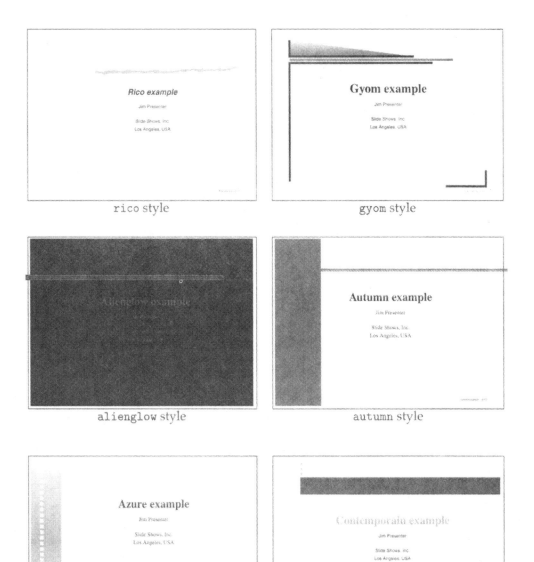

rico style

gyom style

alienglow style

autumn style

azure style

contemporain style

darkblue style

frames style

lignesbleues style

nuancegris style

troispoints style

6.10 Boxes

When building a wall, we put bricks one on top of the other, and in between them we put mortar to make the wall rigid. Similarly, TEX constructs pages by putting lines one above the other, and in between them it puts a special rubber length called *glue*. Each line consists of words, which, in turn, consist of letters (i.e., glyphs). In Section 1.4, we said that TEX treats every glyph as a "box," so a line is a big box that consists of little boxes (i.e., glyphs). A paragraph is an even bigger box that consists of lines and so on. A box is just a rectangle that has *height*, *width*, and *depth*:

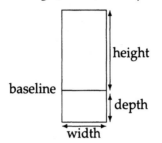

The virtual line where characters sit is called the *baseline*. Here is a little example that demonstrates how TEX perceives two lines of text:

Since boxes are so important, LATEX provides a number of commands that create boxes. Moreover, it provides box variables to which one can store boxes.

The following command can be used to create a box of width equal to *width* that contains an *obj*: positioned at *pos*:

$$\verb|\makebox[|width\verb|][|pos\verb|]{|obj\verb|}|$$

If the *width* is missing, we have to omit the *pos*, so the width of the box is equal to the natural width of the *obj*. The height and the depth of the new box are equal to the natural height and width of the *obj*. The possible values for *pos* are: s (for stretched), l (for flush left), r (for flush right), and c (for centered). The last value is the default one. This shows the difference between these values:

That's all folks!	`\makebox[2\width][l]{...}`
That's all folks!	`\makebox[2\width][r]{...}`
That's all folks!	`\makebox[2\width][c]{...}`
That's all folks!	`\makebox[2\width][s]{...}`

Note that the frames are not produced by the commands that we give on the right-hand side! When LATEX creates a box, it automatically sets the variables \width, \height,

\depth, and \totalheight. These variables contain the width, height, depth, and the total height of the *obj*. The total height is equal to the height plus the depth. These variables are constants, so we cannot alter their values.

The command \mbox{*obj*} is a fast shorthand for the command \makebox{*obj*}. An interesting thing to note is that in a \mbox LATEX uses the main document font. Therefore, if we want to have a word or a short phrase in a mathematical formula that has to be typeset using the main font, we simply can use a \mbox:

$$x > 0 \quad \text{iff} \quad y > 0 \qquad \text{\$x>0\textbackslash quad\textbackslash mbox\{iff\}\textbackslash quad y>0\$}$$

The command \framebox is like the command \makebox except that it puts a frame around the box. Similarly, the command \fbox{*obj*} is shorthand for the command \framebox{*obj*}. The frame is made of lines of thickness \fboxrule, separated by space \fboxsep from the text. Both \fboxrule and \fboxsep are length variables.

➤ **Exercise 6.7** Write down the code that produces the framed *"That's all folks!"* example. □

A box variable can be declared with the command \newsavebox{*cmd*}, provided of course that *cmd* has not been used before. We can define a box variable with the command

$$\text{\textbackslash savebox\{\textbackslash \textit{cmd}\}[\textit{width}][\textit{pos}]\{\textit{obj}\}}$$

The arguments of this command are exactly the same as the corresponding arguments of the \makebox command. Once we have defined a new box variable, we can use it with the command \usebox. The command \sbox{*cmd*}{*obj*} is actually a fast shorthand version of the command \savebox{*cmd*}{*obj*}.

We can draw ruled lines with the command \rule[*raised*]{*width*}{*height*}. The command draws a *width* × *height* line, raised from the baseline by *raised*. In particular, a ruled line with width equal to 0 pt is called a *strut*. Here is a simple example that shows the use of struts:

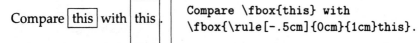

```
Compare \fbox{this} with
\fbox{\rule[-.5cm]{0cm}{1cm}this}.
```

Normally, boxes are put on the baseline, but it is possible to raise boxes with the command \raisebox{*distance*}[*height*][*depth*]{*box*}. This command raises the *box* up by *distance* (or down if *distance* is negative). Moreover, if the optional arguments are present, it makes TEX believe that the box has height and depth equal to *height* and *depth*, respectively. For example, the following piece of code produces the TEX logo:

$$\text{T\textbackslash hspace\{-.1667em\}\textbackslash raisebox\{-.5ex\}\{E\}\textbackslash hspace\{-.125em\}X}$$

➤ **Exercise 6.8** The A in the LATEX logo is in \scriptsize and is raised up by 0.2 em. Moreover, the distance between the L and the A is   0.36 em and the distance between the A and the T is   0.15 em. Write down the commands that generate the LATEX logo. □

➤ **Exercise 6.9** Define a box variable that contains a rectangle that is 3 ex long and 4 ex wide. ☐

If one wants to create a box that contains verbatim text, then one can use the lrbox environment. Here is a simple example:

The characters	`\newsavebox{\temp}`
~ # $ % ^ & _	`\begin{lrbox}{\temp}`
are special.	`The characters \verb|~ # $ % ^ & _|`
	`are special.`
	`\end{lrbox}`
	`\usebox{\temp}`

Note that the space before or after the text in the body of an lrbox is ignored. Moreover, we cannot have more than one paragraph in such a box. If we want to have more than one paragraph, we have to use either a \parbox or a minipage.

The command \parbox[*pos*][*height*][*inner-pos*]{*width*}{*text*} makes a box with width equal to *width*, positioned by *pos* relative to the baseline. The possible values of *pos* are: t (the first line of the box is placed on the baseline), b (the last line of the box is placed on the baseline), c (the box is centered vertically on the baseline), and s (the same as before, but here the box is somehow squeezed). Here is a simple example:

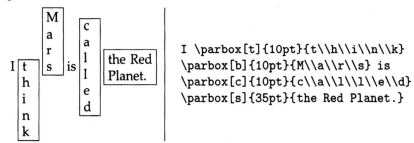

```
I \parbox[t]{10pt}{t\\h\\i\\n\\k}
\parbox[b]{10pt}{M\\a\\r\\s} is
\parbox[c]{10pt}{c\\a\\l\\l\\e\\d}
\parbox[s]{35pt}{the Red Planet.}
```

The optional arguments *height* and *inner-pos* can be used to specify the height of the box and the placement of the *text* inside the box. The possible values are identical to the value that *pos* may take. The following example clearly shows the functionality of the *inner-pos* argument (the horizontal line is the baseline):

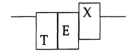

```
\parbox[t][20pt][b]{10pt}{T}%
\parbox[t][20pt][c]{10pt}{E}%
\parbox[t][20pt][t]{10pt}{X}%
```

A minipage is a box that looks like a page by setting the \textwidth and the \columnwidth equal to the width of the box. In addition, one can even have footnotes in a minipage. Minipages can be created with the minipage environment:

```
\begin{minipage}[pos][height][inner-pos]{width}{text}
    text of the minipage
\end{minipage}
```

For an explanation of the meaning of the various arguments that this environment takes, see the discussion on the corresponding arguments of the \parbox command. Here is an example:

An aardvark[a] is a burrowing mammal of southern Africa, having a stocky, hairy body, large ears, a long tubular snout, and powerful digging claws.	`\begin{minipage}{150pt}`
	`An aardvark%`
	`\footnote{Orycteropus`
	`afer} is a ........`
	`\end{minipage}`

---
[a]Orycteropus afer

6.10.1 Fancy Boxes

The package fancybox (by Timothy P. Van Zandt) provides commands that create framed boxes. The package provides four commands that create framed boxes (see Figure 6.11).

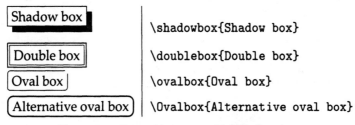

Shadow box	`\shadowbox{Shadow box}`
Double box	`\doublebox{Double box}`
Oval box	`\ovalbox{Oval box}`
Alternative oval box	`\Ovalbox{Alternative oval box}`

Figure 6.11: Framed boxes provided by fancybox.

Note that the frame is separated by space \fboxsep from the text. The width of the shadow is stored in the length variable \shadowsize. In a double box the width of the frames, as well as the distance between the two frames, is controlled by \fboxrule. The width of the frame in a \ovalbox and a \Ovalbox is set by the \thinlines and \thicklines declarations, respectively. The diameter of the corner arcs is set with the \cornersize command. For example, the command

$$\text{\cornersize}\{num\}$$

sets the diameter of the corner arcs to *num* times the lessor of the width or height of the box. Similarly, we can set the diameter with the command \cornersize{*dim*}, where *dim* is just a length.

If we want to create a framed minipage, we can just have the minipage as the argument of a \fbox command. However, if we want to define an environment that will produce a framed minipage, then there is no obvious solution. The package fancybox provides the Sbox environment; this is a variant of the \sbox command that saves the body of the environment in an internal box variable. The contents of this box variable can be retrieved with the command \TheSbox. Here is a simple example:

| hoi polloi (hoi-puh-LOI), noun: The common people generally; the masses. Lizzie insisted that her children distinguish themselves from the hoi polloi by scrupulous honesty. –Kate Buford, Burt Lancaster: An American Life | `\newenvironment{fminipage}%`
`  {\begin{Sbox}\begin{minipage}}%`
`  {\end{minipage}%`
`  \end{Sbox}\fbox{\TheSbox}}`
`...`
`\begin{fminipage}{130pt}`
`hoi polloi (hoi-puh-LOI), noun:`
`...........................`
`\end{fminipage}` |

Note that hoi polloi is actually a transliteration of the Greek term Οἱ πολλοί, which means "the many."

➤ **Exercise 6.10** Define an environment that will be equivalent to a framed equation environment. (Hint: Use the length variables \abovedisplayskip and \belowdisplayskip, which are used to set the white space before and after a math display.) ☐

In some cases, one may want to display a table or a figure in landscape mode just because it is too wide to fit in portrait mode. The environment landfloat is useful in such cases. It takes two arguments: the name of a float object (e.g., a table) and a rotation command. In the example that follows, we use the command \rotatebox:

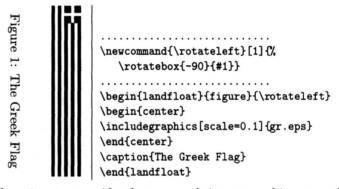

Figure 1: The Greek Flag

```
...........................
\newcommand{\rotateleft}[1]{%
    \rotatebox{-90}{#1}}
...........................
\begin{landfloat}{figure}{\rotateleft}
\begin{center}
\includegraphics[scale=0.1]{gr.eps}
\end{center}
\caption{The Greek Flag}
\end{landfloat}
```

The package fancybox provides the commands \boxput and \boxput* that can put a box behind or in front of another box, respectively. The command has three arguments: the first box, the second box and, just after the name of the command, an optional pair of coordinates that determine where the center of the first box is positioned. For example, (0,0) puts it in the center of the second box, (0,1) places it in the center-top, and (-1,-1) puts it in the bottom-left corner. Thus, the coordinates are always numbers, with one exception: if the second coordinate is b or B, the first box is positioned vertically at the baseline of the second box. Here is a rather complicated example of its use:

The process or method of
selecting one or more in-
dividuals from a group,
as for a service or duty:
a candidate who did not
pursue the nomination,
*but accepted a **draft** by the*
party convention.

```
\boxput{\makebox(0,0){%
Huge\rotatebox{45}{%
\scalebox{2}{%
color[gray]{0.7}{Draft}}}}}{%
\parbox{4cm}{The process or
.........................}}
```

Framing the body of the text and even a whole page is easy with this package. The commands \fancypage and \thisfancypage frame either a complete document or a particular page, respectively. Both commands have two arguments: the first one is used to frame the body of the text and the second the whole page, including the header and the footer. For example, the page in Figure reffancy:page:1 has been framed with the following commands:

```
\thisfancypage{%
  \setlength{\fboxsep}{8pt}%
  \setlength{\shadowsize}{8pt}%
  \shadowbox}{%
  \setlength{\fboxsep}{8pt}\Ovalbox}
```

If we want to frame a flushleft, a flushright, or center environment, we can use the environments Bflushleft, Bflushright, or Bcenter, respectively. Similarly, we can get framed lists with the environments Benumerate, Bitemize, and Bdescription. Naturally, the unframed versions of these environments are the environments enumerate, itemize, and description, respectively.

If we want to produce a framed verbatim environment, we can use the environments that the package provides:

Verbatim This works almost like the corresponding standard LATEX environment.

LVerbatim Just like Verbatim except that the text is indented.

BVerbatim Produces a box with the same width as the longest line of the verbatim text. It has an optional argument that specifies the baseline of the box: b for alignment with the bottom line and t for alignment with the top line. By default, the baseline is assumed to be at the center of the box.

VerbatimOut Has an argument that must be a filename where the body of the environment is written.

SaveVerbatim Stores the body of the environment in a command that is the only argument of the environment. Note that the package does not check whether this command is in use, so it may overwrite the definition of an existing command.

There are some other useful commands for short pieces of verbatim text:

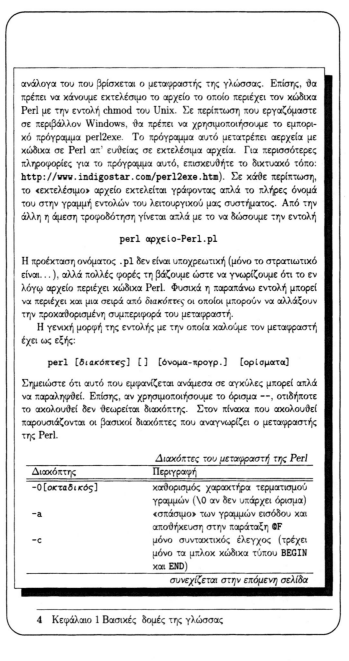

ανάλογα του που βρίσκεται ο μεταφραστής της γλώσσας. Επίσης, θα πρέπει να κάνουμε εκτελέσιμο το αρχείο το οποίο περιέχει τον κώδικα Perl με την εντολή chmod του Unix. Σε περίπτωση που εργαζόμαστε σε περιβάλλον Windows, θα πρέπει να χρησιμοποιήσουμε το εμπορικό πρόγραμμα perl2exe. Το πρόγραμμα αυτό μετατρέπει αρχεία με κώδικα σε Perl απ' ευθείας σε εκτελέσιμα αρχεία. Για περισσότερες πληροφορίες για το πρόγραμμα αυτό, επισκευθήτε το δικτυακό τόπο: http://www.indigostar.com/perl2exe.htm). Σε κάθε περίπτωση, το «εκτελέσιμο» αρχείο εκτελείται γράφοντας απλά το πλήρες όνομά του στην γραμμή εντολών του λειτουργικού μας συστήματος. Από την άλλη η άμεση τροφοδότηση γίνεται απλά με το να δώσουμε την εντολή

                    perl αρχείο-Perl.pl

Η προέκταση ονόματος .pl δεν είναι υποχρεωτική (μόνο το στρατιωτικό είναι...), αλλά πολλές φορές τη βάζουμε ώστε να γνωρίζουμε ότι το εν λόγω αρχείο περιέχει κώδικα Perl. Φυσικά η παραπάνω εντολή μπορεί να περιέχει και μια σειρά από διακόπτες οι οποίοι μπορούν να αλλάξουν την προκαθορισμένη συμπεριφορά του μεταφραστή.

   Η γενική μορφή της εντολής με την οποία καλούμε τον μεταφραστή έχει ως εξής:

    perl [διακόπτες] [] [όνομα-προγρ.] [ορίσματα]

Σημειώστε ότι αυτό που εμφανίζεται ανάμεσα σε αγκύλες μπορεί απλά να παραληφθεί. Επίσης, αν χρησιμοποιήσουμε το όρισμα --, οτιδήποτε το ακολουθεί δεν θεωρείται διακόπτης. Στον πίνακα που ακολουθεί παρουσιάζονται οι βασικοί διακόπτες που αναγνωρίζει ο μεταφραστής της Perl.

	Διακόπτες του μεταφραστή της Perl
Διακόπτης	Περιγραφή
-0[οκταδικός]	καθορισμός χαρακτήρα τερματισμού γραμμών (\0 αν δεν υπάρχει όρισμα)
-a	«σπάσιμο» των γραμμών εισόδου και αποθήκευση στην παράταξη @F
-c	μόνο συντακτικός έλεγχος (τρέχει μόνο τα μπλοκ κώδικα τύπου BEGIN και END)
	συνεχίζεται στην επόμενη σελίδα

4   Κεφάλαιο 1 Βασικές δομές της γλώσσας

Figure 6.12: A page framed with the commands provided by fancybox.

\SaveVerb This command works like the \verb command but has an argument that is a command name in which we store the verbatim text.

\UseVerb This command has one argument (a command name) and is used for including short pieces of verbatim text saved with \SaveVerb.

6.11   New Commands

Although LATEX and the many packages that exist provide commands that can tackle a great many problems, there are still situations where one may want to be able to define a command that will solve a particular problem that no package designer had thought of. For this reason, LATEX provides a facility for the creation of new commands. New commands are defined with the \newcommand command. This command has two required arguments: the name of the new command and the "code" that will be executed every time we use this new command. The name of our new command must start with a backslash and must be followed by one or more letters.[1] Suppose that we want to create a command that will print the expression $\{x_1, \ldots, x_n\}$. Then, the following code will create a new command, \seq, that will do the job every time it is used:

$$\newcommand{\seq}{\{x\_1,\ldots,x\_n\}}$$

Now, we can write $\seq$ to get the expression above. Obviously, the definition above has a big drawback: one cannot use it to print the expression $\{x_1, \ldots, x_k\}$, so one has to define a new command to do the job.

➤ **Exercise 6.11** Define this new command.   □

Now, suppose that we want to be able to define *parametric* commands (i.e., commands that have arguments such as most LATEX commands). The good news is that we can actually create such commands, but the bad news is that we can have at most nine parameters! Here is how we can define a parametric command:

$$\newcommand{\cmd}[n]{commands}$$

n denotes the number of arguments that \cmd has. In order to refer to an individual argument inside the *commands*, we have to use the character # followed by a number. So, if we write #3 in *commands*, we are actually referring to the third argument of \cmd. We now give a concrete example that is a general solution to the problem above:

$$\newcommand{\seq}[1]{\{x\_1,\ldots,x\_{#1}\}}$$

Now, we can write $\seq{k}$ to get $\{x_1, \ldots, x_k\}$. It is important to note that the arguments of a command must be surrounded by curly brackets. Although the definition above is quite useful, whenever we use it we must make sure that it occurs in math mode. A simple solution to this problem is to have the *commands* as an argument of the command \ensuremath. Here is how we can rewrite the definition above:

---

1. Actually, it must be followed by characters having a category code equal to 11.

```
\newcommand{\seq}[1]{%
    \ensuremath{\{x_1,\ldots,x_{#1}\}}}
```

Now, the commands \seq{k} and $\seq{k}$ produce identical output.

▶ **Exercise 6.12** Why do you think we had to use curly brackets to surround the parameter of the command in the expression x_{#1}? ☐

Suppose now that we want to have a command that will also parameterize the "*x*." In order to do that, we must introduce a second parameter:

```
\newcommand{\seq}[2]{%
    \ensuremath{\{#1_1,\ldots,#1_{#2}\}}}
```

Although we have parameterized *x*, in most cases people will use the letter *x* in such expressions, so it could be useful to have a default value for the first parameter. In other words, we want "*x*" to be an optional argument with a default value. LATEX supports command definitions with optional arguments: the default value is surrounded by square brackets and is placed after the number of arguments. Here is the new definition:

```
\newcommand{\seq}[2][x]{%
    \ensuremath{\{#1_1,\ldots,#1_{#2}\}}}
```

Now, the command \seq{n} will print $\{x_1, \ldots, x_n\}$, and the command \seq[y]{k} will print $\{y_1, \ldots, y_k\}$.

▶ **Exercise 6.13** Define a command that will print the phrase *I love to fly!* one hundred times. ☐

If we define a new command with \newcommand, LATEX checks whether this command has already been defined. This is a way to ensure that existing commands will not be redefined accidentally. But there are cases where we need to redefine a command, so LATEX provides the command \renewcommand, which can be used to redefine an existing command. This command has exactly the same arguments as the command \newcommand.

The commands \newcommand and \renewcommand are also available in starred versions. The only difference between a starred and an unstarred command is that the former introduces new commands whose arguments cannot contain more than one paragraph. Consider the following document fragment:

```
. . . . . . . . . . . . . . . . . . . . . . . . . . . . . . . . . . . .
\newcommand*{\bdface}[1]{{\bfseries #1}}
\newcommand{\bldface}[1]{{\bfseries #1}}
. . . . . . . . . . . . . . . . . . . . . . . . . . . . . . . . . . . .
\bldface{text text text \par text   text}
\bdface{text text text \par text   text }
```

Remember that the \par command forces LATEX to start a new paragraph. Now, when LATEX will processes the command \bdface, it will stop and complain with the following message:

```
Runaway argument?
{this is text
! Paragraph ended before \bdface was complete.
<to be read again>
                \par
1.14 \bdface{this is text\par
                        and more text}

?
```

Naturally, the line number refers to a hypothetical input file, and of course LATEX had no problem processing the \bldface command.

Creating simple commands with LATEX is not a difficult task. However, when one wants to create commands that have two optional arguments or optional arguments surrounded by parentheses, one has to know the LATEX internals in order to create such complicated commands. To assist people in creating really complicated commands, Scott Pakin has created a Python script called newcommand.py. Suppose that you want to create the first version of the \seq command. Then, you have to invoke the program and type the following (the % character is the program prompt):

```
$ newcommand.py
% Prototype: MACRO seq
\newcommand{\seq}{%
    % Put your code here.
}
```

As we see, we just type the word MACRO and then the name of the new command. The program responds by generating a "skeleton" command definition. This definition has to be copied to a file. Then, we must fill in the code that actually implements our command. Let us now see how we can define a command with one parameter:

```
% Prototype: MACRO seq #1
\newcommand{\seq}[1]{%
    % Put your code here.
}
```

Suppose that we want to create a command with an optional argument surrounded by parentheses:

```
% Prototype: MACRO seq OPT(#1={x}) #2
\makeatletter
\def\seq{%
```

```
     \@ifnextchar({\seq@i}{\seq@i(x)}%)
}
\def\seq@i(#1)#2{%
   % Put your code here.
}
\makeatother
```

Here, we use the keyword OPT to denote an optional argument. The parentheses are used to denote the tokens that should surround the optional argument. The expression #1={x} is used to specify the default value of the first argument, which is optional. Note that the code above uses the "unknown" command \def. This command is used to create commands with TeX, and its functionality will not concern us here. Now, it is very easy to make the second argument an optional argument:

```
% Prototype: MACRO seq #1 OPT[#2={x}]
\makeatletter
\newcommand{\seq}[1]{%
   \@ifnextchar[{\seq@i#1}{\seq@i#1[x]}%]
}
\def\seq@i#1[#2]{%
   % Put your code here.
}
\makeatother
```

As we see, the optional argument must be surrounded by square brackets. If we want a command with two optional arguments, the following prototype will do the job:

```
% Prototype: MACRO seq #1 OPT[#2={n}] OPT[#3={1}]
\makeatletter
\newcommand{\seq}[1]{%
   \@ifnextchar[{\seq@i#1}{\seq@i#1[n]}%]
}
\def\seq@i#1[#2]{%
   \@ifnextchar[{\seq@ii#1[#2]}{\seq@ii#1[#2][1]}%]
}
\def\seq@ii#1[#2][#3]{%
   % Put your code here.
}
```

Another interesting example is the way one can create a command that has two required arguments surrounded by parentheses and separated by acomma:

```
% Prototype: MACRO seq {(}#1,#2{)}
\def\seq(#1,#2){%
   % Put your code here.
}
```

Note that here we have to put the surrounding parentheses in curly brackets.

6.12 New Environments

Up to now, we have presented many environments, so we hope that the reader will have an understanding of how an environment operates. A new environment is introduced with the command \newenvironment:

\newenvironment{Bar}[n]{opening commands}{closing commands}

Bar is the name of the environment. The opening commands and the closing commands are the code that will be executed before and after the processing of the body of the environment, respectively; n is the number of arguments. In other words, the definition above is equivalent to the following definitions:

\newcommand{\Bar}[n]{opening commands}
\newcommand{\endBar}{closing commands}

As is evident, a new environment can have at most nine arguments and one of them can be an optional argument. We will now give a few examples. Suppose that we want to define an environment that will produce a quote in boldface. Here is a solution to this problem:

```
\newenvironment{bfquote}{%
  \begin{quote}\bfseries}{%
  \end{quote}}
```

Here is an example:

Text before	`Text before`
A short quote.	`\begin{bfquote}`
	`A short quote.`
Text after.	`\end{bfquote}`
	`Text after.`

A more interesting example is the creation of the following environment:

Text before

Quotation 1 (*Hamlet*) To be, or not to be: that is the question

Text after.

```
Text before
\begin{cquote}{Hamlet}{2pt}
To be, or not to be:
that is the question
\end{cquote}
Text after.
```

The environment above has two arguments: the first is the name of the person to whom the quotation is attributed, and the second is the width of the horizontal rule. Moreover, it is evident that the quotes are numbered, so in order to create this new environment, we definitely need a new counter. However, it is important to stress that the arguments of an environment are not visible to the closing commands. This means that we must store the second argument to a (global) length variable in order to make it visible to the *closing commands*. Here is the complete definition of the new environment above:

```
\newcounter{qcounter}
\newlength{\qrulewidth}
\newenvironment{cquote}[2]{%
    \setlength{\qrulewidth}{#2}
    \begin{quote}\rule{\linewidth}{#2}
    \refstepcounter{qcounter}%
    \textbf{Quotation \theqcounter}
    (\textit{#1})}{%
    \newline\rule{\linewidth}{\qrulewidth}
    \end{quote}}
```

The command \newline is actually a "verbose" form of the \\, but it does not accept the optional arguments that the latter command does.

▶ **Exercise 6.14** Define a framed quote environment.   □

If we want to redefine an existing environment, we have to use the command \renewenvironment. This command can have exactly the same arguments that the command \newenvironment can have.

6.13 New Lists

Most of the environments presented in Chapter 4 are defined in terms of two generic list-creation environments: \list and \trivlist. The second environment is actually a special case of the first one, so we will briefly present its functionality at the end of this section. The first environment has two arguments:

```
\begin{list}{label}{commands}
...body...
\end{list}
```

label specifies item labeling and *commands* contains commands for changing, among other things, the horizontal and vertical spacing parameters. Each item of the environment starts with the command \item[*itemlabel*]. This command produces an item labeled by *itemlabel*. If the argument of \item is missing, the *label* is used as the item label. The label is formed by putting the output of the command \makelabel{*itemlabel*} in a box whose width is either its natural width or equal to \labelwidth, whichever is longer. Moreover, if the width of the label is less than \labelwidth, the label is put flush right separated by \labelsep from the item's text (see Figure 6.13 on page 210 for a complete list of all length variables associated with lists). If we define a command that customizes the appearance of the item, we can redefine the command

`\makelabel` in the *commands* part to change the default behavior of the environment. The following code produces an environment like the `description` environment, which, however, frames the item labels:

```
\newcommand{\flabel}[1]{\hfil\fbox{\textbf{#1}}}
\newenvironment{fdescription}{\begin{list}{}{%
    \renewcommand{\makelabel}{\flabel}}}{\end{list}}
```

Note that we put a `\hfil` before the actual label to flush right the label. The following is an example of this new environment:

God morgon	in Swedish.
Aloha kakahiaka	in Hawaiian.
Καλημέρα	in Greek.

```
\begin{fdescription}
\item[God morgon] in Swedish.
\item[Aloha kakahiaka] in
Hawaiian.
\item[Καλημέρα] in Greek.
\end{fdescription}
```

If we want to create numbered lists, we have to include the command

```
\usecounter{cnt}
```

in the *commands*; *cnt* is a new counter and is stepped every time a `\item` command is encountered. In order to show the use of this command, we will design an `enumeration` environment that will display the item labels in a `\fbox`:

```
\newcounter{oenumi}
\renewcommand{\theoenumi}{\roman{oenumi}}
\newcommand{\labeloenumi}{\fbox{\theoenumi}}
\newenvironment{oenum}{\begin{list}{\labeloenumi}{%
                \usecounter{oenumi}%
            \renewcommand{\makelabel}[1]{%
                {\hfil ##1}}}}{\end{list}}
```

Because the redefinition of `\makelabel` occurs inside the first argument of the `newenvironment` environment, we refer to the argument of the command with `##1`. This is necessary to distinguish between arguments in a definition occurring inside another definition. Let us see what we have done:

i one	
ii two	
iii three	

```
\begin{oenum}
\item one
\item two
\item three
\end{oenum}
```

If you have come to the conclusion that the standard `enumerate` environment is defined in a similar way, then you are absolutely right. However, the definition of `enumerate` is far more tricky compared to the definition above. To continue with the definition above, one problem is that the text that follows each item label is not justified. In order to fix this problem, we need to introduce a new length variable, which will be used to measure the width of the box containing the label. Moreover, we must make the \labelwidth wide enough to hold any (reasonably) wide label. Now, we compare the two length variables and create boxes whose width depends on the comparison above.

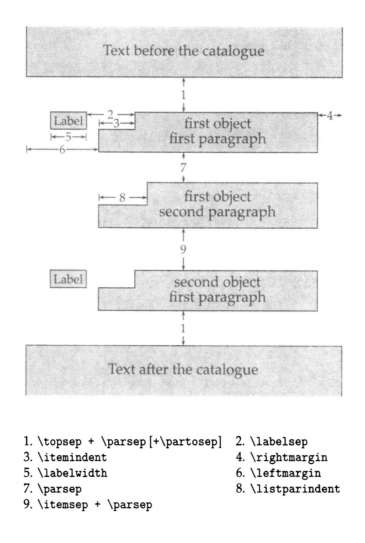

1. \topsep + \parsep [+\partosep]   2. \labelsep
3. \itemindent   4. \rightmargin
5. \labelwidth   6. \leftmargin
7. \parsep   8. \listparindent
9. \itemsep + \parsep

Figure 6.13: Length variables associated with list structures.

➤ **Exercise 6.15** After you read Section 7.2, implement the description above.   □

Environments such as quote are actually lists with a single unlabeled item. A problem that must be tackled is the case where the first character of the environment is a left square bracket. If such a character occurs, TEX will complain of a runaway argument. The solution to this problem is to put the command \relax immediately after the \item command. The following is the standard definition of the quote environment:

```
\newenvironment{quote}{\begin{list}{}{%
  \setlength{\rightmargin}{\leftmargin}}%
  \item\relax}
  {\endlist}
```

For the purposes of this book, the trivlist environment may be considered to be defined as follows:

```
\newenvironment{trivlist}{\begin{list}{}{%
  \setlength{\labelwidth}{0pt}%
  \setlength{\leftmargin}{0pt}%
  \setlength{\itemindent}{0pt}}}{\end{list}}
```

When using this environment, one must always specify a label, even the empty label (i.e., \item[]).

6.14   File Input

When preparing a long document (e.g., a book), it is a good practice to keep individual units in separate files. For example, if you prepare a book, you can keep each chapter in a separate file and have a *master* file that will be used to include all chapters. In order to facilitate this process, LATEX provides the command \include and its friends.

The command \include has one argument—a filename—and forces LATEX to process this file as if it was part of the current file. Suppose that we are preparing a long report that consists of four chapters. Then, the master file will look like this:

```
\documentclass[a4paper,12pt]{report}
.......... preamble commands ........
\begin{document}
\include{chap1}
\include{chap2}
\include{chap3}
\include{chap4}
\end{document}
```

Note that LATEX assumes that the files have the .tex filename extension. If we want to \include only some files, we can put the command \includeonly{$F_1, \ldots, F_n$} in the preamble of the master file to include only the files specified in the argument list of the command. For example, if we process the following file, the output will consist of Chapters 1 and 4 only:

```
\documentclass[a4paper,12pt]{report}
\includeonly{chap1,chap4}
\begin{document}
\include{chap1}
\include{chap2}
\include{chap3}
\include{chap4}
\end{document}
```

The command \input{*file*} is also used to include the *file* in the master file. Moreover, this command is useful when one wants to include a file that does not have the .tex filename extension. For example, the command

$$\input{chap5.ltx}$$

will include the file chap5.ltx.

In many instances, people create large documents that contain lots of figures that are stored in separate files. In such cases, it is useful to be able to store all files associated with a particular document unit in a directory. In order to do this we can use the command \graphicspath (provided by the graphics package) to specify the path to individual graphics files. The command has as arguments either full or relative paths, which must be surrounded by curly brackets:

$$\graphicspath{/path/to/files}{pics/files}$$

Now, we can use any file inclusion command to input both document and graphics files:

$$\include{/path/to/files/myfile.tex}$$

As is evident, for ordinary files we must give the full path.

If we want to see which files LATEX inputs while processing our document, we simply put the command \listfiles in the preamble of our document. On the other hand, if we want to stop LATEX from reading all of the auxiliary files, the files related to indices, bibliographies, and so on, we simply put the command \nofiles in our document's preamble.

6.15 LATEX à l'interactive

TEX is not only an excellent typesetting engine but also a real programming language. Without any doubt, it is a programming language that lacks certain features found in almost all programming languages (e.g., real number manipulation, repetitive constructs, arrays, etc.).[2] On the other hand, it provides facilities for data input and output. This is also true when it comes to terminal input and output, so we can create TEX/LATEX "programs" that can interact with their users. In the next chapter, we will see how to introduce conditional branches and looping using the ifthen package and so be able to create real LATEX programs.

The command \typeout prints its argument to the computer console. For example, consider the following LATEX file:

```
\documentclass{article}
\begin{document}
\typeout{*****************}
\typeout{Hello from LaTeX!}
\typeout{*****************}
\end{document}
```

Here is what we get if we feed this file to LATEX:

```
$ latex test
This is TeX, Version 3.14159 (Web2C 7.3.3.1)
(./test.tex
LaTeX2e <2000/06/01>
Babel <v3.7h> and hyphenation patterns for american,
english, greek, loaded.
(/usr/local/teTeX/share/texmf/tex/latex/base/article.cls
Document Class: article 2000/05/19 v1.4b Standard LaTeX
document class
(/usr/local/teTeX/share/texmf/tex/latex/base/size10.clo))
(./test.aux)
*****************
Hello from LaTeX!
*****************
(./test.aux) )
No pages of output.
Transcript written on test.log.
```

Skeptical readers may think that this ability is actually useless unless, of course, we can provide input to our "programs." The command \typein can be used to input data from our computer console. The command has two arguments: a required text string, which

---

2. TEX is a Turing complete programming langauge, so one can essentially implement any algorithm in TEX.

is used to prompt the user to enter data, and an optional command name. If we specify the optional argument, LATEX stores what the user types to this command; otherwise, it includes the text into the file verbatim. Let us see a trivial, although complete, example:

```
\documentclass{article}
\begin{document}
\typein[\name]{enter your name...}
\textbf{\name}
\end{document}
```

If we feed the file above to LATEX, we get:

```
$ latex example
This is TeX, Version 3.14159 (Web2C 7.3.3.1)
(./tt.tex
LaTeX2e <2000/06/01>
Babel <v3.7h> and hyphenation patterns for american,
english, greek, loaded.
(/usr/local/teTeX/share/texmf/tex/latex/base/article.cls
Document Class: article 2000/05/19 v1.4b Standard LaTeX
document class
(/usr/local/teTeX/share/texmf/tex/latex/base/size10.clo))
(./tt.aux)
enter your name...
\name=Apostolos Syropoulos
[1] (./example.aux) )
Output written on example.dvi (1 page, 284 bytes).
```

Now, if we view the resulting file, it will contain the name of the first author in boldface. Here is another example that is useful if we want to prepare many identical letters interactively:

```
\documentclass{letter}
\address{Spartan \TeX\ Users Group\\
         1, Central Plaza.\\
         SPARTA}
\signature{Dr. Euclid\\President of STUG}
\begin{document}
\typein[\Pname]{enter participant's name}
\typein[\Paddress]{enter participant's address}
\typein[\Ptown]{enter participant's town}
\begin{letter}{\Pname\\ \Paddress\\ \Ptown}
\opening{Dear \Pname}
This is to acknowledge the acceptance of your paper.
\closing{Yours truly}
\end{letter}
\end{document}
```

7

MISCELLANEOUS PACKAGES

In this chapter, we present some packages that do not comfortably fit anywhere else.

7.1 The **calc** Package

The calc package by Kresten Krab Thorup, Frank Jensen, and Chris Rowley re-implements the LATEX commands \setcounter, \addtocounter, \setlength, and \add-tolength in such a way that these commands accept an infix mathematical notation instead of a number. TEX provides commands such as \advance and \multiply in order to do arithmetic, but the calc package provides additional functionality. One can use standard notation for the four arithmetic operations with the symbols +, -, * and /, so if you write 44 / 10, you will get 4. Note that the result is 4 and not 4.4, as the division performed is integer division. Arithmetic with dimensions is also allowed, so you can say, for example, 7cm+2in, but the arithmetic must be well-defined (thus 7cm+2 makes no sense). The \real command allows the use of real numbers in calculations. Thus, if we want to multiply 99 by 2.1, we will use 99 * \real{2.1}.

Let us see an example. Suppose that we want to print the time and date on the documents we create in order to keep track of the newest versions. The counter \time holds the number of minutes since last midnight. Thus, we will have to define two counters, the hour counter and the minutes counter. The hour counter will be set to \time / 60 and the minutes counter to \time - \value{hours} * 60, suggesting the following code:

```
\newcounter{hours} \newcounter{minutes}
\newcommand{\printtime}{%
   \setcounter{hours}{\time / 60}%
   \setcounter{minutes}{\time - \value{hours} * 60}%
   \thehours:\theminutes\ \today}
```

Now, the command \printtime will give: "23:29 March 6, 2001." The package de-fines additional commands for specifying a length using some text. The commands

\widthof{*some text*}, \heightof{*some text*}, and \depthof{*some text*} can be used to store the width, height, and depth, respectively, of *some text* to a length variable. Now one can use commands such as

```
\setlength{\parskip}{\heightof{b} * \real{1.8}}
```

7.2 The **ifthen** Package

The ifthen package by David Carlisle, based on earlier work of Leslie Lamport, provides the command \ifthenelse with the following syntax:

```
\ifthenelse{condition}{''then'' commands}{''else'' commands}
```

The command computes the value of the condition. If the value is true, it executes the *''then'' commands*; otherwise, the value is false and it executes the *''else'' commands*. The condition is either checked on the spot (it could say 1>0 or it can use >, <, or = for the obvious comparisons) or it can be the value set for a Boolean variable. For this task, the package provides the commands \newboolean and \setboolean. For example, \newboolean{BoolVar} defines a new Boolean variable, and the commands \setboolean{BoolVar}{true} and \setboolean{BoolVar}{false} set the variable to true and false, respectively. The truth value of this variable is read by \ifthenelse with the help of the \boolean command. Thus, the commands

```
\newboolean{Boolvar}
\setboolean{BoolVar}{true}
\ifthenelse{\boolean{BoolVar}}{1}{2}
```

will produce the number 1.

Let us continue with the example of the previous section and improve it to give the time in the am-pm format.

```
\newcommand{\printtime}{%
  \setcounter{hours}{\time / 60}%
  \setcounter{minutes}{\time - \value{hours} * 60}%
  \ifthenelse{\value{hours}>12}%
    {\setcounter{hours}{\value{hours} - 12}%
     \thehours:\theminutes\,pm\ \today}%
    {\thehours:\theminutes\,am\ \today}
```

Now, the \printtime command will give "11:29 pm March 6, 2001." Two other commands that help us to check the condition are \isodd{m} (where m is an integer) and \lengthtest. The first checks if the number m is odd, and the second command is used for length comparisons. The code

```
\ifthenelse{\lengthtest{2cm > 2in}}{cm}{in}
```

will print the unit in since 2cm > 2in is false. Another command is \equal{w1}{w2}, which checks whether the character string w1 is the same as the character string w2. For example,

$$\ifthenelse{\equal{TeX}{TEX}}{yes}{no}$$

will print the word no since TeX and TEX are not the same. If the arguments of the \equal command are themselves commands, then these are expanded before they are compared. Thus, the commands

```
\newcommand{\one}{one}
\newcommand{\One}{one}
\ifthenelse{\equal{\one}{\One}}{Yes}{No}
```

will produce Yes since although \one and \One are not the same, after they are expanded they are both equal to one. We close this section by adding that the package provides the command \whiledo that gives the possibility of repeating commands until a condition fails. The syntax is \whiledo{condition}{Commands}. Here is an example that produces the phrase "I must not cheat" 100 times:

```
\newcounter{Pcounter}
\newcommand{\mustnot}{I must not cheat.\ }
\newcommand{\punishment}[1]{%
    \setcounter{Pcounter}{#1}%
    \whiledo{\value{Pcounter}>0}{%
    \mustnot%
    \addtocounter{Pcounter}{-1}%
    }}
\punishment{100}
```

Finally, one more command, the \ifundefined\command command, is available, which checks if the \command is already defined or not.

7.3 Syntax Checking

There are a few ways to check for errors in a LaTeX file without really running the file through the standard LaTeX procedure. One of them is provided by the syntonly package by Frank Mittelbach and Reiner Schöpf. The package implements the command \syntaxonly, which should be placed in the preamble of the document and has the effect of allowing LaTeX to check our code for errors but produces no output. Consequently, since no DVI file is produced, the parsing of the code is much faster.

Another way for syntax-only checking is provided by the LaCHECK program by Kresten Krab Thorup with modifications by Per Abrahamsen. The program checks for common errors in a file and is very helpful for beginners. It checks for mismatched groups (braces), environments and math mode delimiters, bad spacing such as \␣ after an abbreviation, or a missing \@ before the period when a sentence ends with a capital letter. It also checks for bad choices of dots, wrong or absent italic corrections, badly placed punctuation, and poor use of quotation marks. Let us see how LaCHECK performs when it is supplied with an erroneous LaTeX file. Suppose that we feed LaCHECK with a file that contains the following:

```
\documentclass[a4paper]{article
\begin{documnent}
This is a simple \LoTeX\ file.
\emd{document}
```

Then, if we run LaCHECK on this file, we will get the following:

```
$ lacheck test
"test.tex", line 5: <- unmatched "end of file test.tex"
"test.tex", line 2: -> unmatched "\begin{documnent}"
"test.tex", line 5: <- unmatched "end of file test.tex"
"test.tex", line 1: -> unmatched "{"
```

The last message informs us that there is an unmatched left curly bracket. Indeed, we have to place a right curly bracket after the keyword article. The second line informs us that there is an unmatched command. Indeed, if we replace the last line of the input file with

```
\end{documnent}
```

and rerun LaCHECK on the resulting file, we will get the following:

```
$ lacheck test
"test.tex", line 4: <- unmatched "\end{documnet}"
"test.tex", line 2: -> unmatched "\begin{documnent}"
"test.tex", line 5: <- unmatched "end of file test.tex"
"test.tex", line 1: -> unmatched "{"
```

As is obvious now, LaCHECK has detected that our file does not contain the commands that begin and end a "document." The reader may wonder why LaCHECK has failed to recognize that the environment document has not been defined. In order to be able to do this, we must have a system that can process both LaTeX and TeX commands and moreover has the capability to expand macros. Naturally, such a system exists, and it is TeX itself... and this is the reason why the system is not able to check whether the command \LoTeX has been defined or not.

7.4 Typesetting CD Covers

The package cd-cover by Christian Holm provides an easy interface to typeset compact disc covers, especially now that CD burners are widely available.

The package provides three environments rightpagebooklet, leftpagebooklet, and backpage. Moreover, it provides the commands \bookletsheet and \backsheet, which give easy access to the environments above. One difficulty for creating CD covers is that the pages should be output in such a way that, when the booklet is folded, the pages will have the correct order. The DVI file should be processed in landscape mode; that is by using the command

```
dvips -t landscape file.dvi
```

The code that follows provides an example. Here, we use the landscape option of the \documentclass command that swaps the values of the predefined LaTeX length variables \textwidth and \textheight. The output is shown in Figure 7.1 on page 221 scaled to fit in one page.

```
\documentclass[landscape,11pt]{article}
\usepackage{cd-cover}
\renewcommand{\labelenumi}{\oldstylenums{\theenumi}.}

\begin{document}
\bookletsheet{%

{\Large \sc Singer --- Title (A)}\\

\begin{center}
\begin{enumerate}
\setlength{\itemsep}{-4pt}
\item Song 1 (Creator) (B)        \item Song 2 (Creator)
\item Song 3 (Creator)            \item Song 4 (Creator)
\item Song 5 (Creator)            \item Song 6 (Creator)
\item Song 7 (Creator)            \item Song 8 (Creator)
\item Song 9 (Creator)
\end{enumerate}
\end{center}
}{
\begin{center}
{\huge\sc Singer (C)}\\[16pt]
{\huge\sc Title (C)}
\end{center}}

\backsheet{\sc Singer --- Title (D)}{
  {\Large \sc Singer --- Title (E)}

\begin{center}
\begin{enumerate}
```

```
\setlength{\itemsep}{-4pt}
\item Song 1 (Creator) (F)        \item Song 2 (Creator)
\item Song 3 (Creator)            \item Song 4 (Creator)
\item Song 5 (Creator)            \item Song 6 (Creator)
\item Song 7 (Creator)            \item Song 8 (Creator)
\item Song 9 (Creator)
\end{enumerate}
\end{center}
}
\end{document}
```

7.5 Drop Capitals

L ETTRINES are "dropped capitals"such as the letter L that started this paragraph, and we see them very often in magazines and books. It is very easy to produce them using the package lettrine by Daniel Flipo. The package provides the command \lettrine, which is used with the following syntax

$$\text{\lettrine[}options\text{]\{}letter\text{\}\{}paragraph\text{\}}$$

Here, letter is a letter or a word that will be "dropped" and paragraph the text that starts a paragraph. Several parameters are provided for customizing the layout of the dropped capital. These are entered as optional arguments in a comma-separated sequence. The parameters are:

lines=integer Sets how many lines will be occupied by the dropped capital. The default is 2.

lhang=decimal The decimal is any number between zero and one and controls how much the dropped capital will hang out in the margin. The default is zero, and if it is set to one, the dropped capital will be entirely in the margin.

loversize=decimal The decimal is used to enlarge the capital's height and can vary from 1 to 1. The default is zero and if, for example, it is set to 0.1, the height of the dropped capital will be enlarged by 10%, rising above the top of the paragraph.

lraise=decimal This controls the vertical position of the dropped capital. It moves it up or down without changing its height. It is useful for letters such as J or Q that have a positive depth. The default value is zero.

findent=length This parameter controls the horizontal gap between the dropped capital and the indented block of text. The default value is 0 pt.

nindent=length This parameter is used to horizontally shift all of the indented lines after the first one by length.

slope=length This is used for letters such as A or V whose geometry requires that the amount of indentation of each of the lines after the second one changes progressively. The default value is zero, and the effect works after the third line.

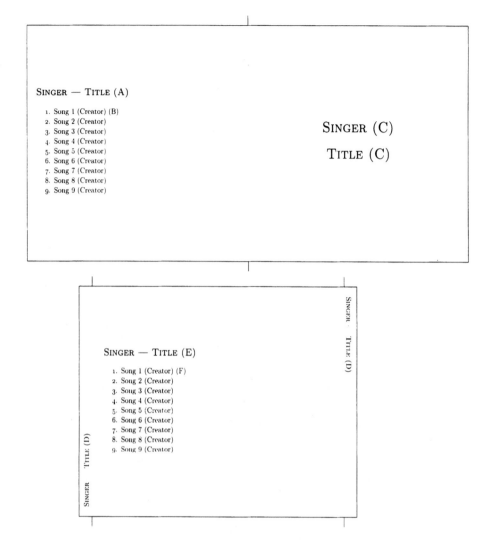

Figure 7.1: cd-cover output

ante=_text_ This is used for entering text to be set before the dropped capital. It is mainly used with guillemots, quotation marks, and so forth, that start a paragraph.

Figure 7.2 shows a usage example of the package. The default values of the above options can easily be changed by putting the following commands in the preamble (here we give them with their default values):

```
\setcounter{DefaultLines}{2}
\renewcommand{\DefaultLoversize}{0}
\renewcommand{\DefaultLraise}{0}
```

«ΠΑΤΙΑ» was a mathematician and a philosopher, daughter of the mathematician Θέων. She studied in the Academy of Athens. Then she returned to Alexandria, Egypt, where she was teaching philosophy and mathematics. She was extremely beautiful and the most important scientist in her era. She was brutally murdered in 415 AD by a Christian mob.

```
\lettrine[lines=4,loversize=%
0.1,slope=-6pt,lhang=.4,ante=%
\textgreek{((},findent=4pt]%
{\textgreek{U}}{\textgreek%
{patia))}} was a mathematician
and a philosopher, daughter
of the...
...by a Christian mob.
```

Figure 7.2: Using the lettrine package.

```
\renewcommand{\DefaultLhang}{0}
\setlength{DefaultFindent}{0pt}
```

The package provides additional functionality. It is possible to use a PostScript file in the position of a dropped capital. Note that this paragraph is not a regular dangerous paragraph. We are using the warning symbol that designates dangerous paragraphs and taking advantage of the extra functionality of the package lettrine to give slope to the text. The PostScript file should be named with *one* letter (in our case, the dangerous symbol is called d.ps). The package provides the commands \LettrineFont and \LettrineFontEPS, which control whether we use text or a PostScript file (that should be loaded) as a dropped capital instead. To load a PostScript file in the position of the dropped capital, we must redefine \LettrineFont to be \LettrineFontEPS. The code that follows shows how this paragraph started.

```
\renewcommand{\LettrineFont}{\LettrineFontEPS}
\lettrine[lines=4,slope=.85em,findent=-1.3em,nindent=1em]{d}{}The
package provides additional functionality...
```

7.6 Preparing a Curriculum Vitae

A curriculum vitae is a very personal issue. Consequently, some may argue that it does not make sense to give a recipe. However, it is also true that style customization in LaTeX can be done only by a minority of its users. The great majority would benefit from such a recipe in the sense that they will at least have a starting point.

The package currvita by Axel Reichert provides a nice and easy interface for creating curriculum vitae. There are two package options: LabelsAligned and TextAligned. The default is TextAligned and uses space generously. If a more compact form is preferred, then we have to use the LabelsAligned option. The package provides two environments: cv and cvlist. The whole curriculum vitae must be in the body of the cv environment:

```
\begin{cv}{CV Heading} CV text \end{cv}
```

The `cvlist` is used to create the curriculum vitae's items; for example,

```
\begin{cvlist}{Personal Information}
  \item Steve Worker\\
    ...

  Sparta
  \item ...
\end{cvlist}
```

On page 224, we show the curriculum vitae of a hypothetical person called Steve Worker.

➤ **Exercise 7.1** Write the code for your curriculum vitae based on the one given in figure 7.3. ☐

Other options of the package are

`ManyBibs` This is for documents with multiple bibliographies using multibbl (see Section 8.2.4). It is particularly useful for multilingual documents.

`NoDate` This suppresses the date printed at the end of the document.

`openbib` This is for cases where you like the "open" format of bibliographies.

The package also provides the command `\cvplace{location}` for printing the `location` where the CV was written together with the date at the end of it. Moreover, the command `\date{date}` can be used to override the default date printed at the end of the CV with the `\today` command.

One can customize several variables of the package. The command `\cvheadingfont` controls the font to be used for the heading of the CV. The default is to use `\Large` and `\bfseries` and if we want to change this, we must *redefine* it. For instance, the redefinition

```
\renewcommand{\cvheadingfont}{\Large\scshape}
```

will change to `\Large` small caps. Similarly, we may change the default values of the following commands:

`\cvlistheadingfont` This controls the font used for the headings of the lists.

`\cvlabelfont` This controls the font of the lists' items.

`\cvlabelwidth` This is a length variable and controls the label's width in the `cvlist` environments.

`\cvlabelskip` This is another length variable that controls the vertical space between the first `cvlist` item and the `cvlist` label.

`\cvlabelsep` A length variable that controls the horizontal space between the items' labels and the items.

`\cvbibname` This controls the bibliography label (whose default is "Publications") and if we want to change the label, we simply redefine it. For example,

```
\renewcommand{\cvbibname}{List of Papers}.
```

Curriculum vitae of Steve Worker

Personal Information

Steve Worker
University of Sparta
Department of Fine Arts
Sparta

Email: `sworker@fn.sparta.edu`

Born in Thebes

Studies

June, 300   B.Sc. in Mathematics from University of Athens.

September, 305   Ph.D. in Mathematics from the University of Pella. Title: *Symmetrizations and Convolutions of Convex Bodies*. Thesis advisor: Euclid.

Research Interests

Geometry of Convex Bodies, Functional Analysis, Geometry of numbers.

Publications List

1. *On a Conjecture by Aristarchus* Annals of the Athenean Mathematical Society 60:187–206, 306.

2. *Is Hippasus's Theorem Correct?* (to appear).

January 4, 310

Figure 7.3: An example of the currvita package.

7.7 Multicolumn Typesetting

The multicol package by Frank Mittelbach provides the multicols environment, which can be used to typeset text in many columns (up to ten). The environment has a required argument, which is the number of columns. If we want to include some long single-column text, we can do this by specifying the text in square brackets after the number of columns. Also, we can specify that the space the long text should occupy:

```
\begin{multicols}{3}
[\section{People}][5cm]
```

The space between columns is controlled by the length variable \columnsep. Moreover, we can add a vertical line that will separate columns by setting the length variable \columnseprule. In the body of the multicols environment, all columns are balanced. However, the environment multicols* disables this feature. The counters unbalance and collectmore can be used to fine-tune the balancing algorithm. The first one should be used to make all but the rightmost column longer by the value of the counter. If set, the value of the second counter is used by the typesetting engine to find proper page breaks.

7.8 Hyphenatable Letter Spacing

Emphasizing text by spacing it out is not considered a good typographic practice by many professionals. However, there are cases (other than emphasizing) where one would want to use this technique. For example, a little letter spacing will increase the readability of small caps or huge titles (e.g., in posters). For small caps, this may not be necessary, as a well-thought out font (such as the default fonts by Knuth) contains glyphs that have some extra horizontal space in both sides of the character. For most fonts, however, this is not the case. Probably, the best way to space out text is provided by the soul package by Melchior Franz. The package can also be used to underline text. The package name derives from the names of the commands \so (space out) and \ul(underline) that it provides. One of the big advantages of the package is that it does not inhibit hyphenation. Table 7.1 shows a summary of the available commands and their result.

The \so command is the basic command for letter spacing. It uses a certain amount of *inter-letter space* between every two characters, *inner space* between the words, and *outer space* before and after the spaced-out text. You can use the command \sodef to modify the default spacing as follows:

$$\sodef\textit{\spaceit}\{\textit{font}\}\{\textit{inter-letter space}\}$$
$$\{\textit{inner space}\}\{\textit{outer space}\}$$

Here, *\spaceit* is a new letterspacing command. All of the arguments above are mandatory. If the new command refers to the default fonts, then the first argument can

Table 7.1: soul package command summary.

\so{spaced{\hyphen}out text}	s p a c e d - o u t t e x t
\caps{CAPITALS, Small Capitals}	CAPITALS, SMALL CAPITALS
\ul{underlined text}	underlined text
\st{strike out text}	~~strike out text~~

\sodef\cs{1em}{2em}{3em}	define the new spacing command \cs
\resetso	resets the \so dimensions
\capsreset	clears the \caps dataset
\capsdef{////}{1em}{2em}{3em}	defines the default \caps data entry
\capssave\cs	save \caps dataset under the name \cs
\setul{1ex}{2ex}	sets the \ul dimensions
\resetul	resets the \ul dimensions
\setuldepth{y}	sets the underline depth 1 point beneath the depth of y

be empty (i.e., {}). The lengths are actually glue (see page 21) and should be given in em units. Here is an example:

```
\sodef\spaceit{}{.2em}{1em plus1em}{2em plus.2em minus.3em}
```

By using the \sodef command, it is possible to redefine the meaning of the \so command instead of defining a new letter spacing command. To reset the \so command to the default values, use \resetso.

A useful variation of the \so command is the \caps command, which in addition to spacing out the text changes lowercase letters to small caps. This command can also be redefined with the \capsdef command. The syntax of this command is as follows:

```
\capsdef\spaceit{font}{command}
        {inter-letter space}
        {inner space}{outer space}
```

Here, the font declaration is of the form

encoding / family / series / shape / size

For example, the font declaration T1/ppl/m/n/6-14 should be used when we want to have the Palatino font family with the T1 encoding, the medium series, normal shape, and all sizes from 6 pt to 14 pt. The *command* can be any font-switching command such as \scshape. The *size* entry may contain a size range as in our example; if we omit the lower bound of the range, then it is assumed to be equal to zero; otherwise, if we omit the upper bound, it is assumed to be a very large number. The \capsdef command redefines \caps, and we can reset the default values with \capsreset. If we want to save our custom values, we should use the \capssave command to create a new command that when invoked will set the values associated with it. For example,

```
\capsdef{/cmr///}{\scshape}{10pt}{20pt}{30pt}
\capssave\widecaps
\resetcaps
\capsdef{/cmr///}{\scshape}{.1pt}{.2pt}{.3pt}
\capssave\narrowcaps
```

defines the two commands \widecaps and \narrowcaps with different spacing-out values.

There are cases when inside a spaced-out phrase there are some parts that should not be spaced out, such as the case of accented letters that would get decomposed. To prevent spacing out an accented letter, we enclose it in braces like this "\so{th{\'e}{\^a}tre}" in order to get "t h é â t r e." In addition to this, anything set in a box or in two pairs of braces ({{}}) will not be spaced out.

A few rules of thumb follow. Punctuation marks should be spaced out except the period (sometimes commas, too). Quotes and numbers are not spaced out. Consequently, all of these should be put outside the \so command or the *inter-letter space* can be temporally canceled with the command \<. For instance, we can write "\so{2\<3 June {{2002}}}" to get "23 J u n e 2002." The hyphen, the en dash, the em dash, and the slash should be entered with the commands \hyphen, \endash, \em-dash and \slash, respectively, instead of writing -, --, ---, and /. An unbreakable space should be followed by a space like this \so{Mrs.\~␣F}. The \\ command works as usual, but without additional arguments.

If the rules of thumb above are not observed, LATEX may complain that "Reconstruction failed."

We will now discuss how it is possible to typeset narrow columns using this package. In many newspapers, magazines, and so forth, when we need to typeset in a narrow column, we usually space out a few words to avoid overfull boxes and bad breaks. Typographically, this is not an acceptable practice, as it severely disturbs the page color. This fact has made the author of the package exclude commands that would provide this feature. The package documentation provides all of the necessary information for those users who are in absolute need of it. Figure 7.4 shows an example from the package's documentation. For multilingual documents, the only problem seems to be that accented letters should be put in braces. The reason is that multilingual LATEX documents use the inputenc package (see Chapter 10), which will decompose accented letters to a sequence of accents followed by the letter to be accented. Thus, in order to space out the word ἔρως, we have to type \so{{>έ}ρως}. However, there are exceptions to this rule. One does not need these extra braces for simpler accents. For example, \so{έρως} will work out fine. Note that if we use Λ with a real Unicode font, then this problem simply does not exist!

We conclude this section with a few words about underlining. The \ul command, which is used to produce underlined text, can be modified with the \setul command

Some maga-   Some magazines
zines and news- and newspapers
papers pre-   prefer this kind
fer this kind   of spacing because
of spacing be- it reduces hyphen-
cause it reduces ation and over-
hyphenation   full problems to
and overfull   a minimum.
problems to a
minimum.

Figure 7.4: Typesetting narrow columns.

\setul{*underline depth*}{*underline thickness*}

where both arguments are lengths or length variables. If we do not want to change one of the parameters, we simply pass an empty argument. Both lengths should be expressed in ex units; they can be restored to their default values with the \reset-ul command. The \setuldepth command sets the underline depth 1 pt beneath its argument's deepest depth. For example, the command \setuldepth{ag} will set the underline depth 1 pt beneath the depth of the letter g. Finally, if we use underlining, it is better to use the package with the overlap option. This option extends the underline segment for each of the underlined letters by 0.5 pt, which helps get rid of little gaps in the underlining that may appear.

8

BIBLIOGRAPHY AND INDEX

The bibliography and, in particular, the index are two parts of a document that usually get neglected during the initial stages of the document creation process. This leads to problems later on, so it is wise to plan ahead, as these two parts are very important for a document to be considered complete and with easily accessible material.

8.1 Preparing the Bibliography

A bibliography is a list of writings used or considered by an author in preparing a particular work. Therefore, a typesetting system must provide the ability for users to refer to bibliographic items that will be typeset according to prespecified rules, usually at the end of a document. For referencing a bibliographic item, LaTeX provides the command \cite. The syntax of this command is the same as that of the command \ref. The argument of the command can be any letters, numbers, or symbols of the *English* language, so for referring to a book that is labeled as knuth:1, we use the command \cite{knuth:1}. Moreover, this command has an optional argument that is a piece of text that will be printed together with the text generated by the \cite command. But how do we label a book (or article) reference? There are two ways of completing this task. The simpler one is the following: LaTeX provides the environment thebibliography, which is a list environment. This list uses the \bibitem command for each bibliographic entry. Each \bibitem command has an argument that is a label used to denote a bibliographic entry. The same label is used by the \cite commands to refer to a particular item in the bibliography. This command also takes an optional argument that is used to set the way the reference appears in the text. Finally, the environment thebibliography has an argument that specifies the number of characters allowed in the referencing names. Here is a complete example:

```
\documentclass{article}
\begin{document}
Donald Knuth has written several important books. In \cite{knuth:2} he
```

```
covers topics ranging from random number generators to floating point
operations and other optimized arithmetic algorithms.
In \cite[Chapter~1]{knuth:1} he describes the game of the name.

\begin{thebibliography}{999}
\bibitem{knuth:1} D. Knuth, \textit{The \TeX book}.
\bibitem[Kn2]{knuth:2} D. Knuth, \textit{The Art of Computer
Programming}, vol. 2, Seminumerical Algorithms.
\end{thebibliography}
\end{document}
```

> Donald Knuth has written several important books. In [Kn2]
> he covers topics ranging from random number generators to float-
> ing point operations and other optimized arithmetic algorithms. In
> [1, Chapter 1] he describes the game of the name.
>
> ## References
>
> [1] D. Knuth, *The TEXbook.*
>
> [Kn2] D. Knuth, *The Art of Computer Programming*, vol. 2, Seminu-
> merical Algorithms.

Figure 8.1: A first example with bibliography.

The formatted output is seen in Figure 8.1. We have asked from the environment
thebibliography to allow a maximum of three characters for referencing (this is what
999 stands for; of course, it could be 637, having the same effect).

As we see, in this example, the references in the text appear with a number or a
sequence of letters in brackets. It is customary though in areas such as literature or
philosophy that references appear as superscripts:

>   . . . as Aristotle has explained in his work.[21]

If we want such a referencing mechanism, we have to redefine the \@cite command.
Here is the standard definition:

```
\newcommand\@cite[2]{%
    [{#1\ifthenelse{\boolean{@tempswa}}{, #2}{}}]}
```

As is evident, the whole body of the definition is enclosed in square brackets. Thus, the
change we want is achieved by writing the following code in the preamble of a LATEX
file:

```
\renewcommand\@cite[2]{\textsuperscript{#1
\ifthenelse{\boolean{@tempswa}}{, #2}{}}}
```

▶ **Exercise 8.1** If we put the code above in the preamble of a LATEX document, LATEX will complain as follows:

```
! LaTeX Error: Missing \begin{document}.
```

Can you suggest a remedy to this problem? □

What if we have several references in the same place? It is typographically correct for these references to appear in one group of brackets with labels separated by commas. For example, [2, 5, 6] is to be preferred over [2], [5], [6] (it also conserves more space). Moreover, if we have consecutive numbers, it is better to use a range. Thus, [2–5] should be preferred over [2, 3, 4, 5]. This mechanism is provided by the package cite by Donald Arseneau. The corresponding package for superscript referencing is the overcite package by the same author. With these packages, a reference to the bibliographic entries 2, 3, 4, 5, 7, 9, 6 (\cite{2,3,4,5,7,9,6}) will become [2–7, 9] and [2–7, 9], respectively. The cite package provides a modified version of the command \cite, which is \cinten. This command removes the square brackets from the citation, allowing further formatting. Note that the package sorted the numbers before typesetting them (the number 6 was given after the number 9). However, older versions did not provide sorting. If you are working on an installation with an older version, you may additionally use the citesort package (by Ian Green) to sort the references before they are typeset.

8.2 Using BibTeX

The method for typesetting the bibliography described in the previous section is useful when we have a few bibliographic references. When we have a lot of them, we would prefer the automatic bibliographic generation available through the use of BibTeX, as this provides a flexible mechanism for dealing with bibliographic entries. BibTeX is a program designed by Oren Patashnik.

The preparation of a bibliography using BibTeX involves two steps: the preparation of a bibliographic database and the choice of a bibliographic style (i.e., we have to choose a program written in the internal language of BibTeX that will format the bibliography according to certain stylistic rules). We now present these steps.

Prepare and store the bibliographic records in a .bib file. Assume that our work is in the file file.tex. We create the file bibfile.bib (the choice of name is irrelevant), which consists of records having the general form

```
@DOC{key,
      author   = "Martin, William Ted",
      publisher = "Springer-Verlag",
```

```
title      = "From \LaTeX\ to $\Lambda$ An {I}ntroduction
             {T}o {D}igital {Typography}",
year       = 2001
}
```

where DOC is the kind of bibliographic reference and can be book, article, booklet, and so on, (see next section). As we see from the previous example, each bibliographic record starts with a left curly bracket and ends with a right curly bracket. However, one may opt to start each bibliographic record with a left parenthesis and end it with a right parenthesis. Note that the case of the field names or the bibliographic records does not matter. In multilingual environments, some additional possibilities are usually provided. The general form of the *fields* is

$$\textit{field-name} = \texttt{"value"} \quad \text{or}$$
$$\textit{field-name} = \texttt{\{value\}}$$

The only exception is the year field—we do not need to enclose the year in quotation marks or curly brackets unless it is a complete date or a year with some attribution (e.g., B.C. and so on). Also, note that fields must be separated by commas.

The key is the label that we use to refer to this bibliographic entry in file.tex through the \cite command. For the author field, we must say that we first write the family name, a comma, and then the first name. Alternatively, we can write

```
author = "William Ted Martin"
```

without a comma, and BibTeX assumes that the last word is the family name. There are cases where the author is a company. For example, the author of the official books for the PostScript language is Adobe Systems, Inc. In such cases, we must inform BibTeX that it should not expect a first name and a family name but, instead, it should treat the author field as one entity. We do this by writing the author like this:

```
author = "{Adobe Systems Inc.}"
```

If the name uses the word Junior, we write

```
Martin, Jr., William
```

Finally, when we have many authors, we can use the keywords and or and others. For example,

```
de Bakker, John and others
```

will give de Bakker, John et al.

For the publisher field, a publisher may appear many times. That is why it may be useful to create a string shorthand. If we write

```
@string{sv = "Springer-Verlag"}
```

then whenever we want to write Springer-Verlag as the publisher, we can just write

```
publisher = sv
```

We note here that when we want to add something to such a string shorthand, we can use the special character #. So, if we want to write Springer-Verlag GmbH, we can use

```
publisher = sv # GmbH
```

The command @preamble has a syntax that is similar to the syntax of the @string command, except that there is no name or equal sign, just the string. This command is useful to add verbatim text to the generated bibliography file. For example, if we want to add an entry to the table of contents we can use the following command:

```
@preamble("\addcontentsline{toc}{chapter}{Bibliography}")
```

For the title field, it is a common problem that the words after the first one are set in lowercase; this holds even if the first letter is capital. To override this, we put either the capital letter or the whole word in curly braces:

```
title = "On the {Meaning} of {N}umbers"
```

Note that the following characters cannot be part of the value of the field:

```
" # % ' ( ) = { }
```

If such a character is needed, we can *escape* it by putting a backslash in front of it and by placing the entire accented character in braces:

```
author = G{\"o}del
```

Note that it is good practice to avoid further nesting of accented characters or else we may not get the expected results for certain bibliographic styles. The records in a bibliographic database can have many more fields, which we will present in the next section. We will close this section by showing the working cycle of BIBTEX.

Suppose that we are preparing a document that we store in a file called text.tex. Our bibliographic database is stored in a file called biblio.bib. Note that files containing bibliographic databases that can be processed by BIBTEX by default have the .bib filename extension. In order to be able to successfully generate a bibliography, we must specify the bibliographic style that will be used. Therefore, we place the command

```
\bibliographystyle{style}
```

somewhere in file text.tex. Note that *style* is the name of a bibliographic style. The next thing that we must do is to put the command

```
\bibliography{biblio1, biblio2, ..., biblioN}
```

somewhere in the file text.tex (usually at the end of the file). The arguments are the names of the bibliographic databases that will be used to create the bibliography. Now, we are ready to process our LATEX file:

```
$ latex text
This is TeX, Version 3.14159 (Web2C 7.3.3.1)
(./text.tex
LaTeX2e <2000/06/01>
(/usr/local/teTeX/share/texmf/tex/latex/base/article.cls
Document Class: article 2000/05/19 v1.4b Standard LaTeX
document class
(/usr/local/teTeX/share/texmf/tex/latex/base/size10.clo))
No file text.aux.
LaTeX Warning: Citation 'man' on page 1 undefined on
input line 4.
LaTeX Warning: Citation 'yager' on page 1 undefined on
input line 4.
LaTeX Warning: Citation 'knuth' on page 1 undefined on
input line 4.
No file text.bbl.
[1] (./text.aux)
LaTeX Warning: There were undefined references.)
Output written on text.dvi (1 page, 292 bytes).
Transcript written on text.log.
```

As is evident LaTeX does not recognize the various bibliography-related labels, so we have to run BibTeX:

```
$ bibtex text
Transcript written on text.log.
This is BibTeX, Version 0.99c (Web2C 7.3.3.1)
The top-level auxiliary file: text.aux
The style file: alpha.bst
Database file #1: biblio.bib
```

Note that we always supply to BibTeX the name of the auxiliary file produced by LaTeX so that BibTeX will resolve the undefined references that are stored in the auxiliary file. Let us now rerun LaTeX:

```
$ latex text
This is TeX, Version 3.14159 (Web2C 7.3.3.1)
(./text.tex
LaTeX2e <2000/06/01>
(/usr/local/teTeX/share/texmf/tex/latex/base/article.cls
Document Class: article 2000/05/19 v1.4b
Standard LaTeX document class
(/usr/local/teTeX/share/texmf/tex/latex/base/size10.clo))
(./text.aux)
LaTeX Warning: Citation 'man' on page 1 undefined on
```

```
input line 4.
LaTeX Warning: Citation 'yager' on page 1 undefined on
input line 4.
LaTeX Warning: Citation 'knuth' on page 1 undefined on
input line 4.
(./text.bbl) [1] (./text.aux)
LaTeX Warning: There were undefined references.
LaTeX Warning: Label(s) may have changed. Rerun to get
cross-references right.)
Output written on text.dvi (1 page, 960 bytes).
Transcript written on text.log.
```

Now, we have to rerun LATEX one more time to have everything in order:

```
$ latex text
This is TeX, Version 3.14159 (Web2C 7.3.3.1)
(./text.tex
LaTeX2e <2000/06/01>
(/usr/local/teTeX/share/texmf/tex/latex/base/article.cls
Document Class: article 2000/05/19 v1.4b
Standard LaTeX document class
(/usr/local/teTeX/share/texmf/tex/latex/base/size10.clo))
(./text.aux) (./text.bbl) [1] (./text.aux) )
Output written on text.dvi (1 page, 932 bytes).
Transcript written on text.log.
```

That's all! We now briefly present the basic bibliography styles:

plain This style file uses numbers in brackets for tags, and the entries are sorted alphabetically.

alpha This style uses alphanumeric tags using the first letters of the author name(s) and the last two digits of the publication year of the record (e.g., [PS00]). Records are sorted alphabetically.

unstr This style uses numeric tags and the entries appear in the bibliography in the order that they are referenced.

abbrv This style is the same as with the plain style but uses a compact form wherever possible (e.g., for month names).

amsplain This style is similar to plain except that document titles are typeset in italics.

amsalpha This style is also similar to alpha except that document titles are typeset in italics.

acm This style is similar to plain except that author names are typeset in small caps.

8.2.1 The BIBTEX Fields

We return now to the issue of what a BIBTEX entry can be and what fields it can contain. The entries of BIBTEX can be: article, book, booklet, inbook (for referring to a part inside a book), incollection (a part of a book with its own title), inproceedings, manual, masterthesis, phdthesis, proceedings, techreport, unpublished, and misc, if nothing else fits.

The above are the common entries, and it is not a complete list. For example, these entries will not work for a language other than English. One should have special entries for one's language that take care of the language selection. Such entries are either provided by a custom bibliographic style for the specific language or one must define them. As an example, Apostolos Syropoulos has designed the bibliographic style hellas that defines entries such as gr-book, gr-article, and so on. If one wants either to define a new bibliography style or to improve or change the functionality of an existing style file, one is advised to study the document contained in file btxhak.tex, which is included in every TEX installation. In addition, one can study the code of an existing style file. All that we can say here is that BIBTEX style files are written in a language that manipulates a stack and uses the so-called *postfix* notation. Languages similar to this language are the PostScript language, the Forth programming language, and the RPL language used in the Hewlett-Packard calculators.

Each bibliographic record can have several fields. The following are the most common: address, annote, author, booktitle, chapter, crossref, edition, editor, howpublished, institution (for technical reports), journal, key, month, note, number, organization (for a sponsor), pages, publisher, school (for a thesis), series, title, type, volume, and year. Most of these entries are self evident but the crossref entry must be explained. Suppose that we have the following two bibliographic entries

```
@inproceedings(objs,
               crossref="pldi",
               author="Julian Dolby",
               title="Automatic Incline",
               pages="7--17")
@proceedings(pldi,
               title="Conference on Programming",
               year=2778,
               organization="ACM SIGPLAN")
```

Then, the following things happen: if the first entry is used, it inherits all of the fields of the second entry, and the second entry will automatically appear in the bibliography, even if it is not called explicitly. In each bibliographic record, we can have three kinds of fields:

Required fields that is, fields that must be present in the record. However, if we either forget or do not have enough data to specify the field, BIBTEX will warn about a missing field.

Optional fields that will be used by BIBTEX only if they are present.
Other fields that are usually ignored by BIBTEX.

Consequently, if we use a particular field in the bibliographic record and the information stored in this field does not appear in the formatted bibliography, this means that the field has been ignored by BIBTEX.

8.2.2 Typesetting a Bibliographic Database

If we want to typeset the whole bibliographic database contained in some file, then one way is to cite each record. Since this is cumbersome, an easier way is to use the \nocite{*} command. The following file will typeset all of the bibliographic records contained in the file.bib file:

```
\documentclass{article}
\begin{document}
\bibliographicstyle{alpha}
\nocite{*}
\bibliography{file}
\end{document}
```

Here, we used the alpha style, but one can use any other style instead.

This simple file can be further customized by loading special packages that will affect the appearance of the bibliography. Here is an example:

```
\documentclass[twocolumn]{article}
\usepackage{bibmods,showtags}
\begin{document}
\bibliographicstyle{alpha}
\nocite{*}
\bibliography{file}
\end{document}
```

Figure 8.2 shows an example output.

8.2.3 Multiple Bibliographies in One Document

There are books (e.g., collections of papers) that need a different bibliography for each chapter or section. The solution for this problem is provided by the package chapterbib by Niel Kempson and Donald Arseneau. The package makes it possible to have one bibliography for each *included* file in a main document (through the \include command) despite the obvious connotation of its name. BIBTEX should be run on each included file separately instead of on the main document file ("root" file). Each of these included files must have its own \bibliographystyle and \bibliography commands. Note that if you are using the babel package, then chapterbib must be loaded before

References

pldi

[ACM78] ACM SIGPLAN. *Conference on Programming*, 2778.

objs

[Dol78] Julian Dolby. Automatic incline. [ACM78], pages 7–17.

Hobby:MetaPost

[Hob92] John Hobby. *A user's manual for MetaPost*. AT&T Bell Laboratories, Murray Hill, NJ, 1992.

Lamport

[Lam94] Leslie Lamport. *LATEX: A Document Preparation System*. Addison Wesley Publ. Co., 2 edition, 1994.

latex

[Mar01] William Ted Martin. *From LATEX to Λ An Introduction To Digital Typography*. Springer Verlag, 2001.

Figure 8.2: Bibliography typesetting using bibmods and showtags.

babel. Naturally, one can use the combine document class to solve this problem in a more general way (see Section 2.9 on page 34).

8.2.4 Bibliography in a Multilingual Environment

A new version of BibTEX, called BibTEX8, capable of handling bibliographic databases written in an extended ASCII, has been developed by Niel Kempson and Alejandro Aguilar-Sierra. The program has a number of switches, but the most important is the -c switch (or its equivalent: --csfile) by which we can specify the so-called csfile to use. The csfile should be used to define how BibTEX8 should treat an extended ASCII (e.g., whether an accented letter should have the same letter ordering as the unaccented one). Each csfile has a number of sections, each having the form of a LATEX command:

```
\section-name{
    section-definitions
}
```

In a csfile, one can have, at most, four sections:

\lowupcase This section is used to define the lower/uppercase relationship of pairs of specified characters. The syntax of this section is

```
\lowupcase{
    LC-1 UC-1 % comment
    LC-2 UC-2 % to use % inside a definition
    ......    % use ^^25 instead
    LC-N UC-N }
```

One is not allowed to redefine the lowercase and uppercase equivalents of a normal ASCII character.

\lowercase This section is used to define the lowercase equivalent of specific characters.

\uppercase This section is used to define the uppercase equivalent of specific characters.

\order This section is used to define the sorting order of the characters. The syntax of the \order section is (CH-N denotes a single character):

```
\order{
  CH-1
  CH-2 CH-3
  CH-4 _ CH-5
  CH-6 - CH-7
  . . . . . . . . . . .
  CH-N        }
```

All characters on the same line have the same sorting "weight." In order to define that a range of characters has the same sorting "weight," we use the construct CH-4 _CH-5 (e.g., A _ Z denotes that characters A through Z have the same sorting "weight"). The construct CH-6 - CH-7 is used to denote that all characters in the range CH-6 to CH-7 should have ascending sorting "weights" starting with CH-6 and ending with CH-7. The position of characters in the file from top to bottom denote their sorting "weights" in increasing order. All characters not present in this section (including ASCII characters) are given the same very high sorting "weight" to ensure that they come last when sorting alphabetically. Therefore, it is a good idea to include the ASCII characters to ensure proper sorting of "mixed" bibliographies.

Let us see an example. Assume that we have a bibliography database in the file greek.bib that includes two citations in Greek and one in English, and assume that we want to print it. We set up a file, say, text.tex, according to Section 8.2.2 slightly modified to support the Greek language (see Section 10.4):

```
\documentclass[a4paper]{article}
\usepackage[iso-8859-7]{inputenc}
\usepackage[greek]{babel}
\bibliographystyle{hellas}
\begin{document}
  \nocite{*}
  \bibliography{greek}
\end{document}
```

The file iso8859-7.csf, by Apostolos Syropoulos, defines the correspondence of lowercase to uppercase Greek letters as well as their order. For example, the \lowupcase part looks like

```
\lowupcase{                                              \order{
    α  A                                                    0-9
    ά  A                                                    A  α
    β  B        and the \order command looks like
    γ  Γ                                                    · · ·
    ...}                                                    Ω  ω  Ὠ  ώ}
```

Now, we run the `text.tex` file through LATEX, then we run `bibtex8 text`, and finally we run LATEX twice. The result looks like this:

Αναφορές

[1] Donald E. Knuth. *The TEX book*. Addison-Wesley, 2000.

[2] Απόστολος Συρόπουλος. *LATEX*. Παρατηρητής, Θεσσαλονίκη, 1998.

[3] Δημήτριος Φιλίππου. *Τα πρώτα βήματα στο TEX*. Παρατηρητής, Θεσσαλονίκη, 1999.

Suppose now that one is preparing a document that uses at least two different scripts (e.g., the Latin and the Hebrew); then it may be necessary to have at least two bibliographic sections: one for the Latin script and one for the Hebrew script. This particular problem can be handled by using the multibbl package by Apostolos Syropoulos. This package redefines most of the commands related to the LATEX user interface for the creation of bibliographies. The command `\newbibliography` is used to create a new auxiliary file, which, in turn, will be used to create a new bibliography section. The command has one argument: the name of an auxiliary file that will be used to create the new bibliography section. The new version of the `\cite` command has two required arguments and one optional one. The first required argument is the name of the bibliography, and the second retains the functionality of the original command. The same design principle applies to the commands `\bibliographystyle` and `\no-cite`. The command `\bibliography` has three required arguments, the first being the name of the auxiliary file, the second the argument that would be used if we had only one bibliography, and the third a string that will be used to typeset the title of the bibliography section and the running heads of this section (the first and second arguments are usually the same). Here is a sample input file that shows how to use the package:

```
\documentclass{article}
  \usepackage{multibbl}
\begin{document}
  \newbibliography{books}
  \newbibliography{papers}
  \bibliographystyle{books}{alpha}
  \bibliographystyle{papers}{unsrt}
```

```
      text text text text text text text
      text text~\cite{papers}{euclid,pythagoras}
      \nocite{books}{*}
      \bibliography{papers}{papers}{Paper List}
      \bibliography{books}{books}{Book List}
   \end{document}
```

8.3 Preparing the Index

An index is something that serves to guide, point out, or otherwise facilitate reference, especially an alphabetized list of names, places, and subjects treated in a printed work, giving the page or pages on which each item is mentioned. In LaTeX, we include a word in the index by using the command \index, so if the word perl should be included in the index, we should use the command

\index{perl}

If the word is to be printed in bold, we use

\index{perl@\textbf{perl}}

The special character @ is used to denote that what appears on its left side must be typeset as it appears on its right side. Thus, the first occurrence of perl will also be used by the sorting algorithm. This is very useful since what is used for sorting and what will be printed may be different! For example, we may want to have the name "Donald Knuth" under the letter K. Then, we should write

\index{Knuth@{Donald Knuth}}

Another thing we may want to change is the way that the page number is typeset. If we want, for example, to have the page number in bold, we would write \index{perl|textbf}. Notice that we wrote textbf without the backslash. Of course, the above can be combined. The command

\index{perl@\textbf{perl}|textit}

will print the word perl in the index (the entry will be typeset in boldface type) sorted as "perl," and its page number will be italic. A common application of this is through the command \see. If we want to send the reader to another index entry, say, to send the reader from the ω to the Ω command, we can write

\index{omega@ω|see{Ω}}

Here, we ask for the entry to be sorted according to the word omega and, in its place, the program must use ω|see{Ω}.

If a word is used repeatedly in a range of pages and we want to have this range in the index, we do not write the relative \index command all of the time. Instead, we write \index{convex|(} at the place where we have the first occurrence and \index{convex|)} at the place where we have the last occurrence. This will produce a page range in the index for the word convex.

Subindices are produced using an exclamation mark. If we want the word "Zeus" to appear in the category of "Greek" which is in the category of "Gods," we will write

$$\text{\textbackslash index\{Gods!Greek!Zeus\}}$$

Let us see the first example. In order to produce an index, we need to load the package makeidx and immediately issue the command \makeindex. At the place where we want the index to be printed, we use \printindex. We create a document with name indtest.tex with the following lines:

```
\documentclass{article}
\usepackage{makeidx}
\makeindex
\begin{document}
This is page 1. \index{perl|(} \index{Java|textbf}
\index{Omega@$\Omega$|textbf} \index{language!formal!lotos}
\index{language!Greek}
\newpage
This is page 2.
\index{omega@$\omega$|see{$\Omega$}}
\index{language!programming!self}\index{"@@\textit{at} symbol}
\index{java}\index{language!Spanish}\index{Goeteborg@Gothenburg}
\newpage
This is page 3.
\index{Goeteborg@G"\"{o}teborg} \index{Java}
\index{omega@$\omega$|see{$\Omega$}}
\index{language!programming!oberon} \index{language!formal}
\newpage
This is page 4.
\index{perl|)} \index{Goeteborg@Gothenburg}
\index{"@@\texttt{@} symbol}
\printindex
\end{document}
```

We now run LaTeX. This first run will create the file indtest.idx, which contains the Index entries as they are read by LaTeX together with their page number information. Now, they must be typeset and sorted. This is done by the program makeindex. To prepare the index, we have to use the program MAKEINDEX by Pehong Chen:

```
$ makeindex indtest
This is makeindex, version 2.13 [07-Mar-1997] (using kpathsea).
Scanning input file indtest.idx....done (17 entries accepted,
```

```
0 rejected).
Sorting entries....done (68 comparisons).
Generating output file indtest.ind....done (36 lines written,
1 warning).
Output written in indtest.ind.
Transcript written in indtest.ilg.
```

As is evident, this process creates the file indtest.ind, which contains the sorted index. It is time to rerun LaTeX. The \printindex command will load the index information from the file indtest.ind into the main document. In Figure 8.3, we can see the result of the example above.

In this example, we observe the use of quotes (") in front of special characters such as @. This is a general principle and is how we can have a special character in the index. For the word "Göteborg" above, we had to type "\ so that the makeindex program will leave a backslash in the indtest.ind file; we also had to type "" so that MAKEINDEX will leave a double quote " in the indtest.ind file. So now, when we rerun LaTeX, it will find in the indtest.ind file the sequence \", thus producing an umlaut above the o in Göteborg. Another example that often appears in this document comes from the commands that we want to have in the index and start with a backslash. Following the point above, if we want the command \alpha in our index (not the character α but the command itself), we must write

$$\index\{alpha@"\verb|"\alpha|\}$$

Better results are achieved by defining a command \PP

$$\newcommand\{\PP\}[1]\{\texttt\{\textbackslash#1\}\}$$

and using it like this

$$\index\{alpha@\PP\{alpha\}\}$$

at symbol, 2	Z, 4
@ symbol, 4	Greek, 1
	programming
Göteborg, 3	oberon, 3
Gothenburg, 2, 4	self, 2
	Spanish, 2
Java, **1**, 2, 3	
	Ω, **1**
language	ω, *see* Ω
formal, 3	
lotos, 1	perl, 1–4

Figure 8.3: A standard index.

8.4   ᴍᴀᴋᴇɪɴᴅᴇx in a Multilingual Environment

In a multilingual document, we have a problem that needs to be addressed. Different alphabets have different orderings of letters, or they are entirely different from the Latin alphabet. Consequently, the sorting algorithm will not produce any satisfactory results. The nice solution would be to produce one index for every different language and sort the entries of each of them with the language letter-order. This is achieved using the package multind by F.W. Long. The package modifies the commands of the makeindex package to accept an additional required argument; this is just the name of the index file. Thus, in a document with mixed English and Greek, we would use:

```
\usepackage{multind}
\makeindex{english}
\makeindex{greek}
```

Then, when we want an English word in the index, say, the word love, we write \index{english}{love}. Similarly, if ἔρως should appear in its index, we will write \index{greek}{ἔρως} This way, the first LATEX run will produce two index files. These will be english.idx and greek.idx. Now, we have to sort the entries and typeset them. The ᴍᴀᴋᴇɪɴᴅᴇx program will work fine for the English language, but it will fail for other languages. There is no global solution to this problem. Each language may or may not have its own program for doing this. The support for the Greek language includes a Perl script written by Apostolos Syropoulos with name ᴍᴋɪɴᴅᴇx that can do for the Greek index file what makeindex does for the English index. After the script is run on the file greek.idx, we rerun LATEX so that the indices get incorporated into the main document. This is done by writing the commands

```
\printindex{greek}{Ευρετήριο ελληνικών όρων}
{\usefont{OT1}{cmr}{m}{n}
\printindex{english}{Ευρετήριο ξενόγλωσσων όρων}}
```

at the places where we want to have them. The second argument of the \printindex command is the section header.

Here, we assume that we are actually preparing a Greek language document so we had to enclose the "foreign" index in a local scope that temporally changes the font in use. Naturally, this is not necessary when preparing our document with Λ. The \printindex command also writes an entry in the table of contents. However, this entry corresponds to a section entry (i.e., the author of the package assumed that indices are just sections). In case we want to change this behavior, all we have to do is to change the corresponding command in the definition of the \printindex command.

➤ **Exercise 8.2** Study the definition of the \printindex command and then modify it so that indeces are "treated" as chapters.   □

8.5 Customizing the Index

We can customize the index either by designing a style file or by redefining the environment that is used to typeset the index. A style file may have at most two sections: the first defines the meaning of the various special characters (input specifiers), and the second defines the commands that will be inserted in the output file (output specifiers). Table 8.1 shows specifiers associated with the definition of the input specifiers. For example, if we want the symbol = to be used instead of @, we have to place the following line in a style file:

`actual '='`

Note that characters and strings must be enclosed in single quotes. We now present the various output specifiers. In what follows, the letter s denotes a string that must be enclosed in double quotes and n denotes a number.

preamble s Preamble of the output file (i.e., what will appear at the very beginning of the output file). The default value is \\begin{theindex}\n. The token \n forces MAKEINDEX to change line.

postamble s Postamble of the output file (i.e., what will appear at the end of the output file). The default value is \n\n\\end{theindex}\n.

setpage_prefix s Prefix of command that sets the starting page number. The default value is \n\\setcounter{page}{.

setpage_suffix s Suffix of command that sets the starting page number. The default value is }\n.

group_skip s Vertical space to be inserted before a new group begins. The default value is \n\n\\indexspace\n.

Table 8.1: Input style specifiers. ch denotes a single character and s a string.

Command	Meaning	Default Symbol
actual ch	see Section 8.3	@
arg_close ch	see Section 8.3	}
arg_open ch	see Section 8.3	{
encap ch	see Section 8.3	|
escape ch	Symbol that escapes the following letter, unless its preceding letter is escape	\\
keyword s	Command that tells MAKEINDEX that its argument is an index entry	\\indexentry
level ch	see Section 8.3	!
quote ch	see Section 8.3	"
range_close ch	see Section 8.3	)
range_open ch	see Section 8.3	(

`heading_flag` s Flag indicating the treatment of new group headers, which are inserted before a new group. The possible groups are symbols, numbers, and the 26 letters. A positive (negative) value causes an uppercase (lowercase) letter to be inserted between prefix and suffix. Default value is 0, which produces no header.

`heading_prefix` s Header prefix to be inserted before a new letter begins; the default value is the empty string.

`symhead_positive` s Heading for symbols to be inserted if `heading_flag` is positive; the default value is Symbols.

`symhead_negative` s Heading for symbols to be inserted if `heading_flag` is negative; the default value is symbols.

`numhead_positive` s Heading for symbols to be inserted if `heading_flag` is positive; the default value is Numbers.

`numhead_negative` s Heading for symbols to be inserted if `heading_flag` is negative; the default value is numbers.

`item_0` s Command to be inserted between two primary items; the default value is \n \\item.

`item_1` s Command to be inserted between two secondary items; the default value is \n \\subitem.

`item_2` s Command to be inserted between two level 2 items; the default value is \n \\subsubitem.

`item_01` s Command to be inserted between a primary and a secondary item; the default value is \n \\subitem.

`item_x1` s Command to be inserted between a primary and a secondary item when the primary item does not have associated page numbers; the default value is \n \\subitem.

`item_12` s Command to be inserted between a secondary and a level 2 item; the default value is \n \\subsubitem.

`item_x2` s Command to be inserted between a secondary and a level 2 item when the secondary item does not have associated page numbers; the default value is \n \\subsubitem.

`delim_0` s Delimiter to be inserted between a primary key and its first page number; the default value is ,␣ (i.e., a comma followed by a blank).

`delim_1` s Delimiter to be inserted between a secondary key and its first page number; the default value is ,␣.

`delim_2` s Delimiter to be inserted between a level 2 key and its first page number; the default value is ,␣.

`delim_n` s Delimiter to be inserted between two page numbers for the same key in any level; the default value is ,␣.

`delim_r` s Delimiter to be inserted between the starting and ending page numbers of a range; the default value is --.

`delim_t` s Delimiter to be inserted at the end of a page list. This delimiter has no effect on entries that have no associated page list. The default value is the empty string.

`encap_prefix` s First part of prefix for the command that encapsulates the page number; the default value is \\.

`encap_infix` s Second part of prefix for the command that encapsulates the page number; the default value is {.

`encap_suffix` s Suffix for the command that encapsulates the page number; the default value is }.

`line_max` n Maximum length of a line in the output, beyond which a line wraps; the default length is 72.

`indent_space` s Space to be inserted in front of wrapped line; the default value is \t\t (i.e., two tabs).

`indent_length` n Length of `indent_space`; the default value is 16, which is equivalent to two tabs.

`suffix_2p` s Delimiter that replaces the range delimiter and the second page number of a two-page list. When present, it overrides `delim_r`. The default value is the empty string. Example: f..

`suffix_3p` s Delimiter that replaces the range delimiter and the second page number of a three-page list. When present, it overrides `delim_r` and `suffix_mp`. The default value is the empty string. Example: ff..

`suffix_mp` s Delimiter that replaces the range delimiter and the second page number of a multiple-page list. When present, it overrides `delim_r`. The default value is the empty string. Example: f..

As an application, we will show you how to define a style file suitable for the generation of glossaries. Moreover, we will use the definition of the `theindex` environment to define an environment suitable for the typesetting of a glossary.

8.6 Glossary Preparation

LaTeX provides the command \glossary, which can be used in order to generate a glossary for, say, a book. However, one must first define a suitable package that can be used by MAKEINDEX to process the generated glossary file. Naturally, in a glossary, the entries do not need to have associated page numbers. However, each \glossary command will print to the glossary file a line of the form

$$\text{\glossaryentry}\{\textit{glossary-text}\}\{\textit{page-number}\}$$

where *page-number* is just a number. A good solution is to ignore the page numbers by instructing MAKEINDEX to make them the arguments of a command that just ignores its arguments! Actually, we will have only one page number, as it makes no sense to have multiple entries for the same key. Moreover, we must let MAKEINDEX know the name of the new command that will appear in the glossary file. Below is the code for a style file that implements all of these features:

```
actual '='
keyword "\\glossaryentry"
preamble
    "\\newcommand{\\Ignore}[1]{}\n
    \\begin{theglossary}\n"
postamble "\n\\end{theglossary}\n"
delim_0 "\\Ignore{"
delim_t "}"
```

Note that we have chosen the = sign instead of the @ sign since we are supposed to explain the meaning of a term. Now, we have to create a little package that will define a theglossary environment. To do this, we use the definition of the theindex environment. We create a package, which we store in file glossary.sty (line numbers are included for future reference):

```
%% Package ''gloss''
1  \RequirePackage{ifthen}
2  \newcommand{\glossaryname}{Glossary}
3  \newenvironment{theglossary}{%
4    \ifthenelse{\boolean{@twocolumn}}{%
5      \setboolean{@restonecol}{false}}{%
6      \setboolean{@restonecol}{true}}%
7    \setlength{\columnseprule}{0pt}%
8    \setlength{\columnsep}{35pt}%
9    \twocolumn[\section*{\glossaryname}]%
10   \markboth{\MakeUppercase\glossaryname}%
11            {\MakeUppercase\glossaryname}%
12   \thispagestyle{plain}
13   \setlength{\parindent}{0pt}
14   \setlength{\parskip}{0pt plus .3pt}
15   \let\item\@idxitem}
16   {\ifthenelse{\boolean{@restonecol}}{%
17      \onecolumn}{%
18      \clearpage}}
19 \newcommand{\printglossary}{%
20   \InputIfFileExists{\jobname.gld}{}{%
21   \typeout{No file \jobname.gld}}}
```

We will now try to explain what the code above does. The command \RequirePackage is used only inside packages to load another package. When we want to load a package with some options, we have to put the options in square brackets:

$$\RequirePackage[options]{package}$$

On line 2, we define the name of the glossary. Note that the babel package redefines this command so that it produces the correct name for the language in use. In line 3, we define a new environment that will be used to typeset the glossary. The internal Boolean variable @twocolumn is set to true when we typeset our document in two columns; otherwise, it is set to false. So, if this variable is true, we do not have to switch back to one column typesetting. This happens when the glossary will be typeset in two columns. Then, on lines 7 and 8, we set the values of the lengths \columnseprule and \columnsep. The first length holds the width of the rule that usually appears between columns in two-column typesetting. The second length holds the length that separates columns in two-column typesetting. On line 9, we use the command \twocolumn to typeset the body of the environment in two columns. The optional argument is used to produce the header for the glossary. On lines 10 and 11, we use the command \markboth to set the running heads. On line 12, we declare the page style of the first page of the glossary. On lines 13 and 14, we set two lengths: \parindent and \parskip. The second one corresponds to space that is left between paragraphs. When the environment ends, we check the value of the internal Boolean variable \@restonecol. If it is true, we again start one column typesetting. Otherwise, we simply start a new page. On lines 19–21, we define the command that will print the index. It uses the command \InputIfFile-Exists, which checks whether the file we want to include in our file exists. If it exists, it includes the file and performs the action specified in the curly brackets after the filename. If the file does not exist, it performs the actions specified in the third pair of curly brackets. Here, the filename has the name of the main file (that is what is stored in variable \jobname) and extension gld. We take this opportunity to present another similar command: \IfFileExists. This command has three arguments: the name of a file, a *then* part and an *else* part. If the file exists, the *then* is executed; otherwise, the *else* part is executed. Back to our business! Now, it is time to test our packages. We first create a file that contains glossary entries:

```
\documentclass[a4paper]{article}
\setlength{\textwidth}{320pt} % We choose this extremely small
\setlength{\textheight}{80pt} % page size to see the two-column
\usepackage{gloss}            % effect.
\makeglossary
\begin{document}
page 1
\glossary{computer=\textbf{Computer} An electronic device.}
\glossary{Stockholm=\textbf{Stockholm} The capital of Sweden.}
\newpage
page 2
\glossary{Athens=\textbf{Athens} The capital of Greece.}
\glossary{Vienna=\textbf{Vienna} The capital of Austria.}
\printglossary
\end{document}
```

The next thing we do is to run LATEX:

```
$ latex gloss
This is TeX, Version 3.14159 (Web2C 7.3.3.1)
(./gloss.tex
LaTeX2e <2000/06/01>
(/usr/local/teTeX/share/texmf/tex/latex/base/article.cls
Document Class: article 2000/05/19 v1.4b Standard LaTeX
document class
(/usr/local/teTeX/share/texmf/tex/latex/base/size10.clo))
(./glossary.sty
(/usr/local/teTeX/share/texmf/tex/latex/base/ifthen.sty))
Writing glossary file gloss.glo
No file gloss.aux.
[1]
No file gloss.gld
[2] (./gloss.aux) )
Output written on gloss.dvi (2 pages, 320 bytes).
Transcript written on gloss.log.
```

Now, we have to run MAKEINDEX:

```
$ makeindex -s gloss.ist -o gloss.gld gloss.glo
This is makeindex, version 2.13 [07-Mar-1997] (using kpathsea).
Scanning style file ./gloss.ist......done (6 attributes
redefined, 0 ignored).
Scanning input file gloss.glo....done (4 entries accepted,
0 rejected).
Sorting entries....done (10 comparisons).
Generating output file gloss.gld....done (17 lines written,
0 warnings).
Output written in gloss.gld.
Transcript written in gloss.ilg.
```

Here, we have to use two command-line switches: -s and -o. The first one is used to specify the name of the style file that the program will use to process the glossary. The second one is used to specify the name of the output file. The last thing we have to do is to rerun LaTeX to include the glossary in our document. The typeset glossary looks like this:

Glossary

Athens The capital of Greece.

Computer An electronic device.

Stockholm The capital of Sweden.

Vienna The capital of Austria.

3

➤ **Exercise 8.3** Write down the style file that the authors used to typeset the index of this book. □

9

Graphics

The ability to include drawings, pictures, and line art in modern publications is more than necessary for any typesetting system. Although LaTeX by itself can be used to produce drawings, such as those found in mathematics books, it provides facilities to include virtually any kind of graphics file. In this chapter, we describe the `picture` environment, which can be used to create simple drawings. In addition, we discuss how to add graphics to a LaTeX file. We also discuss how one can create graphics with other packages. Graphics inclusion was discussed earlier when we dealt with floats (see Section 6.5), but here we will go into detail.

9.1 Drawing with the `picture` Environment

The LaTeX 2_ε format contains a basic set of commands that can be used to draw illustrations made up from simple components, such as straight lines, arrows, simple curves, and text. An advantage of using the `picture` environment is that no special support is required from the device driver. Some limitations of the `picture` environment include a limited range of slopes for lines or arrows, circles of only a fixed range of sizes, and limitations on the thickness of slanted lines, circles, and oval shapes. These restrictions are to be removed in an enhanced version of the package known as pict2e, but this requires special features of the device driver that are not yet widely available. For this reason, we shall focus on the standard `picture` environment, pointing out the restrictions to the graphical components as we go along.

A diagram or illustration is started with the command

> \begin{picture}(*x-size,y-size*)(*x-origin,y-origin*)

With this command, we specify that the plotting area will be *x-size* across and *y-size* units upwards. The last two arguments should be used to optionally specify the lower-left corner of the plotting area when it is different from the default of $(0, 0)$. For example, the command

$$\verb|\begin{picture}(20,40)|$$

sets up a picture 40 units across by 20 units upwards, while

$$\verb|\begin{picture}(200,110)(-100,-10)|$$

defines a plotting area with 200 horizontal units and 110 vertical units with the lower-left corner not at $(0,0)$ but this time at (-100, -10). The dimensions of the units used in the figure are specified separately using

$$\verb|\setlength{\unitlength}{|\mathit{unit\ dimensions}\verb|}|$$

For example, `\setlength{\unitlength}{5pt}` specifies a unit of 5 pt, and `\set-length{\unitlength}{0.1mm}` indicates a unit of 0.1 mm. This makes drawing more straightforward since one can use convenient dimensions such as those of the original data and then scale the figure by adjusting the value of the unit length. Note that scaling does not affect the thickness of lines or the size of text and symbols. To achieve scaled magnification of all of the items in a picture, the graphicx package is required.

The components of a picture are placed with the command

$$\verb|\put(|\mathit{x\text{-}coordinate,\ y\text{-}coordinate}\verb|){|\mathit{component}\verb|}|$$

which also includes things like text, equations, and symbols from LaTeX anywhere in the picture, for example

$$\verb|\put(10,10){$\diamond$}|$$

places a ⋄ symbol in our picture at the coordinate $(10,10)$, aligned upon its reference point. This is useful for graph plotting and the labeling of a figure. It is worth mentioning that we can use the \put command to place things even outside the specified plotting area. For example, the vertical bar at the right was placed at point $(105,0)$ in a plotting area with zero horizontal and vertical units!

We shall now introduce the basic drawing components before giving some illustrations of how they may be combined to produce diagrams and plots.

9.1.1 Invisible and Framed Boxes

The addition of a \makebox command allows us to position symbols and text within an invisible box and to shift the reference point around. The syntax of the command is

$$\verb|\makebox(|x,y\verb|)[|\mathit{position}\verb|]{|\mathit{text}\verb|}|$$

and it generates an invisible box with dimensions x units by y units. An optional argument for position determines the alignment within the box (i.e., lleft, right, top, and bottom). A useful trick is to enclose text or a symbol in a box of zero width; this has the effect of shifting the reference point to its center, which is handy for plotting. Thus,

$$\verb|\put(10,10){\makebox(0,0){$\diamond$}}|$$

now places our earlier diamond centered over the point $(10, 10)$.

Other box-making commands include

$$\texttt{\textbackslash framebox(}x,y\texttt{)\,[}position\texttt{]\{}text\texttt{\}}$$

which is similar to \makebox but places a frame around the rectangle, and

$$\texttt{\textbackslash dashbox\{}dash\_dimension\texttt{\}(x,y)\,[}position\texttt{]\{}text\texttt{\}}$$

which encloses the rectangle in a dashed frame, with the width of the dashes determined by the size of *dash_dimension*.

➤ **Exercise 9.1** Put a ♡ in a 10×10 frame at point $(10, 20)$. ☐

9.1.2 Lines and Arrows

The command

$$\texttt{\textbackslash line(}x\text{-}amount,\ y\text{-}amount\texttt{)\{}units\_across\texttt{\}}$$

produces a line with a slope of *y-amount*/*x-amount* with its size determined by the number of horizontal *units_across*; for example,

$$\texttt{\textbackslash put(0,0)\{\textbackslash line(1,2)\{10\}\}}$$

will create the following line placed with the bottom-left corner at the point $(0, 0)$.

The slope of lines in the picture environment is restricted in that the values of *x-amount*, *y-amount* must be integers between the range 6 and 6, and, in addition, they can have no common divisor greater than one. Legal values include $(1, 2)$ and $(0, 1)$, and examples of values that are illegal include $(1.5, 2)$ and $(4, 8)$. The smallest line is limited to a length of about 10 pt, or 3.6 mm. There are two predefined thicknesses of (sloping) lines available. The default setting is \thinlines, which looks like ——— , while the \thicklines declaration produces a line like this ——— . Other line thicknesses can be requested with the \linethickness{*breadth*} command (e.g., \linethickness{1mm} will produce a line like this ▬▬ and \linethickness{0.05mm} will produce a line like this ———). The declaration \linethickness only applies to lines that are horizontal or vertical. Lines that are slanted, circles, or oval shapes are not altered.

Arrows are drawn using a similar command

$$\texttt{\textbackslash vector(}x\text{-}amount,\ y\text{-}amount\texttt{)\{}units\_across\texttt{\}}$$

that works in the same way as the \line command but is further restricted to values of *x-amount* and *y-amount* that are integers between 4 and 4. Thus, using units of 1 mm, the command \vector(-1,0){10} gives us this arrow ←—— with its reference point on the right-hand side. An arrow that is specified to have a length of zero is an exception to this in that the reference point is at the tip of the arrow.

9.1.3 Circles and Curved Shapes

Circles are produced with the command \circle{diameter} but have a limited range of sizes, the closest one to the requested diameter being produced. Disks are filled circles produced with the \circle* command. So, with 1 mm units, requesting a \circle{3} produces ◯, and a \circle*{3} request gives us a ● . The reference point of a circle or disk is its center. Note also, that the maximum size of a disk is 15 pt, and the maximum size of a circle is 40 pt.

Two other curved elements can be drawn with the picture environment: ovals and Bezier curves. Ovals consist of rectangles with rounded corners and are invoked with the command

$$\oval(width,height)[part]$$

which produces an oval with rounded corners selected from the quarter circles that are available on the system. These are chosen to make it as curved as possible. The *part* argument is an option that specifies which parts of the oval we would like LATEX to draw: left, right, top, or bottom. A single letter produces the indicated half (e.g., [r] gives us the right half of the oval), and two letters result in a corresponding quarter being drawn (e.g., [lb] produces the lower-left corner). If we try \oval(2,4)[t], then this oval is drawn ⌒.

A quadratic Bezier curve has three control points specified by the command

$$\qbezier[number](x_1,\ y_1)(x_2,y_2)(x_3,\ y_3)$$

and the curve is drawn from (x_1, y_1) to (x_3, y_3) with (x_2, y_2) the guiding point. For example, \qbezier(1,1)(5,20)(10,5) produces a curve like this:

Bezier curves are drawn so that the curve at (x_1, y_1) is tangential to an imaginary line between (x_1, y_1) and (x_2, y_2), while the same curve at (x_3, y_3) is tangential to an imaginary line between (x_3, y_3) and (x_2, y_2). Bearing this in mind, it is possible to make two curves join up smoothly by ensuring that they have the same tangent at the point where they join. The optional argument *number* requests that the curve be made up of the specified number of points (e.g., \qbezier[25](0,0)(20,2)(0,4) produces this curve ⋯⋯➤, which is made up of 25 points).

9.1.4 The Construction of Patterns

Repeated patterns are constructed from a compact specification of regularly occurring picture components. They are produced with

```
\multiput(x-coordinate, y-coordinate)(x-shift,
          y-shift){number}{component}
```

This makes it a simple matter to generate hatched shadings and ruler markings. For example,

```
\put(0,1.5){\line(1,0){20}}
\multiput(0.45,0)(1,0){20}{\makebox(0,0)%
   {{\footnotesize /}}}
```

produces

///

We will now lead you through some examples that illustrate how the simple components available in the picture environment can be combined to produce more elaborate drawings. The examples are all drawn in the default Computer Modern family of fonts.

9.1.5   An Example of the Calculation of the Area of a Square

In this first example, we shall draw a box, add arrows and labels, and add a formula for its area. The basic picture environment is set up with a unit length of 0.15 mm, dimensions of 220 units across and 140 units vertically and with the lower-left corber at (25,0). This is specified with the sequence of commands

```
\setlength{\unitlength}{0.15mm}
\begin{picture}(220,140)(-25,0)
      . . . . . . . . . . . . .
\end{picture}
```

Initially, nothing is drawn, but by adding successive picture-drawing commands in the body of the picture environment and running the LATEX program and a DVI viewer, we can watch the figure build up. By running LATEX each time a component is added or modified, we also catch errors and deal with them as they occur.

Next, we draw the box by inserting the line

```
\put(0,0){\thicklines \framebox(100,100){}}
```

in the body of the environment followed by four lines that produce markings delineating the x dimensions

```
\put(-1.5,105) {\line(0,1){16}}
\put(101.2,105){\line(0,1){16}}
\put(105,101.2){\line(1,0){16}}
\put(105,-1.5) {\line(1,0){16}}
```

Double-headed arrows are drawn as a series of vectors by inserting these commands:

```
\put(50,113) {\vector(1,0){50}}
\put(50,113){\vector(-1,0){50}}
\put(113,50) {\vector(0,1){50}}
\put(113,50){\vector(0,-1){50}}
```

We then insert commands for our labels of x (in \scriptsize)

```
\put(50,126){\scriptsize \makebox(0,0){$x$}}
\put(126,50){\scriptsize \makebox(0,0){$x$}}
```

Finally, we add the equation for the area of a square (set in a \small size font)

```
\put(190,50){\small $\textrm{Area}= x^2$}
```

In combination, these pictorial elements produce this figure.

9.1.6   A Diagram for the Calculation of the Area of a Circle

Following a pattern similar to the last example, we build up a graphic for the area of a circle. Once again, the picture starts with an empty picture environment with a unit length of 0.15 mm

```
\setlength{\unitlength}{0.15mm}
\begin{picture}(220,140)(-25,0)
. . . . . . . . . . . . . . . . .
\end{picture}
```

Into this we insert a command to draw a circle that is as near 90 units in diameter as the system can provide:

```
\put(50,50){{\thicklines \circle{90}}}
```

We build up diameter and radius arrows as a series of vectors by inserting the commands

```
\put(50,50){\vector(3,1){44}}
\put(50,50){\vector(-3,-1){44}}
\put(50,50){\vector(-1,3){14.5}}
```

and then add labels for d and r:

```
\put(45,34){\scriptsize $d$}
\put(47,69){\scriptsize $r$}
```

Finally, we insert statements to add the equations for the area of a circle:

```
\put(190,50){\small $\textrm{Area}=
\frac{\pi d^2}{4}$ or $\pi r^2$}
```

Together, the whole sequence of LaTeX statements generates this drawing:

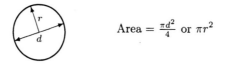

9.1.7 Box-and-Whisker Plots in the Style of John W. Tukey

Box-and-whisker plots, called box plots for short, are a simple and elegant way of summarizing information about the central tendency, range, and shape of the distribution of sample data. They were developed at some length, from earlier ideas, in Tukey's classic (1977) book *Exploratory Data Analysis,* which made them popular. The emphasis is on robust statistics that are easily calculated and insensitive to extreme observations. We shall plot the main features present in data collected on 219 of the world's volcanos and produce a version of Tukey's hand-drawn box plot suitable for reproduction in a document typeset in LaTeX (see [28] pages 40–41, and page 73). We begin with an empty picture environment that contains a declaration that all of the text in the figure will be in a sans serif font.

```
{\sffamily
\setlength{\unitlength}{0.33mm}
\begin{picture}(50,260)(-30,0)
    . . . . . . . . . . . . . . .
\end{picture}}
```

A median line represents the central tendency of the dataset and is larger than 50% of the observations and smaller than the remaining 50%. This is drawn by inserting the command

```
\put(40,65){\line(1,0){20}}
```

We add the box that contains the central 50% of observations, ranging between the upper and lower quartiles. A lower quartile is larger than 25% of the observations, and an upper quartile is larger than 75% of the observations. This is done with the statement

```
\put(40,37){\framebox(20,58){}}
```

The first observations lying within 1.5 times the difference between the lower and upper quartiles are the so-called adjacent values, and they are marked to indicate the limits to the range of typical values.[1]

```
\put(50,185){\circle{2}}
\put(50,6){\circle{2}}
```

We connect the adjacent values to the quartile box with two thin lines, called the whiskers, using these commands:

```
\put(50,7){\line(0,1){30}}
\put(50,95){\line(0,1){89}}
```

Any values external to the range defined by the adjacent values are considered extreme and possibly atypical. We draw these in with these statements:

```
\put(50,200.5){\circle{2}}
\put(50,197){\circle{2}}
\put(47,193){\circle{2}}
\put(53,193){\circle{2}}
\put(50,190){\circle{2}}
\put(50,2){\circle{2}}
```

We draw a floor to the diagram and shade beneath it to indicate the limits of possible measurement with these commands:

```
\put(24,0){\line(1,0){52}}
\multiput(26.5,-4.5)(4,0){13}{%
\makebox(0,0){{\footnotesize /}}}
```

Labels are added for the adjacent values with the stetements

```
\put(54,180){{\tiny Tupungatito}}
\put(54,5){{\tiny Anak Krakatau}}
```

and we label the extreme observations with these additional commands:

```
\put(54,199){{\tiny Guallatiri}}
\put(28,196){{\tiny Lascar}}
\put(12,191){{\tiny Kilimanjaro}}
\put(56,191.5){{\tiny Cotapaxi}}
\put(54,186){{\tiny Misti}}
\put(20,1.5){{\tiny Ilha Nova}}
```

Finally, we put the vertical numerals and a label for the measurement units with this group of statements:

---

1. Note that Tukey's figure connected the whiskers to two values that are close to, although not the actual, adjacent values.

```
\put(-10,-2){\makebox(0,0)[r]{\scriptsize 0}}
\put(-10,48){\makebox(0,0)[r]{\scriptsize 5,000}}
\put(-10,98){\makebox(0,0)[r]{\scriptsize 10,000}}
\put(-10,148){\makebox(0,0)[r]{\scriptsize 15,000}}
\put(-10,198){\makebox(0,0)[r]{\scriptsize 20,000}}
\put(-18, 208){\vector(0,1){15}}
\put(-30,230){{\scriptsize\shortstack{Height\\ (feet)}}}
```

Note that the command \shortstack puts the "words" that make up its argument one above the other. These words must be separated with \\. Back to our example! When the complete sequence of statements is run through LaTeX and viewed or printed, we obtain the following plot.

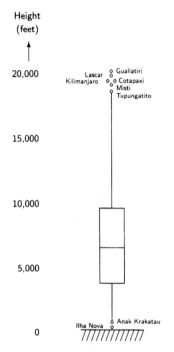

The calculations are most easily carried out in a statistical or spreadsheet software package and the coordinates of the points pasted into the LaTeX document, where they can be inserted into the relevant commands described above.

9.1.8 A Scatter Plot of Temperature

In this example, we demonstrate a simple scatter plot for the exploration of temperature measured during the course of a 24-hour period in Botswana, taken in December 1940 (data from Pearson and Hartley, reproduced in [6], page 352). The figure will show a subsample of the original points suitable for demonstration purposes. The data pairs (time/temperature pairs) that we will use are: $(1,65)$, $(2,69)$, $(3,74)$, $(4,79)$, $(5,83)$, $(6,86)$, $(7,88)$, $(8,90)$, $(9,90)$, $(10,90)$, $(11,89)$, $(12,88)$, $(13,85)$, $(14,80)$, $(15,76)$, $(16,74)$, $(17,73)$, $(18,72)$, $(19,71)$, $(20,69)$, $(21,68)$, $(22,67)$, $(23,67)$, and $(24,66)$. Moreover, we use the formulas $x = t \cdot 1000/25$ and $y = (T - 65) \cdot 1000/(95 - 65)$ to scale our data for plotting.

This time, we begin by defining a plotting symbol to save typing later, and to allow for easy subsequent changes in our plotting symbol just by changing a single statement.

```
\newcommand{\plotsymbol}{%
    \put(0,0){\small \makebox(0,0){+}}}
```

Then, the basic picture environment is defined with a unit length of 0.08 mm

```
{\sffamily
\setlength{\unitlength}{0.08mm}
\begin{picture}(1100,1100)
    . . . . . . . . . . . . .
\end{picture}}
```

and we proceed by inserting picture-drawing statements in the body of the `picture` environment, as before. First, we add some axes

```
\put(0,0){\line(1,0){1000}}
\put(0,0){\line(0,1){1000}}
```

followed by tick marks on the axes

```
\multiput(0,-10)(200,0){6}{\line(0,1){10}}
\multiput(-10,0)(0,189){6}{\line(1,0){10}}
```

We insert the commands for the x-axis numerals, each numeral is centered over its coordinate by enclosing it in a \makebox of zero width

```
\put(0,-40){\makebox(0,0){{\footnotesize 0}}}
\put(200,-40){\makebox(0,0){{\footnotesize 5}}}
\put(400,-40){\makebox(0,0){{\footnotesize 10}}}
\put(600,-40){\makebox(0,0){{\footnotesize 15}}}
\put(800,-40){\makebox(0,0){{\footnotesize 20}}}
\put(1000,-40){\makebox(0,0){{\footnotesize 25}}}
```

and the commands for the y-axis numerals

```
\put(-40,0){\makebox(0,0){{\footnotesize 65}}}
\put(-40,189){\makebox(0,0){{\footnotesize 70}}}
\put(-40,379){\makebox(0,0){{\footnotesize 75}}}
\put(-40,568){\makebox(0,0){{\footnotesize 80}}}
\put(-40,757){\makebox(0,0){{\footnotesize 85}}}
\put(-40,947){\makebox(0,0){{\footnotesize 90}}}
```

We label our axes by inserting the statements

```
\put(500,-110){\makebox(0,0){{%
  \footnotesize Time in hours since midnight}}}
\put(-105,1000){\makebox(0,0){{%
  \footnotesize Temp $^\circ$F}}}
```

and, finally, we plot our data points

```
\put(40,  8){\plotsymbol}
\put(120, 315){\plotsymbol}
\put(200, 591){\plotsymbol}
\put(280, 770){\plotsymbol}
\put(360, 845){\plotsymbol}
\put(440, 797){\plotsymbol}
\put(520, 665){\plotsymbol}
\put(560, 514){\plotsymbol}
\put(600, 372){\plotsymbol}
\put(680, 271){\plotsymbol}
\put(760, 186){\plotsymbol}
\put(840, 104){\plotsymbol}
\put(960, 29){\plotsymbol}
```

Combined together, this sequence of commands draws the following scatter plot.

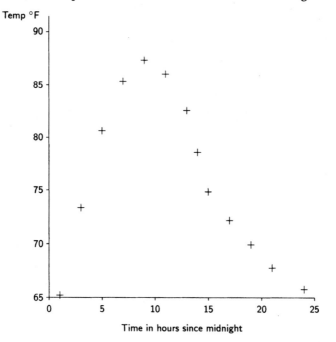

As when drawing a box plot, the reader will often find it convenient to first do the calculation of the coordinates in a statistical or spreadsheet software package and then import the points into their LATEX document for editing into a series of commands for the `picture` environment.

9.1.9 `picture`-Related Packages and Systems

There are a number of packages that are either based on the `picture` environment or extend it. Although there are many such packages, we think that it is better to use a system such as XTEXCAD by Johannes Sixt, which has a graphical user interface and can draw arbitrary circles, lines, and vectors of arbitrary slope. The output of the program is usable once the eepic package is loaded. This package was originally created by Conrad Kwok and was last updated by Piet van Oostrum. An important thing that we must note is that pictures generated with XTEXCAD cannot be used with pdfLATEX.

The bar package by Joachim Bleser and Edmund Lang can be used to draw bar graphs with LATEX. Nowadays, most people use spreadsheets to generate really fancy bar graphs, but nevertheless we believe it is worth the trouble to experiment with this package, so we will briefly present it.

The package defines the environment `barenv`, in which we specify the bars and the axis. The current implementation of the package offers eight different patterns that are used to "color" the bars:

The bars are specified with the \bar command, which has two required and one optional argument:

$$\bar{height}{pattern}[description]$$

`height` is the height of the bar in points and *pattern* a number from one to eight denoting one of the patterns in the figure above. The optional *description* is just the description of an individual bar. Here is a simple example:

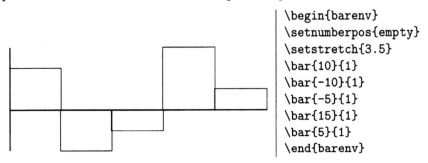

```
\begin{barenv}
\setnumberpos{empty}
\setstretch{3.5}
\bar{10}{1}
\bar{-10}{1}
\bar{-5}{1}
\bar{15}{1}
\bar{5}{1}
\end{barenv}
```

Before describing the meaning of the various commands, we must point out that the example above also shows a limitation of the package: it produces poor results when

lines overlap. The command \setnumberpos controls where the number denoting the
height of the bar will go. The possible values are: empty, axis (under or above the
x-axis), down (under the bar), inside (inside the bar), outside (outside the bar), and
up (on top of the bar). The command \setstretch is used to vertically stretch the bar
graph. Here is another example:

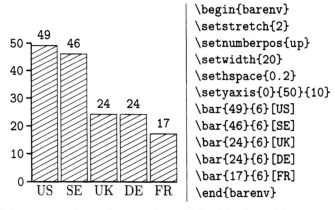

```
\begin{barenv}
\setstretch{2}
\setnumberpos{up}
\setwidth{20}
\sethspace{0.2}
\setyaxis{0}{50}{10}
\bar{49}{6}[US]
\bar{46}{6}[SE]
\bar{24}{6}[UK]
\bar{24}{6}[DE]
\bar{17}{6}[FR]
\end{barenv}
```

The command \setwidth is used to set the width of the bars. Moreover, the command
\sethspace has one argument, which denotes the distance between bars; this number
is multiplied with the actual width of the bars to get the distance between bars. The
commands \setxaxis and \setyaxis are used to draw the x and y axes, respectively.
Both commands have three arguments: the *origin*, the *end*, and a *step*. Furthermore, the
commands \setxname and \setyname have one argument which is the text that will be
displayed on the x and y axes, respectively. We present one more example that shows
some other capabilities of the package:

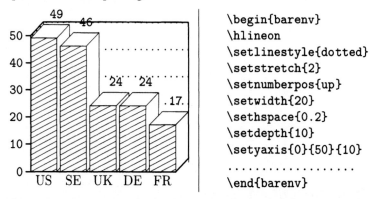

```
\begin{barenv}
\hlineon
\setlinestyle{dotted}
\setstretch{2}
\setnumberpos{up}
\setwidth{20}
\sethspace{0.2}
\setdepth{10}
\setyaxis{0}{50}{10}
 . . . . . . . . . . . . . . . . .
\end{barenv}
```

The command \hlineon activates the background horizontal lines as in the example
above. There are two line styles, solid and dotted, and they can be selected with the
command \setlinestyle. The 3D effect is achieved with the command \setdepth.
This command has one argument, which must be a number greater than or equal to ten.
The command \setprecision is used to define the number of *digits* to be printed

after the decimal sign. We should note at this point that *patterns* 5 and 6 are not drawn correctly, unless we use the eepic package.

9.2 The Gnuplot System

Gnuplot is a free software package that removes much of the tedious calculation required when plotting data, and also makes it possible to plot fitted curves and complex functions by constructing the curves from small line fragments. Since the resulting graphics are generated using standard `picture` commands, they are highly portable. However, they can create large files that may consume LATEX's memory. This figure is a scatter plot of data from a study of child witness testimony, by Graeme, Hutcheson, and colleagues [14]. It displays the empirical relationship between the accuracy and completeness of children's statements in two age groups, adapted for reproduction with kind permission of the authors. An advantage of using Gnuplot over some other graphics software is that, since the output can be entirely in LATEX, the user can easily modify the file (e.g., to change labels or manually move adjacent plotting symbols for increased clarity).

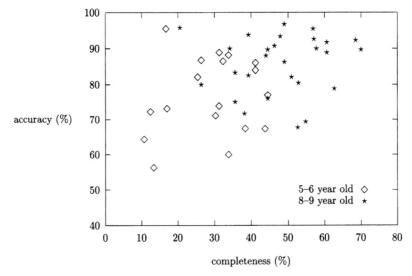

9.3 The **graphicx** Package

In this section, we describe the graphicx package by David Carlisle. The package provides a number of options that should be used only when the appropriate driver program will be used to display, print, or transform the resulting DVI file. The default option is dvips for use with DVIPS. Other useful options include the options pdftex (see Section 9.5), xdvi (for use with the XDVI previewer), dviwin (for use with the DVI-WIN previewer), and dvipdf (should be used when transforming a DVI file using the DVIPDF converter). However, at most installations, the file graphics.cfg is used to automatically load the approriate option when processing an input file with a particular typesetting engine.

The most important command of the package is \includegraphics

$$\includegraphics[keyval-list]\{file\}$$

where *file* is the name of the graphics file (usually a PostScript file) to be included and *keyval-list* is a comma-separated list of parameters in the form parameter=value. The available parameters are:

bb This sets the bounding box and is given in the form bb=a b c d, where (a, b) are the coordinates of the lower-left corner of the graphics and (c, d) the coordinates of the upper-right corner. If these are not set, LaTeX will try to find this information inside the graphics file. If you want to modify it, then GHOSTVIEW shows the coordinates of the current position of the mouse in its graphics window.

bbllx,bblly,bburx,bbury These also set the bounding box and are only here for compatibility reasons with older packages. bbll=a,bblly=b,bburx=c,bbury=d is the same as bb=a b c d.

natwidth,natheight Again, these set the bounding box: natheight=h,natwidth=w is equivalent to bb=0 0 h w.

viewport This modifies the bounding box that is already specified in the graphics file. The four values specify a bounding box *relative* to the original bounding box.

trim The same as above but now the four values specify amounts to be removed from the coordinates of the bounding box specified in the graphics file.

hiresbb This is a Boolean parameter with default value set to true. It causes TeX to look for a %%HiResBoundingBox comment rather than the standard %%BoundingBox.

angle Rotation angle.

origin Rotation origin.

width This is a length that asks for the graphic to be scaled to this width.

height Same as above, for height.

totalheight Same as above, but includes the depth (i.e., it is equal to height plus depth).

keepaspectratio This is a Boolean value key; if set to true, it makes sure that the graphic is not distorted in the attempt to make both the required width and height but scales so that neither dimension *exceeds* the stated dimensions.

scale Scale factor; it can be a positive rational number. If it is less than one, it shrinks the figure; otherwise, it stretches the figure.

clip Again, a Boolean value key with default value set to true. It clips the graphic to the bounding box or the viewport if it is specified.

draft If set to true, it switches locally to draft mode so that the graphic is not printed but the correct space is reserved and the filename printed.

type Specifies the file type which is normally determined by the file extension.
ext Specifies the file extension. Used *only* with the type option.
read Specifies the "read file," which is used to determine the size of the graphic.
command Specifies the file command. Used *only* with the type option.

Note that the order of key values *is* important. The two options

$$[\texttt{angle=-90,scale=4}] \text{ and } [\texttt{scale=4,angle=-90}]$$

are not the same. The former first executes the scaling and then rotates, while the latter does it the other way around (first rotates and then scales).

➤ **Exercise 9.2** Give the necessary commands to make an image the header of a document. □

The overpic package (by Rolf Niepraschk) provides the overpic environment. This environment can be used to include a graphics file, and it defines a plotting area above the included image that has the dimensions of the graphics file. In addition, all of the picture-related commands can be used to place anything that LaTeX can typeset on the included image. To assist users in the placement of things on the image, the package makes it possible to draw a grid of lines on the image. A grid can be 100 units across and 100 units upwards (enabled with the default percent option) or 1000 units across and 1000 units upwards (enabled with the permil option). If we specify the abs option, then we must set the \unitlength length variable. In this case, the placement over the image is expressed in "absolute" units; otherwise, it is expressed in "relative" units. The overpic environment can have the following optional arguments:

scale=*scale-factor* Scale the included image.
grid Draw a grid above the image (default is percent grid).
ticks=*units* Place ticks on all axes at every *units*.
unit=*length* Implicitly set the length variable \unitlength to *length*.

The example in Figure 9.1 demonstrates all of the capabilities of the overpic environment.

9.3.1 Playing with Words

For TeX, a letter or a word is just a box, so we can easily stretch it or change its position. If we plan to use a PostScript driver, then we can do amazing things to boxes! However, we will go into details gradually. To whet your appetite, we can show you that it is possible to ᴅɪꜱᴛᴏʀᴛ, rotate, and reflect or reflect text.

All of this is done with two commands provided by the graphicx package. These are \rotatebox and \scalebox. The syntax for \rotatebox is

$$\texttt{\textbackslash rotatebox[}\textit{key value list}\texttt{]\{}\textit{angle}\texttt{\}\{}\textit{text}\texttt{\}}$$

and for \scalebox

```
\begin{overpic}%
    [scale=.25,grid,
     tics=20,unit=1mm]%
    {golfer.ps}
\put(5,45){\huge \LaTeX}
\put(55,10){%
    \includegraphics[scale=.07]%
        {golfer.ps}}
\end{overpic}
```

Figure 9.1: A demonstration of the capabilities of the overpic environment.

\scalebox{<i>horizontal scaling</i>}
[<i>vertical scaling</i>]
{<i>text to be scaled</i>}

If the optional argument is not given, the scaling will be uniform with a scale factor corresponding to the factor given in the compulsory argument. The scaling factors can be negative, resulting in reflections. Let us see some examples:

Command	Result
\scalebox{3}[.5]{Hello!}	Hello!
\scalebox{.5}[2]{Hello!}	Hello!
\scalebox{-3}[.5]{Hello!}	Hello!
\scalebox{-.5}[2]{Hello!}	Hello!
\scalebox{-1}[1]{Hello!}	!olleH

The command \reflectbox{ <i>text</i> } is equivalent to the last command of the table above.

➤ **Exercise 9.3** How can we get the AꓭBA logo? □

Another way of achieving similar results with \scalebox is provided by the command \resizebox. The difference is that with \scalebox the dimensions of the resized box are now given not as a factor for scaling but as absolute lengths. They can be, for example, 3 cm or a number multiplied by any of the dimensions of the box that is to be resized. These are \height, \width, \depth, and \totalheight (see Section 6.10). The syntax is

\resizebox{*width*}{*height*}{*text to be resized*}

One can use a ! in place of one of the dimensions, and in this case this dimension will be determined from the other one, which must be given explicitly. The next table clarifies the use of this command:

Command	Result
\resizebox{.5cm}{.1cm}{Hello!}	Hello!
\resizebox{.5\width}{2\height}{Hello!}	Hello!
\resizebox{2\width}{!}{Hello!}	Hello!
\resizebox{!}{-2\height}{Hello!}	¡oʅʅəH

Finally, \resizebox has a starred version for which the height of the box refers to the height *plus* the depth. The next table shows the difference:

Command	Result
\resizebox{\width}{2\height}{Bye!}	Bye!
\resizebox*{\width}{2\height}{Bye!}	Bye!

9.4 Images that Can Be Loaded to a LATEX File

The only picture format that can be directly loaded into a LATEX file with the default dvips driver is the Encapsulated PostScript format (or EPS, for short). Any other picture format must first be converted to PostScript. A nice software application for such conversions is the program JPEG2PS by Thomas Merz, which converts the very common JPEG files to PostScript. Actually, this program puts a wrapper around the JPEG file. Note that this program uses features introduced to PostScript level 2. If we have a JPEG file, then the following command can be used to transform our file to EPS:

$jpeg2ps -r *RES* -o file.eps file.jpg

RES is the required resolution in dots per inch.

For formats such as TIFF, GIF, PIXX, and others, one can use the program BM2FONT by Friedhelm Sowa, but this works only with grayscales. However, we strongly suggest that you use a graphics manipulation program to transform your graphics file either to JPEG or to EPS. The Gnu Image Manipulation Program (or GIMP for short) by Spencer

Kimball and Peter Mattis is an excellent freely available graphics manipulation program that the authors strongly recommend.

9.5 Image Inclusion with pdfLATEX

If you are using pdfLATEX and want to include graphics, the graphicx package must be used with the pdftex option. The graphics formats that pdfLATEX can directly handle are PDF, TIFF, JPEG, and PNG. So, if you want to include a PNG file, you have to enter a command such as the following one:

```
\includegraphics[scale=3]{fil.png}
```

It is a fact that PDF and PostScript are very different document formats, so we cannot embed an EPS file into a PDF file. The situation is similar to a Java program that contains Perl code—naturally, the Java compiler will not be able to handle this peculiar program. So, if we have an EPS file, how can we embed it into a LATEX file that will be processed by pdfLATEX? The simplest solution is to use the program EPSTOPDF (by Sebastian Rahtz). The program has a number of options, which are shown below:

Option	Meaning
--help	Print usage
--outfile=*file*	Write result to *file*
--(no)filter	Read standard input (default: --filter)
--(no)gs	Run Ghostscript (default: --gs)
--(no)compress	Use compression (default: --compress)
--(no)hires	Scan HiResBoundingBox (default: --hires)
--(no)exact	Scan ExactBoundingBox (default: --noexact)
--(no)debug	Debug information (default: --nodebug)

9.6 Images in the Background

In the previous sections, we showed how to incorporate pictures into our documents. But what if we want to have a background picture? This is possible with the package eso-pic by Rolf Niepraschk. The package provides the commands \AddToShipoutPicture, \AddToShipoutPicture*, and \ClearShipoutPicture. All of the arguments of the \AddToShipoutPicture command will be added to an internal macro that is a part of a zero length picture environment with basepoint at the lower-left corner of the page. Calling \ClearShipoutPicture, we cancel the effect of \AddToShipoutPicture. The \AddToShipoutPicture* command functions just like the \AddToShipoutPicture command, but it is used to add material only to the current page.

Thus, suppose that we have an image in the file `picture.eps` and we want to set this as the background of an A4 document. Here, we assume that the image is upside-down, so we must rotate it by 180°. This is the code that achieves the result that we want:

```
\AddToShipoutPicture{%
  \setlength{\unitlength}{1mm}
  \put(0,0){\makebox(210,296)[t]{%
    \includegraphics[height=296mm,angle=180]{%
      picture.eps}}}}
```

Let us explain the code above. First, we set the unit length to 1 mm. Then, we create a box that has the height and the width of the logical page. In this box, we include the `picture.eps` file rotated and scaled so that its height will be the height of our page (this is achieved by the `height` parameter in the `\includegraphics` command). Notice that the height of the box matches that of the picture height (i.e., 296 mm) which is also the height of an A4 page. Finally, the whole graphic is put at coordinates $(0,0)$. All of the commands that makeup the argument of `\AddToShipoutPicture` will be executed every time TEX ships out a page. To stop the inclusion of the picture in the background for consecutive pages, we use `\ClearShipoutPicture`.

The eso-pic package uses a more primitive mechanism provided by the package everyshi by Martin Schröder. This package provides the command `\EveryShipout` analogous to the LATEX command `\AtBeginDocument`, whose argument is executed before every shipout. Another interesting use of this package is as a way to add text at the bottom of each page *below* the footer, as seen on this page where we added the logo "Typeset by LATEX 2ε." One may also use the `\AddToShipoutPicture` command for this.

This capability is exploited by the package prelim2e by Martin Schröder, which adds version control to a document. His package provides the commands `\PrelimText` and `\PrelimWords`. The user adds the version to a document by redefining one of these commands with the contents to be written as the document version. For example, one may say

```
\renewcommand{\PrelimWords}{Version 1.2, last revised \today}
```

If the commands are not redefined, they print something like "Preliminary version – June 27, 2002" centered at the bottom of each page.

9.7 The **rotating** Package

The rotating package by Sebastian Rahtz and Leonor Barroca provides the rotating environment `rotate`. Whatever is included between `\begin{rotate}{`*degrees*`}` and `\end{rotate}` will be rotated counter-clockwise by the angle of degrees (a numerical

—positive or negative— quantity). This environment does not attempt, though, to find the required space for what is rotated, so if no special care is taken, the rotated material may be printed on other surrounding material like this:

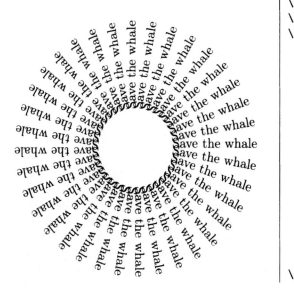

```
\begin{rotate}{-45}
   rotate
\end{rotate}
```

Compare with the rotations in Section 9.3.1. This property allows for more complex tricks like the ones in Figures 9.2 and 9.3 taken from the package documentation. In this example, we use the command \rlap. This command has one argument, and it typesets its argument and backs up as if it has not typeset anything. A similar command is the \llap command. This command creates a box of width zero with its only argument extending to the left of the box. For example, one way to get the symbol ≠ is by using commands \rlap{=}/ and /\llap{=}.

If instead you want to make room for the rotated text, then you may use the environment turn instead of the rotate environment.

In addition to the above, the package provides the sideways environment. This is very useful, for example, for turning tabular material that is wider than the page width.The environments sidewaystable and sidewaysfigure can be used instead of the standard table and figure environments, and they will rotate the table or the

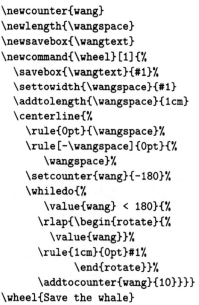

```
\newcounter{wang}
\newlength{\wangspace}
\newsavebox{\wangtext}
\newcommand{\wheel}[1]{%
   \savebox{\wangtext}{#1}%
   \settowidth{\wangspace}{#1}
   \addtolength{\wangspace}{1cm}
   \centerline{%
      \rule{0pt}{\wangspace}%
      \rule[-\wangspace]{0pt}{%
         \wangspace}%
   \setcounter{wang}{-180}%
   \whiledo{%
      \value{wang} < 180}{%
      \rlap{\begin{rotate}{%
         \value{wang}}%
      \rule{1cm}{0pt}#1%
         \end{rotate}}%
      \addtocounter{wang}{10}}}}
\wheel{Save the whale}
```

Figure 9.2: An example of the rotating package.

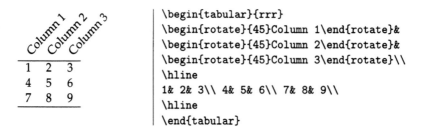

```
\begin{tabular}{rrr}
\begin{rotate}{45}Column 1\end{rotate}&
\begin{rotate}{45}Column 2\end{rotate}&
\begin{rotate}{45}Column 3\end{rotate}\\
\hline
1& 2& 3\\ 4& 5& 6\\ 7& 8& 9\\
\hline
\end{tabular}
```

Figure 9.3: A second example of the rotating package.

figure together *with* its caption. These always take the whole page. In the example of Figure 9.4, we have inserted the table as a figure to easily bypass the problem of sacrificing a whole page for a simple example.

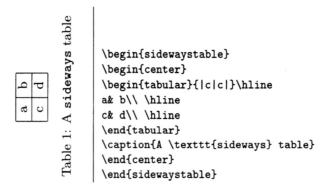

```
\begin{sidewaystable}
\begin{center}
\begin{tabular}{|c|c|}\hline
a& b\\ \hline
c& d\\ \hline
\end{tabular}
\caption{A \texttt{sideways} table}
\end{center}
\end{sidewaystable}
```

Figure 9.4: An example of the sidewaystable environment.

9.8 Mathematics Drawing

Good mathematics drawing is a difficult issue. On a Unix system, the standard tool is XFIG. This tool has the ability to save the file in PICTEX commands, and then the user can modify them accordingly. We discuss this possibility (using PICTEX) in the next section.

A better tool seems to be the program DIA (for DIAgram) available at http://www. lysator.liu.se/~alla/dia/. The native format of DIA is XML compressed with GZIP, but the strong point is that it has the ability to export PSTricks code. After exporting, it is easy for the user to adjust the parameters inside the PSTricks file in order to overcome the inaccuracy of the use of the mouse when drawing. With little work, we can easily get high-quality mathematical drawings. Moreover, it is possible to install DIA on Microsoft Windows. For this and other non-Unix platforms, there are also other drawing tools such as CorelDraw by Corel. We will not discuss these programs except for how to overcome the difficulties they impose.

Most of these programs can save in PostScript format, but the most common problem is that the labels used are not typeset with the same font as the main document font used by LATEX and, even worse, it is very common to have a label in LATEX math mode that these programs cannot typeset. Moreover, the label alignment is usually not as good as LATEX's positioning.

The solution is provided by the package psfrag by Craig Barratt, Michael C. Grant, and David Carlisle. This package provides an easy interface to edit the labels on a PostScript file. It provides the command \psfrag with syntax

> \psfrag{*label to be replaced*}{LATEX *code for the replacement*}

We shall give an example here. The top drawing in Figure 9.5 was generated by XFIG and saved as the EPS file `trigcircle.eps`. The lower drawing in Figure 9.5 is the resulting drawing after modification by the LATEX code given to the right. We have used the commands \raisebox and \hspace in order to achieve better positioning of the labels.

9.9 The PICTEX Package

PICTEX [30] is a collection of macros (by Michael J. Wichura) that have been designed to allow TEX/LATEX users to typeset pictures as a part of their manuscripts. Although PICTEX has been designed for use with plain TEX, LATEX users can still use PICTEX by loading the pictex package by Andreas Schrell. Since PICTEX is a little bit cumbersome to use, we will briefly present the system. Also, we will briefly present MathsPIC, a program that can be used to create PICtures (i.e., drawings made with PICTEX). One thing that the reader must have in mind is that spaces are necessary between keywords and special symbols such as braces and slashes.

Each PICture begins with the command \beginpicture and ends with the command \endpicture. The first thing one has to do is to set the coordinate system with the command

> \setcoordinatesystem units   <*x-units, y-units*>
>   point at *xcoord ycoord*

The units part is used to specify the length of one unit on the x and y axes, respectively. The point part (which is optional) is used to specify the *reference point* of the system. We can (re)set the coordinates system as often as we like. The coordinate system of the following figure

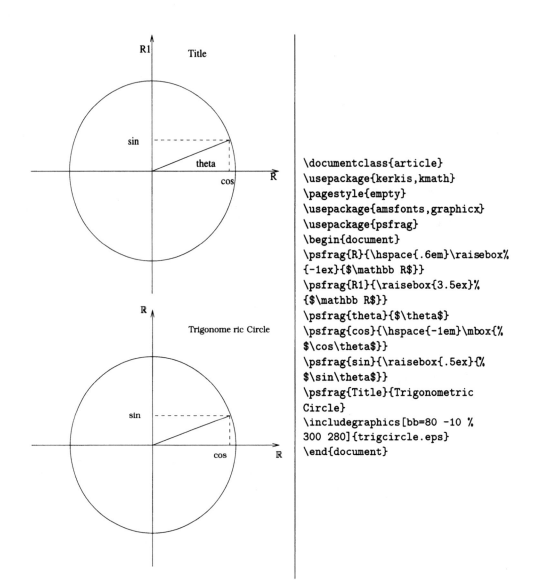

```
\documentclass{article}
\usepackage{kerkis,kmath}
\pagestyle{empty}
\usepackage{amsfonts,graphicx}
\usepackage{psfrag}
\begin{document}
\psfrag{R}{\hspace{.6em}\raisebox%
{-1ex}{$\mathbb R$}}
\psfrag{R1}{\raisebox{3.5ex}%
{$\mathbb R$}}
\psfrag{theta}{$\theta$}
\psfrag{cos}{\hspace{-1em}\mbox{%
$\cos\theta$}}
\psfrag{sin}{\raisebox{.5ex}{%
$\sin\theta$}}
\psfrag{Title}{Trigonometric
Circle}
\includegraphics[bb=80 -10 %
300 280]{trigcircle.eps}
\end{document}
```

Figure 9.5: The original drawing (top), the modified output (bottom), and the LaTeX code that modified the original drawing. Here, we use the experimental packages kerkis and kmath, which support the Kerkis typeface. However, one gets similar results when using either the standard typeface or any other typeface.

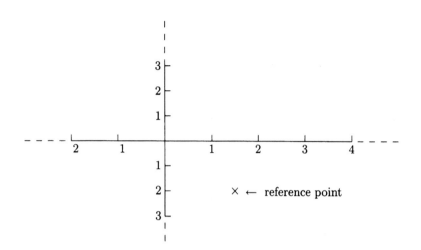

has been established with the command

$\qquad$ \setcoordinatesystem units <.5in,.25in> point at 1.5 -2

We can place a piece of text or another PiCture at a particular point with the command

$\qquad$ \put{$text$}[o_xo_y] at $xcoord$ $ycoord$ <$xshift,yshift$>

This command places the $text$ at ($xcoord$, $ycoord$). The optional [o_xo_y] part is used to place the $text$ inside the resulting box. The valid values are: l(eft), r(ight), t(op), b(ottom), and B(aseline). One can also omit the o_y part. The optional part inside the < > is used to shift the resulting box $xshift$ units right and $yshift$ units up from where it would otherwise go. The command

$\qquad$ \multiput{$text$}[o_xo_y] <$xshift,yshift$> at
... $xcoord$ $ycoord$ ... *n $dxcoord$ $dycoord$ .. /

is used to place the same text at several places in a PiCture. Between "at" and the terminating "/", each occurrence of $xcoord$ $ycoord$ gives the effect of

$\qquad$ \put{$text$}[o_xo_y] <$xshift,yshift$> at $xcoord$ $ycoord$

and each occurrence of *n $dxcoord$ $dycoord$ gives the effect of

$x = x + dxcoord$
$y = y + dycoord$
\put{$text$}[o_xo_y] <$xshift,yshift$> at x y

Here is a simple example:

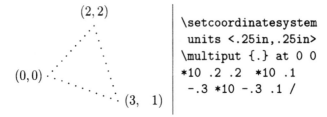

```
\setcoordinatesystem
 units <.25in,.25in>
\multiput {.} at 0 0
*10 .2 .2  *10 .1
 -.3 *10 -.3 .1 /
```

If we use a program that generates the coordinates of a plot, we can use the command

$$\text{\textbackslash multiput } \{text\} \text{ at } "file"$$

which puts the *text* at the coordinates that are specified in the *file*.

If we want to draw axes, we first have to specify the plot area with the command

$$\text{\textbackslash setplotarea x from } xcoord_1 \text{ to } ycoord_1, \text{ y from } xcoord_2 \text{ to } ycoord_2$$

The first pair of coordinates determine its lower-left corner and the second its top right corner. The next step involves the actual drawing of the axes. For this, we use the \axis command. Since the command has many options, we will gradually introduce most of them by giving some examples that demonstrate the features of the command. The drawing

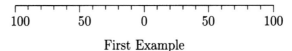

First Example

has been generated by the following code:

```
\setplotarea x from -100 to 100, y from 0 to 0
\axis bottom label {First Example} ticks
    numbered from -100 to 100 by 50
    unlabeled short quantity 21 /
```

Note that bottom specifies that the axis should be placed at the bottom of the plotting area. To draw tick marks, we must specify the ticks keyword. Ticks are usually unlabeled, but they can be numbered. Note that here we request PᴄTᴇX to number the ticks in the range 100 to 100 units by 50 units. Also, we request PᴄTᴇX to put 21 unlabeled equally spaced short ticks. The keyword quantity is followed by the precise number of ticks to be drawn. The \axis command is terminated by ␣/. Ticks they can be short, long, or length <length>: Here is another example

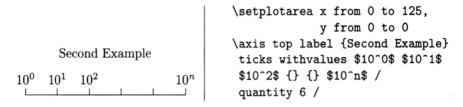

Second Example

10^0 10^1 10^2 10^n

```
\setplotarea x from 0 to 125,
             y from 0 to 0
\axis top label {Second Example}
    ticks withvalues $10^0$ $10^1$
    $10^2$ {} {} $10^n$ /
    quantity 6 /
```

The example above shows how to draw a top axis. As the reader might expect, we can also draw left and right axes. In this example, we ask P̶I̶CTEX to draw six ticks that will be numbered with the "values" specified after the keyword withvalues. Empty values are allowed and must be specified with {}. Note that the withvalues part must be terminated by ⎵/. Ticks can also be put across the plotting area. Let us see two more examples:

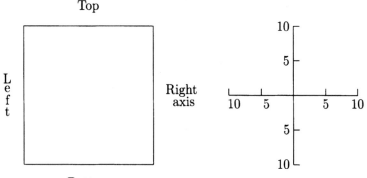

The commands that are necessary to draw the left drawing are:

```
\setplotarea x from 0 to 100, y from 0 to 100
\axis top label {Top} /
\axis bottom label {Bottom} /
\axis left label {\stack{L,e,f,t}} /
\axis right label {\lines{Right\cr axis\cr}} /
```

The \stack command is used to produce a short stack of comma-separated items, and the \lines command produces rows of lines. Since this is a plain TEX command, lines are separated by the \cr command. The right drawing is drawn with the following commands:

```
\setcoordinatesystem units <1pt, 1pt> point at -150 0
\setplotarea x from 0 to 100, y from 0 to 100
\axis bottom shiftedto y=50 ticks
   in withvalues $-10$ $-5$ {} 5 10 / quantity 5 /
\axis left shiftedto x=50 ticks
   in withvalues $-10$ $-5$ {} 5 10 / quantity 5 /
```

The keyword shiftedto is used to shift the axis horizontally or vertically, depending on whether the keyword is followed by x=*units* or y=*units*. Ticks can be placed in(side) or out(side) the plotting area. The last example shows the creation of a logarithmic axis:

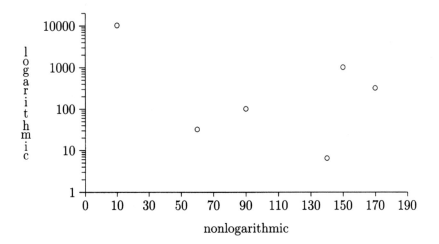

and here is the code that makes this drawing:

```
\setcoordinatesystem units <2.5pt,30pt>
\setplotarea x from 0 to 100, y from 0 to 4.3
\axis left label {\stack{l,o,g,a,r,i,t,h,m,i,c}}
ticks logged numbered at 1 10 100 1000 10000 /
unlabeled short from 2 to 9 by 1
                   from 20 to 90 by 10
                   from 200 to 900 by 100
                   from 2000 to 9000 by 1000
at 20000 / /
\axis bottom label {nonlogarithmic}
ticks out withvalues 0 10 30 50 70 90 110
130 150 170 190  / short unlabeled
quantity 11 /
\put {$\circ$} at 10 4 \put {$\circ$} at 50 2
\put {$\circ$} at 80 3 \put {$\circ$} at 90 2.5
\put {$\circ$} at 75 .8 \put {$\circ$} at 35 1.5
```

The "magic" trick here is accomplished with the keyword logged.

▶ **Exercise 9.4** Draw the picture of Section 9.1.8 using PiCTEX. ☐

One can construct ruled lines with the command \putrule from *xinit yinit* to *xfinal yfinal*. Similarly, we can construct rectangles with the command \putrectangle corners at *xinit yinit* and *xfinal yfinal*. The command \frame [*separation*] {*text*} puts the *text* in a frame separated from the *text* by *separation*.

Another interesting feature of PiCTEX is its ability to create histograms and bar graphs. The commands

```
\sethistograms
\plot xcoord_0 ycoord_0 xcoord_1 ycoord_1
... xcoord_n ycoord_n /
```

produce a histogram composed of rectangles having opposite corners at the points

$(xcoord_0, ycoord_0) - (xcoord_1, ycoord_1)$
$(xcoord_1, ycoord_0) - (xcoord_2, ycoord_2)$
$(xcoord_2, ycoord_0) - (xcoord_3, ycoord_3)$

$\vdots$

Here is the PγCT$_{\! E}$X equivalent of the Figure on page 264:

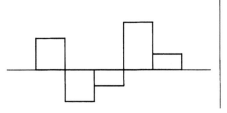

```
\setcoordinatesystem
    units <25pt,25pt>
\putrule from -1 0 to 6 0
\setlength{\linethickness}{.8pt}
\sethistograms
\plot 0 0 1 1 2 -1 3 -.5
      4 1.5 5 .5 /
```

The length variable \linethickness determines the thickness of the lines used. Bar graphs are drawings such as the following one:

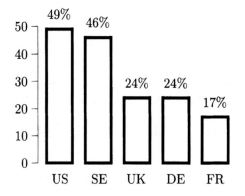

The code that draws the figure above is:

```
\setcoordinatesystem units <2pt,2pt>
\setbars breadth <20pt> baseline at y = 0
        baselabels ([Br] <7pt,-15pt>)
\setlength{\linethickness}{2pt}
\plot 0 49 "US" 15 46 "SE"
30 24 "UK" 45 24 "DE" 60 17 "FR" /
\setbars breadth <20pt> baseline at y = 0
```

```
            endlabels ([tr] <9pt,15pt>)
    \plot 0 49 "49\%" 15 46 "46\%"
    30 24 "24\%" 45 24 "24\%" 60 17 "17\%" /
    \setlength{\linethickness}{.25pt}
    \setplotarea x from -10 to 10, y from 0 to 50
    \axis left ticks numbered from 0 to 50 by 10 /
```

The commands

```
\setbars breadth <β> baseline at z = zcoord
\plot xcoord₁ ycoord₁ xcoord₂ ycoord₂ ... /
```

have the effect of

```
\putrule breadth <β> from xcoord₁ zcoord to xcoord₁ ycoord₁
\putrule breadth <β> from xcoord₂ zcoord to xcoord₂ ycoord₂
```

⋮

when z is the letter y and the effect of

```
\putrule breadth <β> from zcoord ycoord₁ to xcoord₁ ycoord₁
\putrule breadth <β> from zcoord ycoord₁ to xcoord₂ ycoord₂
```

⋮

when z is the letter x. The command

```
\putrule breadth <β> from xcoord_s ycoord_s to xcoord_e ycoord_e
```

draws a rectangle having $(xcoord_s, xcoord_s)$ and $(xcoord_e, xcoord_e)$ as the midpoints of opposite sides of length β. The labels can be attached to the base of the bars by continuing the \setbars with either baselabels or endlabels, respectively. The example above makes full use of the capabilities of this command. The lengths that are surrounded by < and > are used to shift the bars.

To draw a figure or part of a figure that is composed of straight lines, we use the commands

```
\setlinear
\plot xcoord₁ ycoord₁ xcoord₂ ycoord₂ ... /
```

The \plot command connects the points $(xcoord_i, xcoord_i)$ and $(xcoord_{i+1}, xcoord_{i+1})$ with straight lines. Similarly, the commands

```
\setquadratic
\plot xcoord₁ ycoord₁ xcoord₂ ycoord₂ ... /
```

draw quadratic arcs through the points that are specified in the \plot command. Note that the number of points must be odd. The example that follows shows the use of both line-drawing commands:

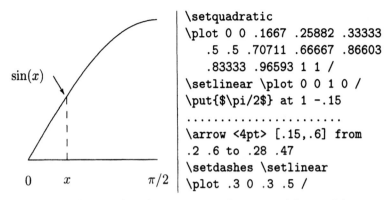

```
\setquadratic
\plot 0 0 .1667 .25882 .33333
      .5 .5 .70711 .66667 .86603
      .83333 .96593 1 1 /
\setlinear \plot 0 0 1 0 /
\put{$\pi/2$} at 1 -.15
. . . . . . . . . . . . . . . . . . . .
\arrow <4pt> [.15,.6] from
.2 .6 to .28 .47
\setdashes \setlinear
\plot .3 0 .3 .5 /
```

The \arrow command is used to draw arrows. The general form of the command is

```
\arrow <ℓ> [β,γ] <xshift,yshift>
    from xcoord_s ycoord_s to xcoord_e ycoord_e
```

The <xshift,yshift> is optional, and the command above draws an arrow of the form

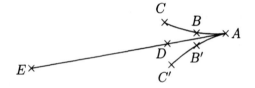

where $E = (xcoord_s, ycoord_s)$, $A = (xcoord_e, ycoord_e)$, ℓ is the distance between A and D, $\beta\ell$ is the distance between B and B', and $\gamma\ell$ is the distance between C and C'. The arrow above was set with

```
\setcoordinatesystem units <1pt,1pt>
\setplotarea x from -160 to 10, y from -30 to 10
\arrow <45pt> [.2,.67] from -150 -25 to 0 0
```

The command \setdashes <ℓ> specifies an interrupted line pattern composed of dashes of length ℓ separated by blank space of length ℓ. The <ℓ> is optional; if it is omitted, PᵢCTEX assumes that the length of the blank space is 5 pt. For dotted lines, we can use the command \setdots, which has the same optional argument as the \setdashes command. If we want a more general pattern, we use the command

$$\setdashpattern <d_1, g_1, d_2, \; g_2, \ldots >$$

which specifies an interrupted line pattern of a dash of length d_1, followed by a gap of length g_1, followed by a dash of length d_2, and so on.

➤ **Exercise 9.5** Express \setdashes in terms of \setdashpattern. □

To revert to nondashed mode, just use the command \setsolid.

Another useful facility that PᵢCTEX offers is the ability to change the plot symbol with the command

\setplotsymbol ({*plot symbol*}[0_x0_y] <*xshift,yshift*>)

The parameters surrounded by square brackets and by < > are optional and have the expected meaning.

If we want to place some text between arrows (i.e., to label something), we can use the \betweenarrows command

\betweenarrows {*text*}[0_x0_y] <*xshift,yshift*>
    from *xcoord$_s$ ycoord$_s$ xcoord$_e$ ycoord$_e$*

As above, the parameters surrounded by square brackets and by < > are optional. The following is a simple example:

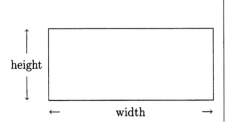

```
\setcoordinatesystem
units <3cm,3cm>
\putrectangle corners
at 0 0 and 1.5 .6 \small
\betweenarrows {width} [t]
<0pt,-5pt> from  0 0 to 1.5 0
\betweenarrows {height} [r] <
-5pt,0pt> from 0 0 to 0 .6
```

PₗCTₑX provides commands that can be used to draw arcs of circles or ellipses. The command

\circulararc θ degrees from *xcoord$_s$ ycoord$_s$*
    center at *xcoord$_c$ ycoord$_c$*

draws an arc of a circle centered at point ($xcoord_c, ycoord_c$); the arc starts from ($xcoord_s, ycoord_s$) and goes counterclockwise by θ degrees. Here is a simple example:

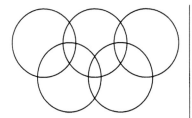

```
\setcoordinatesystem
        units <5pt,5pt>
\multiput {
   \circulararc 360
           degrees from 5 0
        center at 0 0 }
at 0 0 8 0 16 0 4 -5 12 -5 /
```

Similarly, the command

\ellipticalarc axes ratio $\xi : \eta$ θ degrees from *xcoord$_s$ ycoord$_s$*
    center at *xcoord$_c$ ycoord$_c$*

draws an arc of an ellipse whose minor and major axes are parallel to the x and y axes. The numbers ξ and η are proportional to the lengths of the horizontal and vertical axes of the ellipse.

Of course P¡CTEX provides some more facilities, but we will stop our description of P¡CTEX here. The interested reader should consult the user manual. We must also stress that P¡CTEX draws by putting tiny dots next to one another. This means that P¡CTEX is quite memory-demanding, but modern TEX installations usually have no problem with most P¡Ctures.

MathsP¡C is a program that accepts a simple programming notation that is used to specify mathematical drawings. The program yields P¡CTEX code, which then can be processed by LATEX. The program was originally developed by Richard W.D. Nickalls as an MS-DOS utility. The program has been rewritten in Perl by Apostolos Syropoulos in collaboration with the original author. The following is a simple example that shows the code that is necessary to draw the figure on the left:

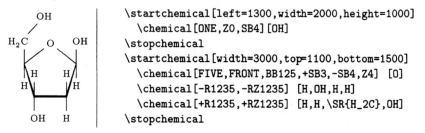

```
paper{units(5mm),xrange(0,10),yrange(0,10)}
point(a){0,0}
point(b){a,polar(5,10deg)}
point(c){a,polar(3,50deg)}
drawpoint(abc)
drawline(abca)
var d=0.5
text($A$){a, shift(-d,0)}
text($B$){b, shift(d,0)}
text($C$){c, shift(-d,0)}
```

9.9.1 The PPCH$_{TEX}$ Package

PPCH$_{TEX}$ is a module originally developed by Hans Hagen for use with CONTEXT[2] to typeset chemical formulas and structures. The module can be used as a normal LATEX package by putting the following code in the preamble of a LATEX file:

```
\usepackage{m-pictex,m-ch-en}
```

The module is actually a P¡CTEX application. The following example shows how to typeset the deoxyribose molecule with PPCH$_{TEX}$ and is due to Ton Otten:

```
\startchemical[left=1300,width=2000,height=1000]
  \chemical[ONE,ZO,SB4][OH]
\stopchemical
\startchemical[width=3000,top=1100,bottom=1500]
  \chemical[FIVE,FRONT,BB125,+SB3,-SB4,Z4] [O]
  \chemical[-R1235,-RZ1235] [H,OH,H,H]
  \chemical[+R1235,+RZ1235] [H,H,\SR{H_2C},OH]
\stopchemical
```

The reader is invited to consult the package's manual for details on the use of the package.

---

2. CONTEXT is another TEX format that has been designed by Hans Hagen. This format is gaining steadily wider acceptance as an alternative to LATEX.

9.9.2 The PSTricks Packages

PSTricks by Timothy Van Zandt is a set of packages that provides a TEX-like interface to access PostScript commands. The packages are now maintained by Denis Girou and extended by him and several other people. The capabilities of these packages are quite extensive, and it would not make sense to cover these in detail in this book. Documentation for these packages is available with standard installations and, of course, in CTAN.

The main package is the pstricks package and should always be loaded if any of the other packages in the suite is loaded.

Figures 9.6–9.12 present a small gallery of graphics that can be done with the PSTricks packages. It is by no means exhaustive, as the packages provide a huge amount of possibilities. However, at least the reader may get a feeling of what PSTricks can do.

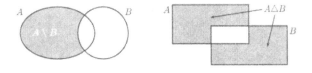

Figure 9.6: An example of pstricks functionality.

Figure 9.7: pstricks, pst-blur shadows (by Martin Giese) , and pst-char.

Figure 9.8: pstricks, pst-3d, and pst-lens by Denis Girou and Manuel Luque.

Although PSTricks are quite useful, one cannot use them with pdfLATEX. For this reason the pdftricks package has been written by C.V. Radhakrishnan and C.V. Rajagopal.

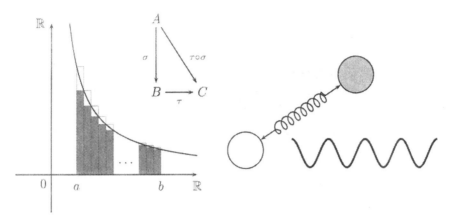

Figure 9.9: pstricks with pst-plot for the graph, with pst-node for the commutative diagram, and with pst-node and pst-coil for the rest.

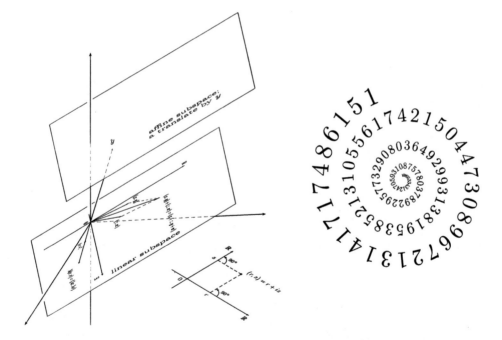

Figure 9.10: pstricks with pst-3d for the graphs and with pst-plot and pst-text for the number spiral (Courtesy of Denis Girou).

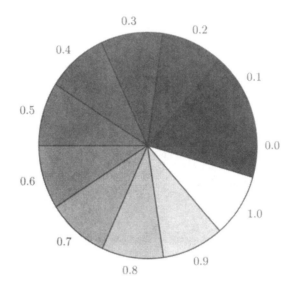

Figure 9.11: pstricks and multido (from the pstricks documentation).

Figure 9.12: pstricks, pst-text, pst-grad, pst-char, and pst-eps (example by Stephan Lehmke, here scaled to fit).

The package uses a special feature found in most TeX implementations (in particular, those that are based on the web2c implementation) that permits the LaTeX processor to escape to a shell, execute some commands and after this to resume processing of the input file and finish its job. Since escape to the shell is used, the package works best on Unix systems. This feature is disabled by default since it dangerous for the security of the system. Although it can be enabled by default it is suggested to pass it to pdfLaTeX as a command-line option like this:

```
$pdflatex -shell-escape filename.tex
```

Since, we have not explained how one can create graphics with the PStricks, we will not go into the details of how to generate PDF files from LaTeX sources that contains PSTricks code. The interested reader, should consult the package's documentation instead.

9.10 Graphs with METAPOST

As we have already stated METAPOST is a reimplementation of METAFONT that produces EPS files. Certainly, it is beyond the scope of this book to give details of this program. The reader interested in METAPOST, in general, should consult [13]. However, there is a METAPOST package that can be easily used to draw graphs like those that we did with the picture environment and PiCTeX, so we will describe this feature of METAPOST.

First of all, METAPOST assumes that the data we are going to use to plot our graph are stored in some external file. This is very convenient, as most of the time we do have data stored externally. Let us start with a simple example. Suppose that the data for the example of Section 9.1.8 are stored in the file temp.data. Next, we create a text file, say scatterplot.mp, with the following METAPOST code:

```
input graph; % percent is used for
beginfig(1); % comments
draw begingraph(250pt,270pt);
gdraw "temp.data" plot btex $+$ etex;
endgraph;
endfig;   %semicolons terminate
end.      %commands
```

Then, we feed this file to METAPOST to get the EPS file:

```
$ mpost scatterplot.mp
This is MetaPost, Version 0.641 (Web2C 7.3.3.1)
(scatterplot.mp
(/usr/local/teTeX/share/texmf/metapost/base/graph.mp
(/usr/local/teTeX/share/texmf/metapost/base/marith.mp
(/usr/local/teTeX/share/texmf/metapost/base/string.mp))
```

```
(/usr/local/teTeX/share/texmf/metapost/base/format.mp
(/usr/local/teTeX/share/texmf/metapost/base/string.mp)
(/usr/local/teTeX/share/texmf/metapost/base/texnum.mp))) [1] )
1 output file written: scatterplot.1
Transcript written on scatterplot.log.
```

The resulting file can be included in any ordinary LaTeX document with the commands that we described in previous sections. Note that we cannot directly use the resulting file with pdfLaTeX, but we can transform it to PDF by feeding it to the mptopdf format (by Hans Hagen) and then include the resulting PDF file in our document. Another problem is that METAPOST, by default, uses the Computer Modern typefaces. Certainly, it is possible to use any font we like, but the relevant details fall outside the scope of this book. The interested reader should consult the METAPOST documentation. Now, it is time to explain the commands of the METAPOST file.

The command input graph must always be present, as it inputs the METAPOST package that allows us to create graphs. The command beginfig is used to start a figure, and a METAPOST file can contain the code of many figures. Each resulting EPS file has the same name as the input file and a file extension corresponding to the number in parentheses that follows this command. The command draw begingraph is used to define the plotting area, and the lengths in parentheses define the width and the height of the resulting picture. The command gdraw is the command that actually draws our graph. Here, we draw a graph with data stored in the file temp.data. Moreover, we plot the points using the plus sign of the default font. The keywords btex and etex are used to delimit LaTeX code. We will discuss this ability later when we introduce labels to our graph. If we omit the plot part, then METAPOST will connect the points without drawing the individual point symbols.

To add labels to graphs is a little bit tricky, as we will see in a moment. The command

$$\text{glabel.} \textit{label suffix} (\textit{string, location})$$

is used to place labels by the axes. The available *label suffix* includes: top, bot, lft and rt. As for the *location*, it can just be a pair of numbers, but we recommend the use of the keyword OUT. This *location* places a label relative to the whole graph. If we replace gdraw "temp.data" plot btex + etex; with

```
glabel.lft(btex Temp $\mbox{}^\circ\mathrm{F}$ etex, OUT);
glabel.bot(btex Time in hours since midnight etex, OUT);
gdraw "temp.data" plot btex $+$ etex ;
```

then we could, in theory, get a graph with labels. However, METAPOST uses plain TeX to format the *string*, and since plain TeX does not have the \mbox command, the processing will fail. The solution is to add the following lines at the beginning of the file:

```
verbatimtex \documentclass[a4paper]{article}
\begin{document} etex
```

and then set the new system variable TEX to latex. On a Unix system with the bash shell, this can be done with the command

export TEX=latex

and on an MS-Windows system with the command

set TEX=latex

Even if you modify the file above as suggested, the output still will not be satisfactory. The reason is that the coordinate system is not completely correct. To change the coordinate system, METAPOST provides the command

setrange(coordinates, coordinates)

where <i>coordinates</i> is a pair consisting of strings, numbers, or the special keyword whatever. The first <i>coordinates</i> give $(x\text{min}, y\text{min})$ and the second $(x\text{max}, y\text{max})$. If we use the keyword origin for the first <i>coordinates</i>, METAPOST will set $x\text{min} = 0$ and $y\text{min} = 0$. Moreover, if we are not concerned about the value of any of the components of the <i>coordinates</i>, we can use the keyword whatever. If we want to specify a big number as a component of the <i>coordinates</i>, we must type the number in the so-called scientific notation surrounded by quotation marks (i.e., "1e6", but not "1E6"!). This trick is necessary, as METAPOST cannot handle a number greater than 32768. Now, if we add the command setrange(0,65,25,92) just after the draw command, we will get the following graph:

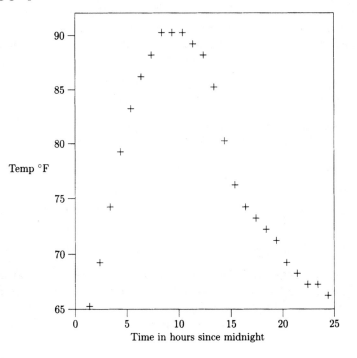

Should we wish to plot a logarithmic or a semilogarithmic graph, we can use the setcoords commands. This command has two arguments, which can have the values: log (for a logarithmic axis) or linear (for a linear axis). If we place a minus sign in front of these keywords, METAPOST makes the x (or y) values run backwards, so the largest value is on the left (or bottom) side.

➤ **Exercise 9.6** Draw the semilogarithmic graph on page 280 using METAPOST. ☐

If we put one or more blank lines between the data in the data file, then META-POST assumes that we have two or more datasets that can be used to plot at least two plots in the same graph. Naturally, the reader may wonder how it is possible to distinguish individual plots in a graph with multiple line plots. There are three ways: a) to draw plots in different colors, b) to draw dashed lines, and c) to draw lines of different widths. In all cases, we simply add some commands just after the file name of the gdraw command. To produce colored lines, we add the keyword withcolor and a color specification. This, in turn, can be either an RGB color or a grayscale color (see the next section). The grayscale shade is specified by writing a decimal number from 0 to 1 in front of the keyword white. The following examples show how to get grayscale and color curves, respectively:

```
gdraw "temp.data" withcolor 0.32white; %gray scale
gdraw "temp.data" withcolor (0.7,0.3.,1); %rgb
```

The simplest way to get a dashed line is to write the keyword dashed followed either by evenly (for a dashed line) or by witdots (for a dotted line). Additionally, we can add a scale factor just after the keywords dashed and withdots. The scale factor starts with the keyword scaled and is followed by a number (the real scale factor). Here are two examples:

```
gdraw "temp.data" dashed scaled 1.5; %dashed line
gdraw "temp.data" widthdots scaled 2; %dotted line
```

If we want a thick line, we just use withpen pencircle scaled *scale-factor*, where *scale-factor* is a length denoting the width of the line. For example,

```
gdraw "temp.data" withpen pencircle scaled 1.5pt
```

draws the plot with a line of width 1.5 pt.

If we want to plot two functions on the same graph, we can use the autogrid command

```
autogrid(label, label)option list
```

Here *label* is grid (for grids), iticks (for ticks inside the graph), and oticks (for ticks outside the graph). The *label* can have a suffix denoting the axis on which the *label* should appear. These suffixes are: .lft, .rt, .top, and .bot. Note that the period is part of the suffix. The *option list* is a command that colors the ticks and is optional.

We note that one should use this command with great care, as our practical experience has shown.

It is even possible to change the appearance of the frame that surrounds the graph by using the command frame:

frame. label option-list

If the <code>label</code> is not specified, then it applies to the whole frame. Other possible values include <code>llft</code> (for the bottom and left sides), <code>lrt</code> (for the left and right sides), <code>ulft</code> (for the left and top sides), and <code>urt</code> (for the top and right sides).

As a final word, we point out that the graph package has some other options, which, however, are useful only if one has a rather deep knowledge of METAPOST. For information about METAPOST, the interested reader should have a look at http://cm.bell-labs.com/who/hobby/MetaPost.html.

9.11 Color Information

There are two main issues concerning the use of color. The first is how to add color to our document, and the second is what one should know for industry-quality color printing.

9.11.1 Color in our Documents

Imagine that you are standing in front of a very beautiful and colorful landscape. You decide to take a picture of it. Later on, you give your photographic film for processing, and when you get the printed photographs you realize that the colors are not those you expected. We are sure that this is not a science fiction scenario but rather a very frequent situation, and most people blame the processing shop for not doing a good job. Fortunately, for the people who run this kind of business, color is a physiological sensation and as such cannot be directly measured or described. So, we cannot blame them, even if we think they are not doing good work! A color model is a mechanism by which we can describe the color formation process in a predictable way. There are two categories of color models: those that are related to device color representation and those related to human visual perception. The first category is directly supported by LATEX, but the second is not supported at all. Grayscale, RGB, HSB, and CMYK are color models of the first category. LATEX supports all of these color models except HSB.

The grayscale color model is used to specify shades of gray. In this color model, black is denoted by 0.0 and white by 1.0, so shades of gray are just numbers between 0 and 1. RGB is the Red-Green-Blue color model. Other colors are derived from combinations of the three *primary* colors and are specified as triplets of numbers from 0.0 to 1.0. For example, purple is defined to be the triplet (0.7, 0.3, 1.0). Obviously, 0.7 is the "amount" of red, 0.3 the "amount" of green, and 1.0 the "amount" of blue (actually, it is blue phosphorus). RGB is an *additive* color model and is used when light is generated. For example, this color model is used in computer monitors and color televisions. HSB is the Hue-Saturation-Brightness color model. This is actually not a color model but rather an alternative convention for specifying colors in the RGB color model, and that is probably the reason why LaTeX does not support this color model. CMYK is the Cyan-Magenta-Yellow-blacK color model and has four primary colors. As in the case of the RGB color model, other colors are derived from combinations of the four primitive colors. For example, purple is defined to be the quadruple $(0.45, 0.86, 0.0, 0.0)$. This is a *subtractive* color model and is used in applications where light is reflected, such as printing.

The terms additive and subtractive refer to the way colors are formed. In the RGB model, each component of a color specification defines the *intensity* of the particular primary color — the larger the number, the higher the intensity. In the CMYK color model, each component of a color specification represents the degree of light absorption — the larger the number, the higher the absorbability.

Color in LaTeX documents can be added in several ways, the most standard one being the color package by David Carlisle. This package supports the color models grayscale, RGB, and CMYK. Additionally, it supports the named color model, which is used to access predefined colors. The set of predefined colors depends on the option one chooses to use. Each option is associated with an external driver program that will be used to transform a DVI file to some format that directly supports color. The most common options are dvips (which is the default option) and pdftex (for use with pdfLaTeX).

Once we have selected the option that is suitable for our purposes, we can define new colors with the \definecolor command:

```
\definecolor{color name}{color model}{color components}
```

The *color name* is the name that we will use in our document to call the color we just defined, *color model* is the model we want to use (it can be either rgb, cmyk, or gray). Finally, *color components* are triplets, quadruplets, or just a single number, depending on the color model used. Here are some typical definitions:

```
\definecolor{RGBpurple}{rgb}{0.7,0.3,1.0}
\definecolor{CMYKpurple}{cmyk}{0.45,0.86,0.0,0.0}
\definecolor{MyWhite}{gray}{1}
```

A typical question that most newcomers ask is: "Where can I get the *color components* of a particular color I want to use in my document?" Since the CMYK color model is used in the printing industry, one must consult the color tables that each company publishes. However, a good source of information for both the CMYK and the RGB color models is the Internet. Just point your favorite Web browser to your favorite search engine and type the necessary search keywords. In addition, for the RGB color model, one can also consult the HTML color tables.

After we have defined the colors that we want to use, we can see how to actually use them. To change the text color, there are two possibilities. The global one is \color{colorname}, and the local one is \textcolor{colorname}{text}. The second command is essentially the same as {\color{*colorname*}*text*}. One can avoid the predefinition of colors and define them "on the fly":

\color[*color model*]{*color components*}   or
\textcolor[*color model*]{*color components*}{*text*}

One can also set the page color. The relative commands are \pagecolor{*colorname*} or if the color is not defined \pagecolor[*color model*]{*color components*}.

Two more commands are available with the color package. These have to do with local coloring of a box. The package provides the commands \colorbox for colored boxes and \fcolorbox for colored framed boxes. Here is an example:

| colored box | \colorbox[gray]{.95}{colored box} |
| colored box | \fcolorbox[gray]{.95}{.8}{colored box} |

Both commands put a color background in the box, while the latter also colors the frame.

➤ **Exercise 9.7** Find a way to demonstrate the difference between an additive and a subtractive color model.   ☐

There are other possibilities for adding color to documents and in much more sophisticated ways. One such example is the color support for the PStricks packages. The main package pstricks already provides more possibilities than the color package, and additional functionality is available. An example is the pst-slpe package by Martin Giese, which adds support for advanced gradients. We show an example from the documentation of the package in Figure 9.13.

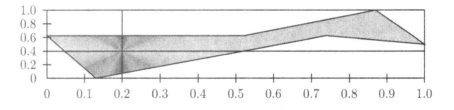

Figure 9.13: pstricks, pst-plot, and pst-slpe.

9.11.2 Coloring Tables

The coloring of tables can be done using the colortbl package by David Carlisle. The package provides commands for coloring rows, columns, cells, and table boundary lines. Figure 9.14 shows some of the capabilities of the package.

Ανθεστηριών						
Σε	Αρ	Ερ	Δι	Αφ	Κρ	Ηλ
			1	2	3	4
5	6	7	8	9	10	11
12	13	14	15	16	17	18
19	20	21	22	23	24	25
26	27	28				

11 13 Ανθεστήρια: Τιμή Διονύσου και Ψυχοπομπού Ερμή
23 Διάσια: Μέγιστη γιορτή Διός και υποδοχή της Άνοιξης

Figure 9.14: An application of the colortbl package: the calendar of the second Athenian month for the year 2000.

Although Figure 9.14 was prepared with the colortbl package, let us note here that Denis Girou is preparing the package pst-cal, which, when released, will considerably ease the creation of calendars. The package should appear as part of the PStricks suite.

Now back to the colortbl package. To color a column, the package provides the \columncolor command with the syntax

```
\columncolor[color model]{color}[left overhang][right overhang]
```

where color model is any of the color models that LaTeX understands, and color is either a named color or the components of a color specification. If we do not specify the color model, LaTeX assumes that we are using a user-defined color. The last two arguments control how close the coloring will get to the boundary of the table columns. The command goes in the definition of the table like this:

```
\begin{tabular}{|>{\columncolor[gray]{0.8}
[1.5pt][5pt]}l|>{\color{white}
\columncolor[gray]{.4}[0pt]}l|}
one & two \\
three & four
\end{tabular}
```

The example above also demonstrates the use of the left/right-overhang parameters. If they are omitted, then the full column width is filled with color. Moreover, we can specify a percentage of the overhang space left around the text of the column, which is the parameter \tabcolsep, like this:

```
\begin{tabular}{|>{\columncolor[gray]{0.8}}l|%
>{\color{white}\columncolor[gray]{.4}[0.5\tabcolsep]}l|}
one & two \\
three & four
\end{tabular}
```

The command \rowcolor is similar to \columncolor and is used to color rows. The \rowcolor command should be positioned at the beginning of a row, and if it crosses a colored column, the row color will overwrite the column color. When overhang arguments are not used, then the overhang information is read from the \columncolor commands, if there are any. Here is an example:

```
\begin{tabular}{|>{\columncolor[gray]{0.2}%
[0.5\tabcolsep]}l|l|}\hline
\rowcolor[gray]{.8} one & two \\ \hline
three & four\\ \hline
\end{tabular}
```

Cell coloring is more tricky and requires the use of the \multicolumn command. We put the cell into a multicolumn and then color it like this:

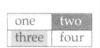

```
\begin{tabular}{|l|l|}\hline
one & \multicolumn{1}{>{\color{white}\columncolor%
[gray]{0.4}}l|}{two} \\ \hline
\multicolumn{1}{|>{\columncolor[gray]{0.8}%
[.5\tabcolsep]}l|}{three} & four\\ \hline
\end{tabular}
```

Let us turn to coloring the ruled lines of a table. For vertical ruled lines, the solution is simple. In the table definitions, instead of using the | character to denote a vertical rule, you can use

$$\{!\color[gray]{.8}\vline\}$$

Coloring horizontal lines is more tricky. The package provides the command \array-rulecolor, which can be given before a table or even inside it. In any case, the command colors the lines that follow it and, if given inside a table, it does not change the color previously specified for the vertical lines:

one   two
three   four

```
\begin{tabular}{|l|l|}%
\arrayrulecolor[gray]{.8} \hline
one & two \\ \hline
three & four\\ \hline
\end{tabular}
```

When using double lines for table separators (using || in the tabular table specification or \hline\hline after a row), one may want to color the white space between them. This is done with the \doublerulesepcolor command, which is used like the \array-rulecolor command:

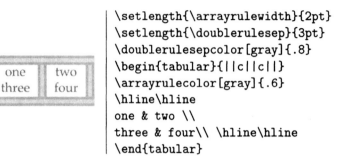

```
\setlength{\arrayrulewidth}{2pt}
\setlength{\doublerulesep}{3pt}
\doublerulesepcolor[gray]{.8}
\begin{tabular}{||c||c||}
\arrayrulecolor[gray]{.6}
\hline\hline
one & two \\
three & four\\ \hline\hline
\end{tabular}
```

We give one more example, where color is used for emphasis. We put the whole table in a colored box using the \colorbox command, and then we emphasize a row using white.

Item	Quantity	Price per Unit	Partial Total
Book A	1	24.99	24.99
Book B	3	29.99	89.97
Book C	2	44.99	89.98
Total			204.94
Tax 8%			16.40
		Grand Total	221.34

```
\colorbox[gray]{0.8}{%
\begin{tabular}{lcrr}
Item & Quantity & Price per Unit
& Partial Total\\[2ex]
Book A & 1 & 24.99 & 24.99\\
Book B & 3 & 29.99 & 89.97\\
Book C & 2 & 44.99 & 89.98\\
\rowcolor{white} Total & \ & \ & 204.94\\
Tax 8\% & \ & \ & 16.40\\
\ &\ &\textbf{Grand Total} & 221.34\\
\end{tabular}}
```

When working with colored tables, it is very convenient to define your own column types, incorporating the color commands in their definition. This saves a lot of typing. For example, for coloring the columns of a table, we may define new column types by

```
\newcolumntype{A}{>{\columncolor[gray]{0.8}[.5\tabcolsep]}c}
```

and then use \begin{tabular}{AAA}. You can do something similar for cells:

```
\newcommand{\cellcol}[2]{\multicolumn{1}{>{\columncolor{#1}}#2}}
```

Provided that you have defined the colors you want to use with \definecolor (see Section 9.11.1), you can say

```
\cellcol{color name}{table alignment}{text of the cell}
```

for every cell you want to color (where the *table alignment* is a character of r, l, c, or a column type that you may have defined previously as above.

9.11.3 Color and the Printing Industry

All of the material of the previous section should be enough for printing colored documents on desktop color printers. However, there is an additional step needed to prepare colored documents for a professional printer. The professional printer will actually perform color separation. This means that the printer will print a color plate four times. Each separation layer will print the corresponding color component of the picture elements. The overprinting of the color components on the page will create the final colors of the color plate. Thus, it is clear that we must somehow perform the necessary step of color separation. The easiest way to do this is by using the aurora package by Graham Freeman. Actually, this is not a LATEX package but rather it consists of PostScript "header" files that should be used to generate four different PostScript files, one for each color. Let us now describe the procedure. First, we run our document through LATEX. Next, we create the four PostScript files with the commands:

```
dvips file -h aurora.pro -h cyan.pro -o file-cyan.ps
dvips file -h aurora.pro -h magenta.pro -o file-magenta.ps
dvips file -h aurora.pro -h yellow.pro -o file-yellow.ps
dvips file -h aurora.pro -h black.pro -o file-black.ps
```

Finally, we send these files to a PostScript printer, usually in the order in which they are generated. Of course, changing the order will not affect the final output as long as we know which layer corresponds to which color.

Let us note here that the aurora package uses PostScript Level 1 commands and not the full possibilities supported by the PostScript Level 2 colorimage command. It should be clear though that the quality of the separation is the work of the PostScript driver (like DVIPS) and not of LATEX. Privately developed drivers have been shown to be capable of full-scale color work with LATEX (see [22]).

9.12 Printing in Landscape Mode

As we have already seen, it is possible to typeset a whole document in landscape mode. However, there are cases where we simply want to typeset parts of a document in landscape mode. For example, if we have a very long table, it makes sense to typeset it in landscape mode in a document that is otherwise typeset in portrait mode. The lscape package (by David Carlisle) provides the landscape environment; its body is typeset in landscape mode. The environment may span several pages.

➤ **Exercise 9.8** Suppose that you are using a document class that does not provide the landscape option. Create a simple package that implements this option. (Hint: Use the internal LaTeX length variable \@tempdima.) □

10

MULTILINGUAL TYPESETTING

The electronic typesetting of a document written in a language other than English is a problem that has been tackled seriously by the TEX community. A complete solution to this problem involves the solution of two subproblems: the preparation of the LATEX file and the typesetting of this file according to the typographic idiosyncrasies of the (main) language of the document. There are at least three different approaches to this really complicated problem:

- The use of standard LATEX packages (i.e., packages that suppose the use of TEX as the underlying typesetting engine and allow multilingual text processing in a portable way). The main drawback of this approach is that in many cases the user has to type commands that seem unnatural.
- The use of customized typsetting engines, such as pTEX, that are based on TEX and take care of most peculiarities of a particular language. The main drawback of this approach is that it is not adequate for documents written in languages other than the native language of the customized typesetting engine.
- Standard TEX extensions, such as ε-TEX and Ω, which have been designed to allow true multilingual typesetting. The drawback of this approach is that these systems have not gained really wide acceptance, mainly because their documentation is still under development.

In this chapter, we will not present the various approaches to multilingual typesetting but, instead, we will describe the tools that are available for the typesetting of documents written in a particular language. However, since there are some tools that are commonly used for the electronic typesetting of documents written in big groups of languages, we will first present the core of these tools and then will present the solution available either for groups of languages or individual languages (if, for example, such languages do not belong to any language group).

10.1 The **babel** Package

The babel package (by Johannes Braams) is the standard package that allows people to typeset multilingual documents with LaTeX. However, one of the biggest drawbacks of this package is that it does not offer the facilities for typesetting documents written in most Asian languages. The package provides various options that correspond to the language(s) that we want to use in a document.

Each document has a main language, which is the last language (option) specified in the option list. For example, with the command

```
\usepackage[german,english,greek]{babel}
```

we inform LaTeX that the main language of our document is the Greek language. All of the standard LaTeX phrases, such as "chapter," "appendix," and so on, appear in the main language of the document. In some cases, we want to use a language with some additional attributes (e.g., we want to typeset polytonic Greek instead of monotonic Greek, which is the default), so, after the \usepackage command, we have to use the command \languageattribute to "activate" these additional attributes. The command has two arguments: the name of a language and a list of attributes. Currently, the greek option supports the polutoniko attribute, which can be used for polytonic typesetting, and the latin option supports the medieval attribute, useful for Latin texts that follow the rules of medieval Latin. Suppose now that we want to write polytonic Greek. To do this, we must have the following commands in our preamble:

```
\usepackage[latin,greek]{babel}
\languageattribute{greek}{polutoniko}
\usepackage[iso-8850-7]{inputenc}
```

The last command is useful only for people having a Greek keyboard. To switch from one language to another, we can use the command \selectlanguage. This command makes sure that the hyphenation patterns as well as all peculiarities associated with its argument, which is the name of an option (language) declared in the preamble, are enabled. For example, if we have in our preamble

```
\usepackage[basque,czech]{babel}
```

then the command \selectlanguage{basque} switches to the basque language and the command \selectlanguage{czech} switches back to the main language of the document. Another way to switch languages is to use the environment otherlanguage. This environment does what the \selectlanguage is doing, but it is useful when we want to mix languages that use different writing directions. This environment has one argument, which is the name of a language. The environment otherlanguage* differs from the otherlanguage environment in that the standard phrases do not change. Similarly, the command \foreignlanguage is used to locally typeset a piece of text in another language. The command has two arguments: the name of the "foreign" language and the text to be typeset with the typographic conventions of the "foreign" language. Of course, the package provides some more commands that are useful for package developers, and we will not describe them. Interested people should consult the package's documentation for more information.

Given a character set, an encoding arranges the characters in a specified order. The associated *code point* assigned to each character is used as a means for accessing that character. A code point is an integer value that is assigned to a character. Each character receives a unique code point. The standard approach to typesetting LaTeX documents that are prepared in some extended ASCII character set is to use some input encoding file that will map the non-ASCII characters either to commands or to ASCII characters. The inputenc package (by Alan Jeffrey and Frank Mittelbach) is used to perform this mapping. The package provides a number of options that correspond to the extended ASCII character set in which the document is written. In what follows, we present the standard options and their corresponding character sets:

- `ascii`: ASCII encoding for the range 32–127.
- `latin1`: ASCII plus the characters needed for most Western European languages, including Danish, Dutch, English, Faroese, Finnish, Flemish, French, German, Icelandic, Italian, Norwegian, Portuguese, Spanish, and Swedish. Some non-European languages, such as Hawaiian and Indonesian, are also written in this character set.
- `latin2`: ASCII plus the characters needed for most Central European languages, including Croatian, Czech, Hungarian, Polish, Romanian, Slovak, and Slovenian.
- `latin3`: ASCII plus the characters needed for Esperanto, Maltese, Turkish, and Galician. However, `latin5` is the preferred character set for Turkish.
- `latin4`: ASCII plus the characters needed for the Baltic languages (Latvian, Estonian, and Lithuanian), Greenlandic, and Lappish.
- `latin5` is essentially the same as `latin1`, except that some Turkish characters replace less commonly used Icelandic letters.
- `decmulti`: DEC Multinational Character Set encoding.
- `cp850`: IBM 850 code page, almost the same as ISO Latin 1, but character arrangement is not the same.
- `cp852`: IBM 852 code page.
- `cp437`: IBM 437 code page, which is the original American code page and contains letters, digits, mathematical symbols, and some characters useful in the construction of pseudographics.
- `cp437de`: IBM 437 code page (German version).
- `cp865`: IBM 865 code page.
- `applemac`: Macintosh encoding.
- `next`: Next encoding.
- `ansinew`: Windows 3.1 ANSI encoding, extension of Latin-1.
- `cp1252`: Synonym for `ansinew`.
- `cp1250`: Windows 1250 (Central and Eastern Europe) code page.

Some other options (encodings) exist, such as the `iso-8859-7` option, which is used to typeset Greek, but those must be obtained separately. To change the input encoding in a document, we should use the \inputencoding command, which has as its argument the name of an input encoding.

10.2 The Ω Typesetting Engine

Ω has been designed to facilitate the typesetting of multilingual documents without any restriction. A Λ file consists of text written either in some extended ASCII character set or in any Unicode encoding. Unless, we use the UCS-2 input encoding, we have to use some ΩCP (i.e., a binary form of an ΩTP that can be readily used by Ω, which will convert the characters of the extended ASCII character set to their Unicode counterparts). In addition, Ω extends TeX's capabilities by allowing the use of 65,536 fonts that may contain up to 65,536 glyphs (TeX supports 256 fonts with up to 256 glyphs). The same capabilities are available to length variables and counters. Also, Ω introduces some new *primitive* commands that are necessary to properly typeset multilingual documents.

In French typography, quotations start with an opening guillemet and an unbreakable space. After the text, we have an unbreakable space and a closing guillemet. If the quoted text span to more than one line, the guillemets must also appear at the beginning of each line that contains quoted text. It is almost impossible to write a macro that will implement this typographic convention, so Ω offers the commands \localleftbox and \localrightbox to solve problems such as this. Both commands must be used in a local scope and have one argument, which is the "symbol" that will appear on the left or right side, respectively, of each output line. For example, the text

« Jean Calas, âgé de soixante et huit ans, exerçait la profession de
« négociant à Toulouse depuis plus de quarante années, et était
« reconnu de tous ceux qui ont vécu avec lui pour un bon père. Il
« était protestant, ainsi que sa femme et tous ses enfants, excepté un,
« qui avait abjuré l'hérésie, et à qui le père faisait une petite pension.
« Il paraissait si éloigné de cet absurde fanatisme qui rompt tous les
« liens de la société qu'il approuva la conversion de son fils Louis
« Calas, et qu'il avait depuis trente ans chez lui une servante zélée
« catholique, laquelle avait élevé tous ses enfants. »

has been typed in as follows

```
{<<~\localleftbox{<<~}Jean Calas,... tous ses enfants.~>>}
```

TeX is a typesetting system that only supports left-to-right typesetting. Therefore, it is inadequate for many languages such as Hebrew, Arabic,[1] and so on. Ω, on the other hand, provides primitive commands that can be used to specify the direction of pages and paragraphs. The commands \pagedir, \bodydir, \pardir, \textdir, and \mathdir are used to specify the direction of a page, of the main body of text, of a paragraph,

---

1. Actually, this is not strictly true. But it is true that it is quite cumbersome to typeset Arabic text with LaTeX.

of a short text passage or of a mathematical formula, respectively. All commands have an argument consisting of three letters that specify the "top" of each page, the "left" of each page, and the "top" of each character. The command and the argument are separated by at least one space, and the argument must not be enclosed in braces. The letters can be T (for top), B (for bottom), L (for left), and R (for right). The direction specified by the first letter must be orthogonal to the direction specified by the second letter. The third letter may take all possible values. For example, the command \pagedir TLT specifies that the top of the logical page is the top of the physical page, the left of the logical page is the left of the physical page, and the top of each character is the top of the physical page. For a traditional Japanese text, the corresponding command is \pagedir RTR because the top of the logical page is the right of the physical page, the left of the logical page is the top of the physical page, and the top of each character is at the right of the physical page. Finally, for Arabic and Hebrew, the corresponding command is \pagedir TRT just because the left of each logical page is the right of the physical page.

▶ **Exercise 10.1** How can we embed English text in an Arabic document and vice versa? ☐

In Section 6.10, we presented the various box construction and manipulation commands provided by LaTeX. These commands make use of the primitive commands \hbox, \vbox, and \vtop. These commands are used to construct horizontal and vertical boxes. A horizontal box produces material that is appended to the current paragraph. A vertical box produces material that is used to build up the current page. The difference between \vtop and \vbox is demonstrated by the following example:

vbox	\hbox{A
A vtop and a box	\vtop{\hbox{vtop}\hbox{box}} and a
box	\vbox{\hbox{vbox}\hbox{box}}}

For more information on these commands, the reader must consult the TeXbook. In Ω, all of these primitive commands can have an optional argument that is used to specify the writing direction of the contents of the box. Here is how we can specify the writing direction:

$$\left. \begin{array}{l} \texttt{\textbackslash hbox} \\ \texttt{\textbackslash vbox} \\ \texttt{\textbackslash vtop} \end{array} \right\} \texttt{dir } \textit{direction\{material\}}$$

The authors believe that all commands that can be used to specify the writing direction must somehow find their way into future releases of Λ. For example, Apostolos Syropoulos has reimplemented some standard environments with an optional argument that is used to specify the writing direction. Of course, this work is completely experimental, and it will take some time before it is finalized.

We have already mentioned Ω translation processes, and here we will go into the details. An ΩTP is a little "program" that is used to map one character set to another so that Ω can process our documents. For example, if we prepare a document using the ISO-8859-7 extended ASCII character set, we need a mechanism to map the characters of this set to Unicode, as this is the default character set Ω understands. Although this step may seem redundant, it is necessary in cases where people do not prepare their input files with a Unicode editor. Moreover, there are certain problems that are not related at all to Unicode. For example, in Turkish we use both i and ı, so when we type fil (elephant) we want to get fil and not fil[2] (which is not a Turkish word)! This is definitely a problem that can be tackled by ΩTPs.

An ΩTP defines a *finite state automaton* [i.e., an abstract machine consisting of a set of states (including the initial state), a set of input events, a set of output events, and a state transition function]. The function takes the current state and an input event and returns the new set of output events and the next state. Some states may be designated as "terminal states." The state machine can also be viewed as a function that maps an ordered sequence of input events into a corresponding sequence of (sets of) output events. A deterministic finite state automaton is one where the next state is uniquely determined by a single input event. But what's a state? We will explain the notion of a state by means of a simple example. Suppose that we want to write a simple program that will count the words of a text file. A simple strategy is to read the input stream character by character and to advance a counter whenever we see a nonblack character provided a Boolean variable is set to true. To avoid advancing the counter while reading the characters of a word, we set the Boolean variable to false. So, from this example, it is obvious that our program will be in two states: In and Out. In addition, we change state simply by setting the Boolean variable to true or false, respectively. We will now describe the structure of an ΩTP file.

Usually, ΩTPs are stored in files that have the .otp filename extension. An ΩTP file consists of six parts. Some of them have default values and may not be present in an ΩTP file. The six parts are: the input, the output, the tables, the states, the aliases and the expressions. The input and output parts specify the number of bytes occupied by each character of the input and output streams. The input and output parts are specified as follows:

input: *number* ;
output: *number* ;

Here, *number* is either a decimal, an octal (preceded by @'), a hexadecimal (preceded by @"), or an ASCII character enclosed by a grave accent and an apostrophe. Note that for hexadecimal numbers the digits above 9 can be in either uppercase or lowercase form. If we omit either of these parts, Ω assumes that *number* is equal to 2 (the number of bytes each UCS-2 character occupies). The tables part

---

2. The fi ligature of the main font of this book is designed so that the dot above letter i does not vanish in the fi glyph. So, we had to use another font where the dot vanishes.

is used to define arrays that will be referred to later in the expressions part. To make things clear we give a simple example. Suppose that we want to map the characters of some extended ASCII character set to their Unicode counterparts. Then, we define the array ASCII with length equal to the number of characters with code point greater than 127 (in extended ASCII). Next, we assign to each array element the code point of the corresponding Unicode character. Now, it is easy to get the Unicode code point of an extended ASCII character with code point C with the expression ASCII[C-F], where F is the code point of the first non-ASCII character that appears in the extended ASCII character set. Of course, we can use this "trick" only if Unicode preserves the character order of the corresponding extended ASCII character set, which is the case for most character sets. The table part begins with table: and is followed by one or more *table-specs*. Each *table-specs* is terminated with a semicolon (;) and its syntax follows:

$$table\text{-}id\,[table\text{-}length] = \{table\text{-}entries\}$$

Here, *table-id* is the name of the array. The name of a *table-id*, as well as all names appearing in an ΩTP file, must start with a letter and can be followed by zero or more letters, underscores, or digits. The *table-length* is the length of the table (i.e., a number). Finally, *table-entries* is a comma-separated list of numbers. The following is the table part of a real ΩTP:

```
tables: tab8859_7[@"60] = { @"00A0, @"0371, @"0372, ...};
```

The states part is used to define *states* that will be used later in the expressions part. The states part is optional and if present it must be specified as follows:

$$states: \textit{state-list};$$

Here, *state-list* is a comma separated list of state names. Here is a fragment of the states part of a real ΩTP:

```
states: ESCAPE, JISX0208_1978, JISX0208_1983, JISX0212;
```

The aliases part is used to define expressions that are frequently used in the expressions part. The aliases part is also optional and its syntax is

$$aliases: \textit{aliases-list}$$

where the *aliases-list* consists of one or more definitions:

$$aliase\text{-}name = \textit{left};$$

left is defined below. Here is a simple aliases part:

```
aliases: ESC = @"1b; LSO = @"0f;
```

The last part of an ΩTP, namely the expressions part, is the most interesting and important part. This part starts with the keyword expression followed by a colon and a list of expressions. Individual expressions have the form

$$\textit{LeftState TotalLeft Right PushBack RightState};$$

where *LeftState* defines for which state this expression is applicable, *TotalLeft* defines the left-hand side *regular expression*, *Right* defines the characters to be output, *PushBack* declares which characters must be added to the input stream, and *RightState* is used to define the new state. Roughly speaking, a regular expression is a character string that contains *wild-card* characters. For example, a very simple form of regular expression is used in commands that list the files of a directory. Here is how an ΩTP operates: if it is in a given state *LeftState* and the regular expression *TotalLeft* matches the beginning of the input stream, then it skips to the character that immediately follows the substring that matched *TotalLeft*, the characters generated by *Right* are put onto the output stream, the characters generated by the *PushBack* are placed at the beginning of the input stream, and, finally, the system changes its state to *RightState*. The characters that have been put back to the input stream will be looked at upon the next iteration of the automaton. We will now describe the syntax of expressions.

The *LeftState* can be either empty or of the form *<StateName>*. The syntax of *TotalLeft* is

beg: lefts end:

Note that both beg: and end: are optional. If beg: is present, the regular expression will succeed only if it can match the beginning of the input stream. Similarly, if end: is present, the regular expression will succeed only if it can match the end of the input stream. The *lefts* is a list of *left* items separated by vertical bars. Certainly, if the list consists of only one *left*, we must not put a vertical bar at the end of it. A *left* item can be:

- A number, a list of space separated numbers, or a range of numbers specified as n-m, where n and m are numbers, which match any current character that has code point in the specified range.
- The character " . ", which matches anything.
- A parenthesized list of *left* items separated by vertical bars. This list denotes a choice (i.e., the ΩTP will try each *left* from left to right to see if it can be "applied" to the input stream). If we put the symbol ^ in front of the parenthesized list, then this means that if each *left* will fail, the whole expression will succeed.
- An *alias-name* surrounded by curly brackets. In this case, we substitute the *left* item with what the *alias-name* stands for.
- A *left* item followed by <n,m>. Here, both n and m are numbers, and m is optional. Its meaning is that *left* must match between n and m times; if m is missing, then *left* must match at least n times.

The syntax of *Right* is: => *chars*, while the syntax of *PushBack* is: <= *chars*. Here, *chars* is one or more *char* items, which, in turn, can be:

- An ASCII character string enclosed in double quotation marks (e.g., "abc") or a number.
- The expression \n, where n is a number, corresponds to the nth character of the string that matched the *TotalLeft*. For example, \1 is the first character of this string.

- The expression \\$ corresponds to the last character of the string that matched the *TotalLeft*, while the expression \\($-n), where n is a number, corresponds to the nth character from the end of the string that matched the *TotalLeft*.

- The expression * denotes the whole string that has been matched.

- The expressions \\(*-n) and \\(*+n), where n is a number, denote the whole string that has been matched without the last or first n characters, respectively, and the expression \\(*+n-m), where m is also a number, is the matching string without the first n and the last m characters.

- Finally, it can be an arithmetic expression: #*arithmetic-expression*.

For instance, the following is an example of an ΩTP that transforms in a Greek text the letter β to ϐ if it does not occur at the beginning of a word.

```
expressions:
{LETTER}@"03B2@"03B2 =>   \1   @"03D0 @"03D0 ;
{LETTER}@"03B2 =>   \1   @"03D0 ;
.  => \1;
```

The number @"03B2 corresponds to β and the number @"03D0 to ϐ. Of course, LETTER is an alias that corresponds to the (code points of the) Greek letters. Here is another example. Suppose that we want to write an ΩTP that will transform text written in the Latin transcription of the Cherokee syllabary (see Table 10.8) to Unicode. The only real problem is the handling of the syllables: s, sa, se, and so on. The following code fragment shows exactly how we can tackle this particular problem:

```
. . . . . . . . . . . . . . . . . . . . .
's' end: => @"13CD;
's' ^('a'|'e'|'i'|'o'|'u'|'v') => @"13CD <= \2;
's' 'a' => @"13CC;
's' 'e' => @"13CE;
. . . . . . . . . . . . . . . . . . . . .
```

Here is what we actually do: if the input stream contains only the letter s, then we emit the character with code point @"13CD; otherwise, if there is a leading s that is not followed by an a, or an e, and so on, then we emit the same character and push back the character that follows s. Of course, it is now easy to handle the cases where the head of the input stream consists of the letters s and a, or s and e, and so on.

As we have shown with the array "trick," arithmetic expressions are really useful, so it is not surprising that ΩTPs support arithmetic expressions. The calculations performed by an arithmetic expression refer to the string that matched the regular expression. An arithmetic expression can be one of the following:

- A number or the expressions \n, \\$, or \\($-n). All of these expressions have the expected meaning.

- Suppose that a and b are arithmetic expressions. Then $a + b$, $a - b$, $a * b$, a div: b, a mod: b, and (a) are all arithmetic expressions. The symbols +, -, *, and div: are used to perform addition, subtraction, multiplication, and division, respectively. The symbol mod: returns the quotient of the integer division of the two operands. Parentheses are used to override operator precedence (e.g., to perform an addition before a multiplication).

- The expression *table-id*[*arithmetic-expression*] is used to get an element of an array.

We note, again, that an arithmetic expression must be prefixed with the symbol #. Here is how we can implement the array "trick":

```
expressions:
@"00-@"9F        => \1;
@"A0-@"FF        => #(tab8859_7[\1-@"A0]);
                 => @"FFFD;
```

The character @"FFFD is the *replacement character* and is used to replace an incoming character whose value is unknown or unpresentable in Unicode. Readers not familiar with regular expressions are advised to experiment with a real programming language that supports regular expressions, such as Perl [29].

The *RightState* can either be empty or can have one of the following forms: <*state-name*>, <push: *state-name*>, or <pop:>. If it is empty, the ΩTP stays in the same state. If it is of the first form, the ΩTP changes to state *state-name*. The second form changes the ΩTP to state <*state-name*> but saves the previous state into a special data structure. Finally, the third form returns the ΩTP to the state that was previously saved into this special data structure and of course deletes this state from the data structure.

➤ **Exercise 10.2** Unicode contains two different characters for the Greek small letter theta—ϑ (@"03D1) and θ (@"3B8). According to [4], when typesetting Greek text, we should use ϑ only at the beginning of a word. Write an ΩTP that will implement this typographic convention. ☐

➤ **Exercise 10.3** Write an ΩTP that solves the "fi" ligature problem described above. ☐

An Ω compiled translation process (or ΩCP, for short) is the binary equivalent of an ΩTP. The program OTP2OCP transforms an ΩTP to an ΩCP. In addition, the program OUTOCP transforms an ΩCP to human readable form. This program is provided for debugging purposes. If we want an Ω/Λ file to read an ΩCP, we have to use the following command:

\ocp\ *InternalOCPname=RealOCPname*

Here, \ *InternalOCPname* is a new control sequence with which we will refer to the actual ΩCP named *RealOCPname*. An ΩCP list is a mechanism to combine ΩCPs. The

ΩCPs of an ΩCP list are applied one after the other to the input stream. ΩCP lists are like the pipes that are available in most operating systems. Roughly speaking, pipes are sequences of programs where the output of one program is the input of the next program in the pipe. Merging two pipes means that the input to the first program of the second pipe is the output of the last program of the first pipe. However, this is the fundamental difference between pipes and ΩCP lists. An ΩCP list consists of pairs where the first element is a number and the second is an ΩCP. Now, when we merge two ΩCP lists, we get a new ΩCP list whose elements are sorted using the first element of each pair in ascending order.

To build an ΩCP list, we use the five commands \nullocplist, \addbefore-ocplist, \addafterocplist, \removebeforeocplist, and \removeafterocplist. The command \ocplist is actually used to build an ΩCP list:

$$\text{\ocplist } \backslash ListName = ocpList$$

If we are creating an *ocpList* from ΩCPs only, then we must use the \nullocplist command at the end. The effect of this command is to create an empty ΩCP list, which is then populated with ΩCPs. Here is an example:

```
\ocp\OCPutf=inutf8
\ocp\OCParab=uni2cuni
\ocplist\OCPArablistutf=\addbeforeocplist 1000 \OCPutf
                       \addbeforeocplist 1000 \OCParab
                       \nullocplist
```

If the \nullocplist is replaced by an existing ΩCP list, then we get a modified version of the existing ΩCP list. An ΩCP list forms a queue, so we can add or remove ΩCPs from the head or the tail of the queue. Given an ΩCP list ℓ, the command \addbe-foreocplist *n ocp ℓ* adds the *ocp* at the head of the list. The number *n* is used to form the pair that we were talking about before. This number is used to place the ΩCP in a position so that the order is preserved. The command \addafterocplist *n ocp ℓ* adds the *ocp* at the end of the list and takes care so that the order is preserved. The command \removebeforeocplist *n ℓ* removes from the head of ℓ the ΩCP with number *n*. Similarly, the command \removeafterocplist *n ℓ* removes from the tail of ℓ the ΩCP with number *n*. Having defined our ΩCP lists, we must be able to activate and deactivate them. To activate an ΩCP list, we use the command \pushocplist *ocpList*. The command \popocplist deactivates the last ΩCP list that has been activated. To deactivate all ΩCP lists, we use the command \clearocplist. To see which ΩCP is active while Ω processes an input file, set the variable \ocptracelevel to a number greater that zero, such as

$$\text{\ocptracelevel=1}$$

The authors of Ω present in [11] the use of external ΩCPs. An external ΩCP is a program written in some real programming language that reads data from the keyboard (or the standard input in general) and writes data to the screen (or the standard output in general). When using an external ΩCP, Ω actually forms a pipe that sends the *current* input stream to the external ΩCP, which, in turn, processes the input stream and sends the result back to Ω for further processing. The command \externalocp is used to introduce an external ΩCP, while it is activated as a normal ΩCP. Let us give a simple example. Suppose that we have an ΩCP that solves the "fi" ligature problem and an external program that prints the word fifi ten times. When the following Λ file is processed, the output will "contain" the text that precedes the use of the ΩCP list, the "word" fifi ten times (without fi ligatures), and the text that follows the use of the ΩCP list.

```
\documentclass{article}
\begin{document}
\ocp\FIlig=filig
\externalocp\FIgen=figen {}
\ocplist\MyOCP=
  \addbeforeocplist 1 \FIgen
  \addbeforeocplist 2 \FIlig
  \nullocplist
text text text text text text text
{\pushocplist\MyOCP text text }
text text text text text text
\end{document}
```

Note that the name of the external program must be followed by {}. Also, the text inside the local scope is completely ignored, as the external ΩCP does not need input.

➤ **Exercise 10.4** What is the paper size used in the example above? □

In [11], the authors of Ω present a little Perl script that has been designed to detect spelling errors by using ISPELL. However, the script presented in their paper has a couple of errors (remember: to err is human!). Here is a slightly modified version of the external ΩCP that actually works:

```
#!/usr/bin/perl
@IN = <>;
$in = sprintf "%s", @IN;
open OUT, "| ispell -t -l > tmp.spell";
print OUT $in;
close OUT;
open IN, "tmp.spell";
while (<IN>) {
  foreach $mot (split /\n/, $_) {
    $MOTS{$mot}=$mot;
  }
}
```

```
close IN;
foreach $mot (sort keys %MOTS) {
  $in =~ s/$mot/\\textcolor{red}{$mot}/g;
}
print $in;
```

Try the following Λ input file to see what happens:

```
\documentclass[a4paper]{article}
\usepackage{color}
\externalocp\OCPverif=verif.pl {}
\ocplist\verifier=
    \addbeforeocplist 100 \OCPverif
    \nullocplist
\begin{document}
\pushocplist\verifier
This is a semple taxt that has errors.
\end{document}
```

Ω can operate in four different input *modes*. Each mode characterizes the character set used to prepare the input file. The input modes are:

onebyte Each character occupies exactly one byte and includes ASCII, the various ISO-8859-X, and the shifted East Asian character sets.

ebcdic The EBCDIC character set is IBM's 8-bit extension of the 4-bit Binary Coded Decimal encoding of digits 0–9. This mode is useful only on machines that support this character set.

twobyte Assumes that each character occupies exactly two bytes (i.e., the input is encoded with the Unicode UCS-2 encoding).

twobyteLE The same as twobyte, but characters are encoded in *little endian* order. In little endian order, the most significant bits are stored at the end of a byte cluster. The opposite convention is called *big endian* order. To see the difference, consider the character Σ (GREEK LETTER CAPITAL SIGMA). In big endian, this letter is @"03A3, and in little endian it is @"A303.

Here are the primitives for manipulating modes:

\DefaultInputMode *mode* Sets the default input mode to *mode*.

\noDefaultInputMode Ω processes input just like TeX does.

\DefaultOutputMode *mode* Sets the default output mode to *mode*.

\noDefaultOutputMode Ω generates output just like TeX does.

\InputMode *file mode* The input mode for *file* is changed to *mode*, where *file* can be either currentfile, meaning the current input file, or a file number (i.e., a number that is used to identify an external file name). For most purposes, the use of currentfile will be enough.

\noInputMode *file* Now, Ω processes *file* just like TeX does.

\OutputMode *file* *mode* The output mode for *file* is changed to *mode*, where *file*
can be either currentfile, meaning the current input file, or a file number.

\noOutputMode *file* Ω generates output to file *file* just like TeX does.

There are also a number of primitives for manipulating *translations*. These commands
are primarily intended for "technical translations," such as onebyte to twobyte or
from little endian to big endian. However, we warn the reader to avoid using these
commands, as they may change the way commands are expanded. Here is the list of
these commands:

\DefaultInputTranslation *mode* *ocp* Sets the default input translation for *mode* to
ocp.

\noDefaultInputTranslation *mode* There is no longer a default input translation for
mode.

\DefaultOutputTranslation *mode* *ocp* Sets the default output translation for *mode*
to *ocp*.

\InputTranslation *file* *ocp* The input translation for *file* is *ocp*, where *file* can
be either currentfile or a file number.

\noInputTranslation *file* There is no longer an input translation for *file*.

\OutputTranslation *file* *ocp* The output translation for *file* is *ocp*, where *file*
can be either currentfile or a file number.

\noOutputTranslation *file* There is no longer an output translation for *file*.

Currently, one can use the OmegaSerif and OmegaSans typefaces with Ω for truly
multilingual typesetting. In particular, to typeset Arabic text, one should use the OT1
font encoding and the omarb font family. For European text, one should use the OT1
font encoding and the omlgc font family. In addition, one has at his/her disposal the
uctt monospaced font family.

In the following sections, the reader will have the chance to see real applications of
ΩCPs.

10.3   The ε-TeX Typesetting Engine

ε-TeX is actually a successor of both TeX and TeX--XₑT (the bidirectional version of
TeX). Normally, ε-TeX operates in TeX mode. To enter the TeX--XₑT mode, we must set
the state variable \TeXXeTstate. Usually, we set this variable to 1 to enter TeX--XₑT
mode and to 0 to revert to normal TeX mode. The commands \beginR and \endR are
used to typeset text with a right-to-left writing direction. The commands \beginL and
\endL are used to typeset text with a left-to-right writing direction inside a right-to-left
"environment." ε-TeX extends the allowable number of counters and length variables
from 256 to 32767. Another interesting feature of ε-TeX is that it adds syntactic sugar so
arithmetic expressions are written in a natural way. Here is an example:

```
\ifdim\dimexpr (2pt-5pt)*\numexpr 3-3*13/5\relax + 34pt/2<\wd20
         . . . . . . . . . . . . . . . . . . . . . . . . . .
\else
         . . . . . . . . . . . . . . . . . . . . . . . . . .
\fi
```

Note that length expressions are prefixed by the \dimexpr command and numerical expressions are prefixed by the \numexpr command. Note also that we use the "expected" symbols when writing down an arithmetic expression. The constructs \ifdim-\fi and \ifnum-\fi introduce two control constructs like those found in ordinary computer languages. In addition, ε-TeX provides a number of other constructs that are really useful when it comes to macro definitions. All in all, we do believe that neither Ω nor ε-TeX can be a true TeX successor. We must definitely incorporate ideas from both systems when the time comes to develop a universally accepted TeX successor.

10.4 The Greek Language

Although the name of TeX derives from the common root of two Greek words, only recently has it become possible to prepare a LATeX/Λ document and process it with the mainstream tools. babel provides the greek option (by Apostolos Syropoulos[3]) and the polutoniko language attribute. The option allows people to prepare documents in monotonic Greek, while the language attribute is useful for the typesetting of polytonic Greek. When preparing a Greek document, the following transliteration is actually employed:

α	β	γ	δ	ε	ζ	η	ϑ	ι	κ	λ	μ	ν
a	b	g	d	e	z	h	j	i	k	l	m	n
ξ	o	π	ϱ	σ	τ	υ	φ	χ	ψ	ω	ς	
x	o	p	r	s	t	u	f	q	y	w	c	

Claudio Beccari designed the standard Greek fonts in such a way that one does not have to use the character "c" to get the letter ς [1]. Of course, this transliteration may be helpful for people without a Greek keyboard, but for Greeks it is unacceptable to write Greek using a Latin transliteration. So, the first author designed the iso-8859-7 input encoding, which is suitable for both Unix and Microsoft Windows. For MacOS, Dimitrios A. Filippou has designed the macgreek encoding. But, although we can use these two encodings, we still need a mechanism to enter all possible accents and breathing symbols that are needed to correctly typeset polytonic Greek. Here is what we have to type in order to get the correct accent and breathing symbols:

---

3. When we do not mention the author of a babel option, the reader must assume that it is Johannes Braams.

Accent	Symbol	Example	Output
acute	'	g'ata	γάτα
grave	'	dad'i	δαδὶ
circumflex	~	ful~hc	φυλῆς
rough breathing	<	<'otan	ὅταν
smooth breathing	>	>'aneu	ἄνευ
subscript	|	>anate'ilh|	ἀνατείλη
dieresis	"	qa"ide'uh|c	χαϊδεύης

Note that the subscript symbol is placed *after* the letter. The last thing someone must know in order to be able to write normal Greek text is the punctuation marks used in the language:

Punctuation Sign	Symbol	Output
period	.	.
semicolon	;	·
exclamation mark	!	!
comma	,	,
colon	:	:
question mark	?	;
left apostrophe	' '	''
right apostrophe	' '	'
left quotation mark	((	«
right quotation mark	))	»

When typesetting Greek text, the command \textlatin can be used for short passages in some language that uses the Latin alphabet, while the command \latintext changes the base fonts to the ones used by languages that use the Latin alphabet. However, all words will be hyphenated by following the Greek hyphenation rules! Similar commands are available once someone has selected some other language. The commands \textgreek and \greektext behave exactly like their Latin counterparts. For example, the word Μίμης has been produced with the command \textgreek{Mίμης}. Note that one cannot put a circumflex on a vowel using these commands—it is mandatory to enable the polutoniko language attribute.

The greek option offers the commands \Greeknumeral and \greeknumeral, which are used to get a Greek numeral in uppercase or lowercase form:

Command	Output
\Greeknumeral{9999}	‚ΘϠϞΘ'
\greeknumeral{9999}	‚θϡϟθ'

In order to correctly typeset the Greek numerals, the greek option provides the following commands: \qoppa (ϟ), \sampi (ϡ), and \stigma (ϛ). In addition, we can get the so-called Attic (or Athenian) numerals with the command \athnum provided by the athnum package (by Apostolos Syropoulos with assistance from Claudio Beccari, who designed the necessary glyphs in the standard font).

Command	Output
\athnum{9999}	⊠XXXXX⊡HHHH⊡ΔΔΔΔIIIIII
\athnum{2002}	XXII

Note that the package grnumalt (by Apostolos Syropoulos) can be used to get Attic numerals without using the greek option and Greek fonts.

The command \Grtoday typesets the current date using Greek numerals instead of Arabic, so the command \today prints 20 Αὐγούστου 2002 and the command \Grtoday prints Κ' Αὐγούστου ͵ΒΒ'. Finally, the greek option provides the commands \Digamma (F) and \ddigamma (ϝ), which are necessary to typeset archaic Greek texts. The package grtimes [24] (by Apostolos Syropoulos with assistance from Antonis Tsolomitis) allows users to typeset their documents using the Times Roman typeface.

Although Ω is currently our best choice for multilingual typesetting, there is still no single package that can cover at least the European languages. In [9] the authors of Ω describe a Greek option of the omega package, dated 1999/06/01, that can practically process text in Greek, English and French. Recently, Javier Bezos released his lambda package, which is actually a reimplementation of the omega package. The only "real" advantage of this package is that it can also process Spanish text. However, the package is not yet stable, so it is necessary to develop a package that will cover at least the European languages, and the authors of this book are working in this direction.

When using the omega package, we declare the main language of the document with the \background command. Additional languages can be "loaded" with the \load command. The last version of the package supports the greek, the usenglish, and the french options (languages). To switch languages, one can use the command \localLanguage, where Language is any of the three languages. Certain features can be enabled by adding a comma-separated list of key=value pairs. The key accents allows the processing of either monotonic or polytonic Greek text. The only difference is that the symbol = is used instead of ˜. Of course, this is just a convention implemented by an ΩTP. To allow the two forms of the letter β, use the key beta with value twoform.

10.4.1 Writing Greek Philological Texts

Although LATEX comes tuned to write the most demanding mathematical text, this is not the case with philological texts. The capability to work with this kind of text can be added using the package teubner by Claudio Beccari. Using this package, it is easy to typeset most philological documents. It is in a preliminary version now, but hopefully in the future it will provide full support for the most demanding texts. It is supposed to be used with the fonts of Claudio Beccari that have been recently extended to include the italic font used in the past 100 years or so by the Teubner printing company in Lipsia; the font is so well-known in Greece that it is normally referred to with the name of "Lipsiakos."

The Lipsiakos font can be locally selected with the command \textLipsias or by using the declaration \Lipsiakostext, which switches to the Lipsiakos font. For example, to typeset a paragraph with the Lipsiakos typeface, we can use the following construct:

\begin{Lipsiakostext} *text* \end{Lipsiakostext}

The package defines several verse environments, several commands for accessing special symbols used by philologists, and commands for producing metrics. The interested user can find all of the details in documentation that accompanies the package. As a simple demonstration of the capabilities of the package, we give a simple example that shows how numbering on the (left) side of a poem works (we start the verse enumeration from 1 and the "subverse" enumeration from 10):

> 10   Κυανέοις κόλποισιν ἐνημένη, ἀερόμορφε,
> 11   Ἥρα παμβασίλεια, Διὸς σύλλεκτρε μάκαιρα,
> 12   ψυχοτρόφους αὔρας θνητοῖς παρέχουσα προσηνεῖς,
> 13   ὄμβρων μὲν μήτηρ, ἀνέμων τροφέ, παντογένεθλε·
> 5 14 χωρὶς γὰρ σέθεν οὐδὲν ὅλως ζωῆς φύσιν ἔγνω·
> 15   κοινωνεῖς γὰρ ἅπασι κεκραμένη ἠέρι σεμνῶι·
> 16   πάντων γὰρ κρατέεις μούνη πάντεσσί τ᾽ ἀνάσσεις
> 17   ἠερίοις ῥοίζοισι τινασσομένη κατὰ χεῦμα.
> 18   ἀλλά, μάκαιρα θεά, πολυώνυμε, παμβασίλεια,
> 10 19 ἔλθοις εὐμενέουσα καλῶι γήθοντι προσώπωι.

The input code that produces the example above follows:

```
\begin{Lipsiakostext}
  \begin{VERSI}\SubVerso[10]%
    Kuan'eoic k'olpoisin >enhm'enh, >aer'omorfe, \ \\
    ..............................................
    >'eljoic e>umen'eousa kal~wi g'hjonti pros'wpwi.
  \end{VERSI}
\end{Lipsiakostext}
```

10.4.2 Working with Thesaurus Linguae Grecae

Thesaurus Linguae Grecae is a collection of all of the Greek documents from the beginning of the Greek civilization until 1453 A.D. in electronic form. The collection is available in a CD from http://www.tlg.uci.edu and is one of the most important tools for classicists and people with an interest in Greek literature. The texts are written in files using a transliteration similar (but different) to the transliteration of the babel package. For example, they write w)\| instead of babel's >'w| for the character ὢ. There are several interfaces to the CD: programs that read the text written in this transliteration and show them in Greek on screen. However, the best one seems to be the DIOGENES program (by Peter Heslin). It is a Web interface (it also comes with a command-line tool) that is capable of showing the TLG texts through a Web browser in several encodings, among them Unicode and the babel transliteration. This makes it ideal for typesetting a passage or the whole text of a work using babel or Ω. It is available from http://www.durham.ac.uk/p.j.heslin/diogenes.

10.5 The Latin Language

The latin option (by Claudio Beccari) of the babel package is set up in such a way that hyphenation is prohibited between the last two lines of a paragraph. The medieval attribute should be used for Latin text that follows the conventions and rules of medieval Latin. The main difference between "normal" Latin and medieval Latin is that in the latter we systematically use the letter u even in cases where "normal" Latin uses the letter v; quite the opposite happens when we use uppercase letters—the letter V systematically replaces the letter U even when U is used in "normal" Latin. In medieval Latin the following digraphs were used: æ, œ, Æ, and Œ. In classical Latin, these digraphs were actually written as two letters. To access these digraphs, one should just use the commands presented in Chapter 3. The latin option provides the commands, or *shorthands* in babel's terminology, ^ℓ and =ℓ. The former is used to place a breve accent above the vowel ℓ (e.g., the command ^i prints ĭ). The latter is used to place a macron accent above the vowel ℓ (e.g., the input =o is typeset as ō). Note that both commands cannot be used to put accents on digraphs.

10.6 The Dutch Language

The dutch option of the babel package provides a few commands that we will describe now. The "ℓ shorthand, where ℓ is one of the letters a, e, i, o, or u, has the effect of the command \"ℓ, but the former takes special care when the letter i is used and produces ï instead of ī. Also, this shorthand does not disturb the hyphenation process. The commands "y and "Y produce the ligatures ij and IJ. These ligatures are very common in Dutch. The command "| disables ligatures. In modern Dutch typography, it is customary to use apostrophe-quote for quoted text (e.g., 'this is quote'). However, the traditional convention was to use the symbols „ and " to open and close a quotation (e.g., „This is a quote"). These symbols are produced by the commands "' and "', respectively. The Afrikaans language is actually a dialect of the Dutch language spoken in South Africa. This dialect is also supported by babel (option afrikaans). If we do not want to prepare our input file using all of these shorthands, then we simply prepare our input file with an editor that is aware of the ISO-8859-1 character set and use the t1enc package. As far as it regards Λ, one should type a document using a Unicode editor to avoid all of these commands. Otherwise, the appropriate ΩCP must be loaded. And since Ωbabel is still far away, one has to redefine the commands that produce the predefined name strings such as "chapter," "preface," and so on.

10.7 The Esperanto Language

Esperanto is an artificial language intended for use between people who speak different native languages. Esperanto was developed during the period 1877–1885 by L.L. Zamenhof of Warsaw, Poland. The esperanto option defines all of the language-specific macros for the Esperanto language. The language uses the Latin alphabet and the letters Ĉ, ĉ, Ĝ, ĝ, Ĥ, ĥ, Ĵ, ĵ, Ŝ, ŝ, Ŭ, and ŭ. All of these extra letters can be accessed with the command "ℓ, where ℓ is any of the extra letters. For example to typeset the sentence "Mi faris ĝin por vi," we simply type Mi faris "gin por vi. Of course, if we type our document with the aid of an editor that supports the Latin 3 encoding, then our input file will contain the extra characters of this encoding. For example, in the following screen dump, it is clear that we type Latin 3 characters and use the necessary input encoding:

```
emacs@ocean1.ee.duth.gr

Buffers Files Tools Edit Search Mule TeX Help
\documentclass[a4paper]{article}
\usepackage[esperanto]{babel}
\usepackage[latin3]{inputenc}
\begin{document}
Mi faris ĝin por vi.
\end{document}
--3:--   esp.tex              (LaTeX)--L1--All--
(No changes need to be saved)
```

This is in general the mechanism by which we prepare multilingual LATEX input files. If we want to typeset Esperanto language documents with Λ, we must use the in88593 ΩCP. Similar ΩTPs exist for other Latin encodings: in88591 (for Latin 1), in88592 (for Latin 2), and in88594 (for Latin 4).

10.8   The Italian Language

Currently, Λ offers no special tools for typesetting Italian documents, but the babel package offers the italian option (by Claudio Beccari). This option disables hyphenation between the last two lines of a paragraph. The command "" is used to enter the opening American quotation marks. Although the introduction of this command may seem redundant, the reality is that with an Italian keyboard it is not easy at all to enter a back-tick! To insert guillemets, one can either use the command "< and "> or the symbols << and >>, if the t1enc package has been loaded. The various commands that place accents on letters are redefined to avoid elaborate input sequences such as \'{\i} that can be replaced with the much simpler and more readable \'i. Of course, if t1enc has been loaded, these facilities are superfluous. The commands \unit, \ped, and \ap are introduced to enable the correct typesetting of mathematical signs and symbols that are used in the physical sciences and technology (ISO 31–12:1992). More specifically, the \unit command is used to typeset the unit part of a quantity (e.g., the text 3\unit{cm} will be typeset as 3 cm). Note that we get the same output even in math mode! The commands \ap and \ped are used to typeset superscripts and subscripts (e.g., the command $V\ped{min}$ will be typeset as V_{min}).

10.9   The Irish and "British" Languages

The default language of both LaTeX and Λ is American English, so the english option is used to enable hyphenation according to the British English conventions. To use the default American English hyphenation patterns, use the american option. The options welsh, irish, and scottish are used for Welsh, Irish, and Scottish, respectively.

10.10   The German Language

German is an official language in five countries: Germany, Austria, Luxembourg, Lichtenstein, and Switzerland. The babel package provides the options german and austrian (both by Bernd Raichle), which are suitable for German that uses the old orthography. In August 1998, the new orthography of the German language was officially announced. The new orthography is a unified set of rules on German spelling, hyphenation, and pronunciation. The purpose of this reform was to eliminate inconsistencies in the way the German language has been written to date. The options ngerman and naustrian support the new orthography. As far as it regards babel, the only difference between German and Austrian is the spelling of the name of the first month of the year—Januar in German and Jänner in Austrian. To place an umlaut above a vowel, we use the command "ℓ, where ℓ is a vowel. The commands "s and "z produce the letter ß, while the commands "S and "Z produce the digraph SS. In addition, the commands "', "',

"<, and "> produce the German quotation marks, „ and ", and the opening and closing guillemets, « and », respectively. The command "$\ell\ell$, where ℓ is one of the consonants c, f, l, m, n, p, r, or t, will force TEX to hyphenate the double consonant as $\ell\ell$-ℓ. Similarly, the command "ck will force the letters ck to be hyphenated as k-k. Note that these hyphenation conventions belong to the old orthography. As in the case of the Dutch language, we can prepare our input files using the ISO-8859-1 character set if we use the t1enc package.

10.11 The French Language

The frenchb option (by Daniel Flipo) of the babel package implements the typographic conventions of the French language. First of all, this option uses the symbol "–" in all levels of the itemize. If we want to change the symbol used in all levels of the itemize environment, we can use the command \FrenchLabelItem. For example, the command

$$\renewcommand\{FrenchLabelItem\}\{\textemdash\}$$

will change the itemization symbol to an em dash. If we want to change the symbol used in a particular level of the itemize environment, we can use the commands presented in Section 4.3.1. An interesting feature of French typography related to lists is that they occupy less vertical space. This feature can be turned off with the command \French-ListSpacingfalse. The command \FrenchListSpacingtrue turns on this feature. To write quoted text, one can use the commands \og and \fg, which yield the symbols « and », respectively, plus the necessary unbreakable space. If we prepare the input file using the ISO-8859-1 character set, then we have to type << and >> to get the opening and closing guillemets. The command \up is provided to typeset superscripts like Mme (abbreviation for "Madame") or 1er (for "premier"), and so on. The command has one argument, which is the text that appears as a superscript. Since family names must be typeset using small capitals, the command \bsc is used to correctly typeset family names. For example, the input text Antonis~\bsc{Tsolomitis} will be typeset as Antonis Tsolomitis.

The commands \primo, \secundo, \tertio and \quarto print the symbols 1°, 2°, 3°, and 4°, respectively. Similarly, the commands \fprimo), \fsecundo), \ftertio), and \fquarto) print the symbols 1°), 2°), 3°), and 4°), respectively. Note that we must type in the parenthesis. More generally, we can create any ordinal with the command \FrenchEnumerate and \FrenchPopularEnumerate. For example, the command \primo is defined as follows:

$$\newcommand\{\primo\}\{\FrenchEnumerate\{1\}\}$$

Note that the output of the command \FrenchPopularEnumerate{1} is 1°)! So, we really do not understand why the user has to type the parenthesis when using commands such as \fsecundo). The commands \No and \no print N° and n°, which are abbreviations for "Numèro" and "numèro," respectively. The command \degres is used to typeset temperatures and alcohol's strength. To typeset a temperature, use 20~\degres C. To typeset the strength of an alcoholic beverage, use 46\degres. The command \nombre is provided to facilitate the typesetting of numerals. This command functions properly in both text and math mode and prints the number in clusters of three digits separated either by a space or by a comma (if the current language is other than French). The commands \FrenchLayout and \StandardLayout are provided so that either the French or the Anglo-Saxon typographic conventions are applied throughout a document. The frenchb option implements also the typographic rule that specifies that some white space should be added before the punctuation symbols ;, !, ?, and :.

Both the omega and the lambda packages provide support for the French language. In particular, the lambda package provides the options uppercase (with values unaccented and accented) and guillemets (with values line, paragraph, and normal), which have the expected meaning. In addition, we can specify the input character encoding with the option charset, which may assume the value isolat1.

10.12   The Breton Language

The Breton language is a Celtic language spoken in Brittany (NW France). The breton option (by Christian Rolland) can be used to typeset Breton LaTeX documents. The characters :, ;, ! and ? are shorthands that print a little white space and then the corresponding glyph. The commands \kentan, \eil, \trede, \pevare, and \pempvet print the numerals $1^{añ}$, 2^{l}, 3re, 4re, and 5^{vet}, respectively.

10.13   The Nordic Languages

The babel options danish, icelandic, norsk, swedish, finnish, and samin can be used to typeset documents written in Danish, Icelandic, Norwegian, Swedish, Finnish, or the Northern Sámi[4] language. The nynorsk dialect of the norsk option is used when we want to typeset Nynorsk instead of Bokmål.[5]

The danish option defines the commands "', "', "< and "> which print the symbols „, ", «, and », respectively. The norsk option also defines the commands "< and "> plus the command "$\ell\ell$, where ℓ is one of the letters b, d, f, g, l, m, n, p, r, s, or t, and the command "ee. The command "$\ell\ell$ forces TeX to hyphenate the double consonant $\ell\ell$ as $\ell\ell$-ℓ. The command "ee will force TeX to hyphenate ee as é-e. The swedish option

---

4. The Sámi language belongs to the Finnic group of languages that are spoken in Lapland.
5. Bokmål ("book language") is used by about 90% of the population. After 400 years of union with Denmark, the Norwegian written language had been very much influenced by Danish. Nynorsk ("new Norwegian") was constructed as a reaction to this. Based on common spoken language in rural Norway, a new written language was constructed by Ivar Aasen, but it has never gained much popularity outside rural regions.

defines the commands "a, "o, "w, and their uppercase forms. These commands print the letters ä, ö, and å, respectively. The command "$\ell\ell$, where ℓ is one of the letters b, d, f, g, l, m, n, p, r, s, and t, is used to hyphenate $\ell\ell$ in compound words as $\ell\ell$-ℓ. For example, the words sko"ttavla and stra"ffeihet will be hyphenated as skott-tavla and straff-feihet, respectively. In Swedish typography, it is customary to use the English quotation marks so there are no special commands to access quotation marks. The finnish option provides the commands "', "', "<, and ">, which function just like their Danish counterparts. These commands produce the same symbols when used in the icelandic option (by Einar Árnason). In addition, the icelandic option provides the following commands:

Commands:	"o	"O	"ó	"Ó	"e	"E	"é	"É
Result:	ǫ	Ǫ	ǫ́	Ǫ́	ę	Ę	ę́	Ę́

The \tala command is used to typeset numbers. Here are two examples:

\tala{1234567}	1 234 567
\tala{123,4567}	123,456 7

The command \grada produces the symbol °, while the command \gradur is used to typeset temperatures and alcohol's strength, as shown in the following example:

$$5\text{\textbackslash gradur C} \quad | \quad 5\,°C$$

The command \upp functions like the command with the same name provided by the frenchb option. The samin option does not provide any special commands, but we need the letters Ŧ and ŧ, which are not available in most standard fonts. An easy way out is to put the following definitions in the preamble of our input files:

```
\newcommand{\tx}{t\hspace{-.35em}-}
\newcommand{\Tx}{T\hspace{-.5em}-}
```

➤ **Exercise 10.5** The commands above produce correct visual results when we typeset our documents with the Computer Modern typeface. The main font of this book is the Palatino typeface and, of course, the commands above are not suitable for this typeface. Rewrite the commands so that they produce correct visual results with this typeface. We remind you that we set Palatino as the document's main font by using the palatino package. ☐

10.14 The Thai Language

The Thai language is written like all European languages—from left to right and from top to bottom. Thai text must be typed in using the TIS–620 encoding (TIS stands for Thai Industrial Standard). It is interesting to note that ISO-8859-11 is equivalent to TIS–620.

Currently, there are two approaches to typesetting Thai documents using LaTeX—the ThaiTeX system and the unofficial thai option of the babel package.

The ThaiTeX system (by Vuthichai Ampornaramveth) consists of Thai fonts, the thai package, and the program CTTEX, which is a Thai word separator. The Thai script is written continuously without using spaces for breaking between words. A program such as LaTeX, then, needs to know where to break the sentence for a new line. So, CTTEX is a preprocessor that does exactly this thing: it adds spaces between words. All Thai LaTeX files can contain both Thai and English text. The author of the package provides his own commands for font size, series, and shape selection, as the usual commands do not give the expected result. So, the font selection commands are \sptiny, \spscriptsize, and so on. Also, the commands for font series and shape selection are \spbf (for boldface series), \spit (for italic shape), and \sprm (for upright shape). Now we describe how to process a document written in Thai. Initially, we prepare our LaTeX file with our favorite editor (EMACS mule mode is a good choice). Then, we use CTTEX to separate the words in Thai sentences with a space so that TeX can render Thai paragraphs correctly. After that, we can process the resulting file with LaTeX. The following is the output of a simple Thai LaTeX document.

สวัสดีครับ ผมชื่อ อโพสโตโลส

More information on ThaiTeX, Thai fonts, and utilities are available from http://thaigate.rd.nacsis.ac.jp.

Another approach to typesetting Thai with LaTeX is the thai option of the babel package. The thai option was originally developed by Surapant Meknavin and further developed by Theppitak Karoonboonyanan. All LaTeX files that contain Thai text must be preprocessed with SWATH (another Thai word separator). This option also provides the command \textthai and the declaration \thaitext for switching to Thai text mode if the main language of the document is not the Thai language. Unfortunately, the current version of the thai option is not officially part of the babel package. Moreover, a serious drawback of this language option is that it cannot coexist with ThaiTeX on the same installation—there are two files with the same name but with rather different functionality. We believe that this is something that has to be tackled by the maintainers of this language option. The latest version of the thai option is available from ftp://ftp.nectec.or.th/pub/linux.tle/3.0/SOURCES/.

The current version of Ω provides the intis620 ΩCP, which transforms Thai text encoded with TIS–620 to Unicode. In addition, it provides a set of Thai fonts. However, ΩCPs are useless when it comes to word-breaking, and one has to resort to an external ΩCP. Here is a skeleton Λ document that shows how things should be done:

```
. . . . . . . . . . . . . . . . . . . . . . . . . . . .
\ocp\InTIS=tis620
\externalocp\WordBreaker=swath.pl {}
```

```
\ocplist\ThaiProcList=
    \addbeforeocplist 100 \WordBreaker
    \addbeforeocplist 101 \InTIS
    \nullocplist
.........................
\begin{document}
\pushocplist\ThaiProcList
.........................
```

Here is the source of the Perl script that this Λ input file uses:

```
#!/usr/bin/perl
open OUT, "| cttex > tmp.cttex";
while (<STDIN>) {
  print OUT;
}
close OUT;
open IN, "tmp.cttex";
while (<IN>) {
  print STDOUT;
}
close IN;
```

The code is presented without explanation as we do not plan to teach Perl. For this purpose, the reader should consult the Perl bible [29].

10.15 The Bahasa Indonesia Language

The official language of Indonesia is called Bahasa Indonesia. The word bahasa in Indonesian means "language," so Bahasa Indonesia is the Indonesian language! This language uses the basic Latin alphabet (i.e., there are no accented or "strange"-looking letters). The bahasa option (by Jörg Knappen) just provides a translation of the language-dependent fixed words for "chapter," "caption," and so on.

10.16 The Slovenian Language

The slovene option of the babel package can be used to typeset Slovenian language documents with LaTeX. The shorthand "c, where c is one of the letters c, C, s, S, z, or Z, produces the letter č. To get guillemets, one has to use the shorthands "< and ">. The shorthands "' and "' produce the symbols „ and ", respectively. Naturally, by using the inputenc package with the latin2 option, one can avoid using all of these shorthands.

10.17 The Romanian Language

The romanian option of the babel package provides only translations of the various predefined names, so this package is not particularly useful for Romanian language document typesetting. More features are provided by the romanian package (by Adrian Rezus). This package, which cannot coexist with the babel package, allows the typesetting of multilingual documents. The languages supported are Romanian, English, French, and German. The commands \romanianTeX, \originalTeX, \germanTeX, and \frenchTeX are used to switch languages. The shorthands "a and "A produce the symbols ă and â. Note that there are no uppercase forms of these shorthands. In addition, the shorthands "i, "I, "s, "S, "t, and "T produce the symbols î, Î, ş, Ş, ţ, and Ţ, respectively. Note that it is customary nowadays to put a comma under t and T and not a cedilla.

10.18 The Slovak Language

LaTeX documents written in Slovakian can be processed with the slovak option (by Jana Chlebikova) of the babel package. In addition to changing all of the predefined fixed names, the option provides the \q command, which has as argument one of the letters t, d, l, and L and yields the letters ť, ď, ľ, and Ľ, respectively.

10.19 The Czech Language

The czech option of the babel package can be used to typeset Czech language documents. This option provides the \q command, which functions exactly like the \q command of the slovak option. In addition, it provides the \w command, which puts the ° accent over the letters u and U. As is obvious, one can use the latin2 option of the inputenc package to avoid using these commands. Of course, the same thing applies to Slovak language documents as well.

CSLaTeX (by Jaroslav Šnaidr, Zdeněk Wagner, and Jiří Zlatuška) is a nonstandard LaTeX-based format that has been designed to facilitate the typesetting of either Slovak or Czech language documents. The packages czech and slovak are the main tools to typeset Czech or Slovak documents. The options IL2, T1, and OT1 enable the corresponding font encodings. The options split and nosplit turn the splitting of hyphens on and off.

10.20 The Tibetan Language

The documents written in the Tibetan language can be processed with the otibet package for Λ (by Norbert Preining). The package contains all of the necessary files to process documents writen in Tibetan. The Tibetan text is typed in using either the "Wylie's transcription" or the Unicode v.2.0 transcription. Table 10.1 shows the transcription employed. Double entries in the transcription rows denote Wylie/Unicode transcriptions.

Table 10.1: Transcription table for the otibet package.

Transl.	Tib.	nxa/nna		zha	
ka		ta		za	
kha		tha		'a	
ga		da		ya	
nga		na		ta	
ca		pa		la	
cha		pha		sha	
ja		ba		shxa/ssa	
nya		ma		sa	
txa/tta		tsa		ha	
thxa/ttha		tsha		a	
dxa/dda		dza		/	
dxha/ddha		wa		.,	

To allow Λ to distinguish between a prefixed "g" and a main "g," we have to put "g" in curly brackets. Short passages in Tibetan can be written using the \texttb command, while long passages can be written as follows:

```
{\tbfamily \pushocplist\TibetanInputOcpList ...}
```

➤ **Exercise 10.6** Create an environment Tibetan to be used for Tibetan text. □

The package allows the mixing of English and Tibetan text. However, to write, say, Greek and Tibetan text, as in the example

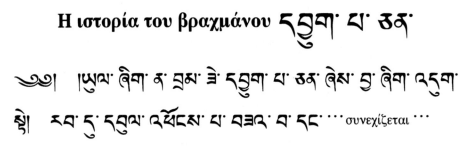

we have to use the following code:

```
\ocp\Greek=in88597
\ocplist\GreekInputOcpList=
\addbeforeocplist 1 \Greek
\nullocplist
\newcommand{\greek}[1]{{\pushocplist\GreekInputOcpList%
\fontfamily{omlgc}\selectfont #1\popocplist}}
```

Now, the heading of the text is coded as follows:

```
\textbf{\greek{H ιστορία του βραχμάνου} \textttb{dbyug pa can }}
```

For more information on Tibetan fonts and software, see http://www.cfynn.dircon.co.uk/Links/bodhsoft_links.html. The otibet package is available from http://www.logic.at/people/preining/tex/tex.html.

10.21 The Japanese Language

The writing system of Japanese is really a very complex one. A regular Japanese document contains a liberal mixture of three separate systems! One system is the kanji, which are the ideographs borrowed from Chinese. Today there are about two thousand kanji ideographs in regular use in Japan. The two other systems, which are generically called kana, are much more simple because they are both syllabic. Katakana, the first syllabary, is more angular and is used mostly for transcribing words of foreign origin. Hiragana is more cursive and can be used for grammatical inflections or for writing native Japanese words where kanji ideographs are not used. Because of the complexity of the Japanese writing system, there are a number of Japanese character set standards, all of which are identified by a code starting with "JIS," which stands for Japanese Industrial Standard. The most popularly used Japanese character set is known as JIS X 0208-1990. It includes 6879 characters, among which are the hiragana and katakana syllables, 6355 kanji ideographs, the Roman, the Greek, and the Cyrillic alphabets, the numerals, and a number of typographic symbols. There are three different ASCII-based encodings that are in common use for Japanese text: the ISO-2022-JP encoding, the EUC (Extended Unix Code) encoding, and the Shift-JIS encoding (usually employed in Microsoft Windows and MacOS).

To typeset a Japanese document using LATEX, one can either use the pLATEX system or Λ. pLATEX sits on top of the pTEX typesetting engine, a modified version of TEX developed for the typesetting of Japanese text by the ASCII Corporation (see http://www.ascii.co.jp/pb/ptex). pLATEX(i.e., LATEX for pTEX) offers modified versions of all standard LATEX document classes: `jarticle` (for articles), `jbook` (for books), and `jreports` (for reports). The figure that follows shows the first few lines of a pLATEX file that describes the new features of a pLATEX release that was available at the second half of the year 2000.

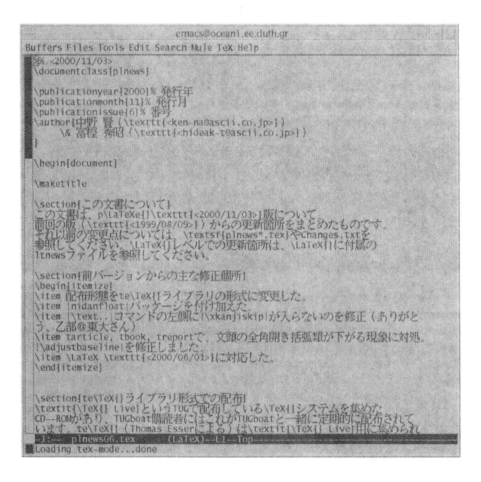

Note that similar files accompany each release of standard LATEX. pTEX also provides modified versions of BIBTEX and MAKEINDEX capable of handling Japanese bibliographies and indices.

By default, pLATEX uses virtual Japanese fonts that are called min10, min9, etc., and goth10, goth9, etc. Therefore, one has to resort to fonts that are available to one's system. For example, on an MS-Windows installation people use TrueType fonts, and on Unix systems the various drivers make use of the VFLIB library, which understands all possible font formats. The situation is better for people working on MacOS. This operating system provides both PostScript Type 1 fonts and TrueType fonts, so they can map min10 to Ryumin-Light and goth10 to GothicBBB-Medium to preview DVI files. Moreover, ASCII Corp. and other publishers who prepare their publications with pLATEX use Japanese PostScript fonts to typeset their products. If you wonder how they manage to do this, the answer is very simple: the resulting PostScript file uses resident PostScript fonts (i.e., fonts that are provided by a PostScript printer). Such PostScript files cannot be directly previewed with Ghostscript, and we need to patch the program to be able to preview our documents. The necessary information is available from http://www.cit.ics.saitama-u.ac.jp/~far/howto/gs-cid.html. Of course, the Omega-j system, described below, has the same font problems, but one can use the Ω virtual fonts bundled with pTEX or the one that is part of Omega-j. Now, for ordinary people, a good solution is to use TrueType fonts. Let us suppose that we have a complete TrueType font at our disposal. Then, we also need the programs TTF2PK and TTF2TFM (see Section 12.6).

The Omega-j system developed by Matt Gushee allows Ω users to typeset Japanese documents written in the ISO-2022-JP encoding. Moreover, Ichiro Matsuda has developed the necessary ΩTP for the two other encodings that are part of the pTEX distribution, while Hideyuki Suzuki has developed an improved version of the original ΩTP for the ISO-2022-JP encoding. The people that have developed pLATEX provide also a Japanese version of Λ that can be used directly to typeset Japanese text, but this version is not portable. A good solution is to create a little package that will be loaded by Λ and that would provide all of the necessary definitions. Here is a little package that the first-named author of this book has created:

```
\ocp\JISInput=injis
\newcommand{\japanese}{\InputTranslation currentfile \JISInput%
    \fontfamily{ommincho}\selectfont}
\endinput
```

Of course, we have to use the new command introduced by this little package at the beginning of the body of a Λ document. Moreover, if we want to write text in another input encoding, we can define similar commands. Naturally, we also need a simple font definition file. Before we present the simple font definition file that we created, we must stress that the author of Omega-j provides an ΩVP file that can be used to create all of the necessary support files, so again we have to find a real font! Of course, one can rename this file and use it also for a boldface font and so on. Anyway, here are the contents of the simple font definition file (see Section 12.4):

```
\ProvidesFile{ot1ommincho.fd}
\DeclareFontFamily{OT1}{ommincho}{}
\DeclareFontShape{OT1}{ommincho}{m}{n}{
  <-> ommincho}{}
\endinput
```

Note that the command \ProvidesFile is used to identify a font definition file. Using these tools, we created a simple Λ file and successfully processed it with Λ. The output of our file follows

私の名前はアポストロスです。

The Omega-j system is available for download from `http://www.havenrock.com/archives/classic/docproc/nihongo/omega-j/index.html`.

10.22   The Spanish Language

Although Spanish is spoken in many areas on the globe, babel still provides only the spanish option (by Javier Bezos) for typesetting Spanish text. If we request Spanish language typesetting with the command

```
\usepackage[activeacute,spanish]{babel}
```

then we can place an acute accent over the letters a, e, i, o, and u, simply by typing ʼℓ, where ℓ is one of these letters. Similar shorthands can be used for uppercase letters. To get the letters ñ and Ñ, use the shorthands ʼn and ʼN, respectively. The shorthands ˜-, ˜--, ˜--- produce a hyphen, an en dash and an em dash, but these commands take care so that a line break does not occur after the dash. The shorthands "u and "c are used to get the letters ü and ç. To get guillemets, we have to use the shorthands "< and ">. The shorthands << and >> behave rather strangely since they place the guillemets, but if they occur inside another pair of << and >> they print the American opening and closing quotation marks. Now, if we have a pair of << and >> that occur inside another pair, which in turn occurs in another pair, then the innermost symbols produce the symbols grave accent and apostrophe. Here is an example that makes things clear:

«El artículo "Estudio de la palabra 'añejo' y sus usos" apenas tenía interés»	`<<El art'iculo <<Estudio de la palabra <<a'nejo>> y sus usos>> apenas ten'ia inter'es>>`

The symbols << and >> are actually shorthands for the commands \begin{quoting} and \end{quoting}. The feature just described is used by default. To deactivate this feature, use the command \deactivatequoting. To reactivate it, use the command \active-quoting. In the past, Spanish grammar dictated that if a word was to be hyphenated before a double r, the double r should be transformed into a single r. To enable this rule, we must use the shorthand "rr, e.g., contra"rreloj. The shorthand "| is used to disable the formation of a ligature, but it is not used at all.

In Spanish mathematical typography, the operators lim, max, and min are accented (e.g., the expression $\lim_{x\to 0} \frac{1}{x} = \infty$ should actually be typeset as $\lím_{x\to 0} \frac{1}{x} = \infty$). To deactivate this feature, use the command \unaccentedoperators. On the other hand, the command \accentedoperators can be used to reactivate this feature. The command \dotlessi is useful to get the letter ı in normal text mode as well as in math mode.

Spanish is also supported by the lambda package. To enable the typesetting of Spanish text, use the spanish option.

10.23 Other Iberian Languages

With the aid of the babel package, it is now possible to typeset LaTeX input files written in Portuguese, Catalan, Galician, and Basque. The corresponding options supporting typesetting in these languages are: portuges, catalan, galician (by Manuel Carriba), and basque (by Juan M. Aguirregabiria).

The portuges option defines the shorthands "|, "<, and ">, which are used to disable ligatures and to produce left and right guillemets. These shorthands are defined in all options described in this section. The catalan option provides a number of shorthands and the following new commands:

Command	Output
\l.l	l·l
\L.L	L·L
\lgem	l·l
\Lgem	L·L

The activeacute option activates the shorthand 'ℓ, where ℓ can be e, i, o, or u, which places an acute accent over ℓ. Similarly, the activeagrave option activates the shorthand 'ℓ, where ℓ can be a, e, or o, which places a grave accent over ℓ. The shorthand "c produces the letter ç, while the shorthands "l and "L produce the same output as the commands \lgem and \Lgem. The galician option provides the following shorthands: 'a (accent over all vowels), 'n (ñ), "u (ü), "a for feminine ordinals as in 2ª, and "o for masculine ordinals as in 2º. The basque option provides the shorthand ~n, which produces the letter ñ.

10.24 The Estonian Language

The typesetting of Estonian input files is currently supported only by babel and the estonian option (by Enn Saar). The shorthand "i, where i is one of the letters a, o, or u, produces the letter ï. The shorthands ~s, ~z, and ~o print the letters š, ž, and õ, respectively. Of course, all of these shorthands can also be used with capital letters. The author of this option recommends that people use the t1enc package to get better hyphenation. Note that this package must be loaded before the babel package.

10.25 The Korean Language

Chinese characters, known as *hanja*, were used to write Korean, in a system called *Ido* (known also as *Ito* and *Idoo*), until the 15th century. However, the system never gained wide acceptance, and its use was very restricted. In 1446, after many years of study and testing by the ruler of the time, King Sejong, and his scholars, a unique Korean alphabet, known as *Hunminjongum*, was introduced. The modern Korean alphabet, *Hangul*, was derived from this earlier form. However, the Chinese characters were not abandoned altogether. For example, Hangul is used almost exclusively in South Korea, but certain newspapers and scholars still use Chinese ideograms, in parentheses, just after words referring to ideas or concepts in general. On the other hand, North Korea has completely abandoned all Chinese characters and uses Hangul exclusively. The KSC-5601 character set, also known as *Korean Wansung*, includes the Hangul, the Roman, the Greek, and the Cyrillic alphabets plus the Chinese ideograms. Note that the official name of KSC-5601 has changed, and it is now known as KSX-1001. Today, most people use the EUC-KR encoding to type their texts.

The HLATEX "system" (by Koaunghi Un) facilitates the typesetting of Korean text with both LATEX and Λ. The basic packages that HLATEX provides are the hfont and the hangul packages. The hfont package is used to activate the use of Korean fonts (which are also part of the HLATEX), but it does not change predefined words such as "chapter," "glossary," and so on. For example, the input file shown below

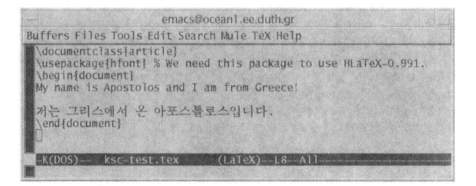

```
emacs@ocean1.ee.duth.gr
Buffers Files Tools Edit Search Mule TeX Help
\documentclass{article}
\usepackage{hfont} % We need this package to use HLaTeX-0.991.
\begin{document}
My name is Apostolos and I am from Greece!

저는 그리스에서 온 아포스톨로스입니다.
\end{document}

-K(DOS)--- ksc-test.tex      (LaTeX)--L8--All
```

will be typeset as follows:

My name is Apostolos and I am from Greece!
저는 그리스에서 온 아포스톨로스입니다.

Note that we will get the same output regardless of what typesetting engine we use (LaTeX, pdfLaTeX, or Λ). If we want to use the Moonhwabu TrueType font, we must use the moonttf package. The hangul package "koreanizes" the predefined words and thus is suitable for real Korean documents. The package provides the following options:

hanja All names (e.g., "chapter," "bibliography," etc.) are typeset in Hanja. Otherwise, they are typeset in Hangul.

hardbold Documents are typeset with real boldface fonts.

softbold Documents are typeset with poor man's bold.

nojosa This option is used to turn off the automatic selection of an appropriate josa. A josa is a functional unit that is used to determine the case of nouns and pronouns. The author of HLaTeX has opted to include the automatic josa selection just because in certain cases the josa has different forms depending on the last syllable of the preceding noun or pronoun. Josa is a linguistic phenomenon of the Korean and Japanese languages.

The commands \textmj, \textgt, and \textttz are used to select a Roman, a sans serif, or a monospaced font family. These commands are "koreanized" versions of the corresponding standard commands. The following shows the default font families supported by HLaTeX:

My name is Apostolos and I am from Greece!
저는 그리스에서 온 아포스톨로스입니다.
My name is Apostolos and I am from Greece!
저는 그리스에서 온 아포스톨로스입니다.
My name is Apostolos and I am from Greece!
저 는 그 리 스 에 서 온 아 포 스 톨 로 스 입 니 다.

Numbering in general can be altered by using the following commands:

Command	Output
\jaso:	ㄴ ㄷ ㄹ ㅁ ㅂ ㅅ ㅇ ㅈ ㅊ
\gana:	ㅏ 나 다 라 마 바 사 아 자 차
\ojaso:	㉡ ㉢ ㉣ ㉤ ㉥ ㉦ ㉧ ㉨ ㉩
\ogana:	㉯ ㉰ ㉱ ㉲ ㉳ ㉴ ㉵ ㉶ ㉷ ㉸
\pjaso:	㈀ ㈁ ㈂ ㈃ ㈄ ㈅ ㈆ ㈇ ㈈
\pgana:	㈎ ㈏ ㈐ ㈑ ㈒ ㈓ ㈔ ㈕ ㈖ ㈗
\onum:	① ② ③ ④ ⑤ ⑥ ⑦ ⑧ ⑨ ⑩
\oeng:	ⓐ ⓑ ⓒ ⓓ ⓔ ⓕ ⓖ ⓗ ⓘ ⓙ
\peng:	(a) (b) (c) (d) (e) (f) (g) (h) (i) (j)

Note that all of these commands expect a counter as their only argument.

➤ **Exercise 10.7** Give the command that will set the page numbering to \gana. ☐

The HLᴬTᴇX system provides also the halpha bibliography style and two style files for the generation of indices and glossaries.

10.26 The Hebrew Language

The Hebrew language is supported through the hebrew option of the babel package (by Boris Lavva). This option assumes that we are actually using ε-TᴇX as our typesetting engine. In addition, text using only consonants is fully supported. Support for vowels (*nikud*) is not yet available through the standard packages, but one may find the necessary files to use vowels at Sivan Toledo's homepage at http://www.math.tau.ac.il/~stoledo. The following is an example from the documentation of his work:

וַיָּשִׂימוּ עָלָיו שָׂרֵי מִיסִים לְמַעַן עַנֹּתוֹ בְּסִבְלֹתָם.

The Hebrew script has a right-to-left writing direction. If we want to type in a Hebrew input file, we need a tool that provides both the fonts and the capability to write text from right to left. A good choice is to use LʏX from http://www.lyx.org, which fully supports bidirectional writing. Information about setting up LʏX to use Hebrew can be found on Toledo's homepage mentioned above. However, one can always use the babel transliteration for the Hebrew letters if a short passage is to be written. The following table shows the transliteration employed:

'	a	b	c	d	e	f	g	h	i	j	k	l	m
א	ב	ג	ד	ה	ו	ז	ח	ט	י	ך	כ	ל	ם
n	o	p	q	r	s	t	u	v	w	x	y	z	
מ	ן	נ	ס	ע	ף	פ	ץ	צ	ק	ר	ש	ת	

As is customary for the babel package, one can write in Hebrew without having a Hebrew-enabled keyboard by using the correspondence of the Latin letters to the Hebrew ones. If a Hebrew keyboard is available, then one may load the inputenc package with the appropriate option: 8859–8 for Unix systems, cp1255 for Microsoft Windows, cp862 for the IBM code page usually found on DOS, and si960 for the "old-code" 7-bit Israeli Standard Hebrew encoding. It should be noted here that up to now only the Microsoft code page (cp1255) supports the input of vowels and dots (nikud).

The command \R{*Hebrew text*} switches to Hebrew and to the right-to-left direction in order to typeset the *Hebrew text*. Similarly, the command \L{*Non-Hebrew text*} is used to switch to another language (left-to-right). These two commands can be used in the middle of a paragraph, thus allowing us to set a Hebrew word in a non-Hebrew paragraph and vice versa. For example, with default language hebrew and with american and greek loaded, the input

```
\R{abg} \L{\textlatin{abc}} \R{abg} \L{\textgreek{abg}}
```

will produce: בגת abc בגת αβγ

The \extrashebrew command changes to Hebrew encoding and to right-to-left writing direction. This is undone by \noextrashebrew. Two useful box commands are \hmbox and \embox. The first one is for creating a right-to-left box, whereas the second one is for the left-to-right direction. These are very useful when writing Hebrew inside math:

$$\int\limits_{אבדה} f(x) = 1$$

was produced by the input

```
$\int\limits_{\hmbox{\scriptsize 'acd}} f(x)=1$
```

The \hebmonth{*month*} command produces month names in Hebrew. The command \hebdate{*day*}{*month*}{*year*} translates a given Gregorian date to Hebrew, while \hebday replaces the \today command in Hebrew documents. The \datehebrew command redefines the command \today to produce Gregorian dates in Hebrew. Here are two examples:

```
\hebdate{26}{12}{2000}     26 בדצמבר 2000
\hebday                    29 באוגוסט 2001
```

The hebrew option of the babel package supports most aspects of document typesetting. This includes a right-to-left table of contents, table of figures, headers, footnotes, sectioning commands, and so on. To create a right-to-left table of contents, list of figures, and list of tables, we use the commands \rltableofcontents, \rllistoffigures, and \rllistoftables, respectively. By reversing the first two letters of these commands, we get the lists in the left-to-right direction. Thus, the following commands are available for left-to-right lists: \lrtableofcontents, \lrlistoffigures, and \lrlistoftables. The command \captionshebrew will change the caption names with the Hebrew equivalents.

The work of Toledo on nikud is using the ligature mechanism of TeX in order to access precomposed accented letters (the accents are used to specify the correct vowel on a syllable). One font that can be used is the Hebrew font provided by the Ω project, which has been properly modified to work with LaTeX.

Ω has most of the files for Hebrew support ready. Fonts are available (that support nikud as well) and an ΩCP for translating the ISO-8859-8 to Unicode (called in88598.ocp). But, the current *virtual* fonts must incorporate support for Hebrew. Support for the Microsoft code page (CP-1255) is still to be added as well as the necessary commands for calling the Hebrew scripts. The following defines a simple environment that permits the typesetting of Hebrew text with Λ:

```
\ocp\NewHebrew=in88598
\ocplist\NewHebrewList=
  \addbeforeocplist 1 \NewHebrew
  \nullocplist
```

```
\newenvironment{newhebrew}{%
  \textdir TRT\pardir TRT\pushoplist\NewHebrewList%
}{\popocplist}
```

10.27 The Cyrillic Script

The Russian, Bulgarian, and Ukrainian languages use the Cyrillic script and are supported by the babel options russianb, bulgarian and ukrainian, respectively. The input encodings supported by the inputenc package are: iso88595 for Unix systems, cp1251 for Microsoft Windows, and maccyr for the Macintosh. Other supported input encodings include: koi8-r, koi8-ru, koi8-u, cp855, and more. Of course, the choice of input encoding depends heavily on the operating system that one uses to prepare an input file. Naturally, sticking to ISO standards is the best solution.

The following code snippet shows what we have to type to typeset a document that conains text fragments that use three different scripts. Note that we have to use the \inputencoding command to (locally) change the input encoding:

```
{\selectlanguage{russian}\inputencoding{iso88595}
\begin{verse} ''Russian text'' \end{verse}}
{\selectlanguage{greek}% Greek text in Latin transliteration
\begin{verse} ''Greek text'' \end{verse}}
\begin{verse} ''English text'' \end{verse}
```

will be typeset as follows:

> Я помню это чудное мгновение
> Когда передо мной явилась ты. (Пушкин)

> Θυμάμαι εκείνη την εκπληκτική στιγμή
> Όταν μπροστά μου εμφανίστηκες εσύ. (Πούσκιν)

> I remember that wonderful moment
> when you appeared in front of me. (Puchkin)

Observe that we had to enter the Greek text using its Latin transliteration, as it is not possible to have three different scripts in an extended ASCII file! Of course, this is possible when typing in the input file using Unicode.

The fonts for Cyrillic script are available in several encodings. The default is the T2A encoding, which contains both Cyrillic and the standard English letters. Other encodings are the X2, OT2, LCY, and LWN, and they are made available if loaded as an option with the fontenc[6] package; for example to use the X2 encoding, we should use

---

6. This package is used to select the font encoding of a particular script. For example, for the Latin script there are two font encodings: the T1 (or Cork encoding) and the OT1 encodings.

```
\usepackage[X2]{fontenc}
```

The different font encodings offer different characters. For example, the characters «
and » are not available with the OT2 encoding but are available with the T2A. Although
the T2A encoding is the default, the X2 encoding is more complete, as it contains no
Latin letters. With X2, the user has more Cyrillic characters available. In addition to
this, users are forced to switch languages using standard (babel-type) commands, and
LaTeX pays back with correct hyphenation for all languages used. To locally switch to
the Cyrillic script, we use the command

```
\textCyrillic{Cyrillic text}
```

For a global switch to the Cyrillic script, we use the \Cyrillictext command instead.
The standard commands \textlatin and \latintext switch back to the Latin alphabet
locally and globally, respectively (all of these commands are not needed for the default
T2A encoding unless a third language other than English and a language that uses
the Cyrillic script is to be used). Of course, to turn on the proper hyphenation for the
languages that use the Cyrillic script we use, for a short phrase, the command

```
\foreignlanguage{bulgarian}{Bulgarian text}
```

should be preferred (here written for Bulgarian).

The \daterussian redefines the \today command to produce Russian dates and
similarly for the commands \datebulgarian and \dateukrainian.

An important issue to be noted for these languages is that the character " is made
active and used as in Table 10.2.

The quotes in Table 10.2 can also be typeset by using the commands in Table 10.3.

The French quotes are also available as ligatures << and >> in Cyrillic font encodings
LCY, X2, T2A, T2B, and as < and > characters in Cyrillic font encodings OT2 and LWN.

Currently, Ω provides the following ΩTPs:

in88595 Suitable for conversion to Unicode from ISO-8859-5 (Latin/Cyrillic).

inav Can be used to convert to Unicode from Cyrillic Alternativnyi Variant.

incp1251 When the input file has been prepared using the Cyrillic MS-DOS encoding
CP-1251.

incp866 Suitable for conversion to Unicode from Cyrillic MS-DOS encoding CP-866.

inkoi8 For conversions to Unicode from Cyrillic KOI-8 (GOST 19769-74).

inov Can be used to convert to Unicode from Cyrillic Osnovnoj Variant.

inucode For conversion to Unicode from Cyrillic U-code.

Also, the current version of Ω fonts does support the Cyrillic alphabet, so it is easy
to prepare documents with Λ. Of course, what remains to be done is to change the
standard names in the predefined documented classes.

Table 10.2: The extra definitions made by russianb, bulgarian, and ukrainian.

"|	disable ligature at this position.
"-	an explicit hyphen sign, allowing hyphenation in the rest of the word.
"---	Cyrillic em dash in plain text.
"--~	Cyrillic em dash in compound names (surnames).
"--*	Cyrillic em dash for denoting direct speech.
""	like "-, but producing no hyphen sign (for compound words with hyphen; e.g. x-""y or some other signs such as as "disable/enable").
"~	for a compound word mark without a breakpoint.
"=	for a compound word mark with a breakpoint, allowing hyphenation in the composing words.
",	thin space for initials with a breakpoint following surname.
"‘	for German left double quotes („).
"›	for German right double quotes (").
"<	for French left double quotes («).
">	for French right double quotes (»).

Table 10.3: More commands that produce dashes and quotes.

\cdash---	Cyrillic emdash in plain text.
\cdash--~	Cyrillic emdash in compound names (surnames).
\cdash--*	Cyrillic emdash for denoting direct speech.
\glqq	for German left double quotes („).
\grqq	for German right double quotes (").
\flqq	for French left double quotes («).
\frqq	for French right double quotes (»).
\dq	the original quotes character (").

10.28   The Armenian Language

The armtex package (by Serguei Dachian, Arnak Dalalyan, and Vardan Hakobian) is an effort to provide a complete solution to the problem of typesetting Armenian documents with LaTeX.

The package is loaded in the usual way, by putting the following command in the document's preamble:

\usepackage[*options*]{armtex}

It accepts several options. The most important one is the option latin. When this option is not used, the document language is changed to Armenian; that is, the entire document (including the main text, the headers, the table of contents, words such as "chapter," "appendix," etc.) is typeset in Armenian. When this option is used, the document language is not changed to Armenian, but, nevertheless, all of the commands described below become available.

The package automatically loads the OT6 font encoding since it uses Armenian fonts conforming to this encoding. So, using the fontenc package to explicitly load the OT6 encoding is unnecessary (and may even cause some LaTeX errors).

If a user has a standard Armenian keyboard to type in an input file, it is necessary to use the ARMSCII8 input encoding: that is, the following command must appear in the document's preamble:

\usepackage[armscii8]{inputenc}

Otherwise, the text must be entered using the following transliteration:

Ա ա	A a	Հ հ	I i	Յ յ	Y y	Ս ս	T t
Բ բ	B b	Լ լ	L l	Ն ն	N n	Ր ր	R r
Գ գ	G g	Խ խ	X x	Շ շ	Sh sh	Ց g	C c
Դ դ	D d	Ծ ծ	C' c'	Ո ո	O o	Ւ ւ	W w
Ե ե	E e	Կ կ	K k	Չ չ	Ch ch	Փ փ	P' p'
Զ զ	Z z	Հ h	H h	Պ պ	P p	Ք ք	Q q
Է է	E' e'	Ձ ձ	Dz dz	Ջ ջ	J j	ու	ev
Ը ը	U' u'	Ղ ղ	Gh gh	Ռ ռ	R' r'	Օ o	O' o'
Թ թ	T' t'	Ճ ճ	J' j'	Ս u	S s	Ֆ ֆ	F f
Ժ ժ	G' g'	Մ մ	M m	Վ վ	V v	ու ու	U u

Most punctuation and general symbols can be accessed with the expected keystrokes. The following table shows the very few exceptions:

!	´	\!	!
?	ˆ	\?	?
|	´	\\|	—

As we have already seen, mixing scripts with LaTeX is a cumbersome task. In the case of the armtex package, there are three types of font-changing commands.

1. Orthogonal commands that work like \itshape or \bfseries:

   \artmfamily, \arssfamily, \armdseries, \arbfseries,
   \arupshape, \aritshape, \arslshape.

2. Orthogonal commands that work like \textit or \textbf:

\armtm, \armss, \armmd, \armbf, \armup, \armit, \armsl.

3. Commands that are not orthogonal, which work like \it or \bf:

\artm, \artmit, \artmsl, \artmbf, \artmbfit, \artmbfsl,
\arss, \arsssl, \arssbf, \arssbfsl.

Besides choosing Armenian fonts, these commands enter Armenian mode (if needed); that is, they switch to Armenian encoding and set up the commands \- and \today to be Armenian. The declaration \aroff is used to leave the Armenian mode. The commands \rm, \bf, and others are redefined to leave the Armenian mode automatically.

▶ **Exercise 10.8** Define the command \noarmtext that can be used in Armenian mode to typeset its argument using the Latin alphabet. □

The declaration \armdate can be used to have \today produce the current date in the Armenian way. The declaration \armdateoff cancels the effect of \armdate and so dates are printed in the language that was in use before we switched to the Armenian language. Here is an example:

\today	19 սեպտեմբերի 2001թ.
\armdateoff \today	19th September 2001

The declaration \armhyph sets up the command \- to be Armenian. The declaration \armhyphoff cancels the effect of \armhyph.

The commands \armnamesoff and \armnames can be used to force LaTeX to print in the output file the words "chapter," "appendix," and so on in either English or Armenian. Basically, one does not need to mess with these commands, but they can become useful when typesetting a multilingual document combining, for example, babel and armtex packages.

If we want to use Armenian letters for mathematical symbols, we can use the following mathematical font selection commands: \mathartm, \mathartmit, \mathartmbf, and \mathartmbfit.

For additional information, the interested reader may refer to the very detailed manual (in Armenian) distributed with the package.

The following shows the names of the authors in Armenian, but which name is which?

Ապոստոլոս Սիրոպուլոս & Անտոնիս Չոլմիդիս.

10.29 The Polish Language

To typeset Polish text, one can use the polish option of the babel package (by Elmar Schalueck and Michael Janich). This option provides a number of shorthands that can be used to print additional Polish letters. More specifically, the shorthands "a, "A, "e, and "E print the letters ą, Ą, ę, and Ę, respectively. In addition, the shorthands "l, "L, "r, "R, "z, and "Z print the letters Ł, ł, ż, Ż, ź, and Ź, respectively. Guillemets can be printed with the shorthands "< and ">, while "' and "' produce the opening („) and closing (") quotation marks.

The PLATEX system (by Mariusz Olko and Marcin Woliński) has been specifically designed to typeset Polish language documents. The system consists of a version of LATEX with the hyphenation patterns of the Polish langauge preloaded, the polski package, and some other support files. The system solves the problem of directly entering Polish language text, much like inputenc does. However, the polski package offers a number of options that make things even easier. plmath should be used to load Polish language fonts for use in mathematical text. nomathsymbols deactivates the use of Polish names for standard mathematical functions. The options T1, OT1, and OT4 activate the corresponding font encodings, and last, the MeX option allows the processing of LAMEX files. LAMEX is a special LATEX format that is based on an obsolete version of LATEX.

10.30 The Georgian Language

The Georgian language uses two scripts the *Mkhedruli* (secular script) and the *Khutsuri* (ecclesiastic script). Mkhedruli is a caseless script (i.e., there is only one form for each letter) and is the official script of Georgia. The Khutsuri script is used mainly by the Georgian Orthodox Church. Currently, only the mxedruli and the xucuri packages (by Johannes Heinecke) provide rudimentary support for typesetting Georgian language documents with LATEX. The two input encodings Georgian-Academy and Georgian-PS have not found their way into the TEX/Ω world so one must type in Georgian text using a transliteration. For the Mkhedruli script, the following transliteration is employed:

ა	a	ი	i	რ	r	შ	+s
ბ	b	კ	.k	ს	s	ჩ	+c
გ	g	ლ	l	ტ	.t	ც	c
დ	d	მ	m	უ	u	ძ	j
ე	e	ნ	n	ფ	p	წ	.c
ვ	v	ო	o	ქ	k	ჭ	.+c
ზ	z	პ	.p	ღ	.g	ხ	x
თ	t	ჟ	+z	ყ	q	ჯ	+j
						ჰ	h

The mxedruli package provides the commands \mxedr, \mxedb, \mxedi, and \mxedc, which select the Roman, the boldface, the italic, and the caps and small caps typefaces, respectively. It is even possible to put accents above Georgian letters (this is a feature of some Kartvelian languages). The commands \^, \=, and \" are used to put a circumflex, a macron, or an umlaut on the letter that follows the command. The following table shows the transliteration employed for the *Khutsuri* script:

Ⴆ ⴒ	A, a	ⴈ ⴈ	I, i	ⴃ dh	R, r	ⴢ y	+S, +s
Ⴢ y	B, b	ⴉ ⴉ	K, .k	ⴑ h	S, s	ⴗ h	+C, +c
Ⴀ ⴀ	G, g	ⴒ ⴍ	L, l	ⴐ p	.T, .t	ⴍ q	C, c
Ⴖ ⴟ	D, d	ⴃ ⴃ	M, m	ⴍ uy	U, u	ⴓ th	J, j
ⴈ ⴋ	E, e	ⴉ ⴌ	N, n	Ⴔ Ⴔ	P, p	ⴓ	.C, .c
Ⴑ ⴑ	V, v	ⴍ ⴍ	O, o	ⴕ ch	K, k	ⴕ q	,+C, .+c
ⴇ ⴈ	Z, z	ⴓ ⴓ	.P, .p	ⴌ n	.G, .g	ⴖ p	X, x
ⴈ ⴋ	T, t	ⴋ ⴗ	+Z, +z	ⴏ y	Q, q	Ⴞ x	+J, +j
						ⴒ ⴒ	H, h

Currently, there is only one upright font available for this script. Note that numbers and punctuation symbols have no special transliteration.

10.31 The Ethiopian Language

The ethiop package (by Berhanu Beyene, Manfred Kudlek, Olaf Kummer, and Jochen Metzinger) is a complete solution to the problem of typesetting documents written in any of the Ethiopian languages with either LATEX or Λ. The package is an extension of EthTEX, and it has been designed as if it were an option of the babel package. This means that if we load both the babel and the ethiop packages, then we can actually use \selectlanguage{ethiop}. Of course, if we want to typeset a monolingual document, there is no reason to load the babel package.

The typing of Ethiopian text is not a straightforward task. Unless we use a Unicode editor to type in our text, we do need to master the transliteration that the authors of ethiop have devised. The transliteration is based on the observation that the letters of the Ethiopian languages do represent syllables that start with a consonant and are followed by a vowel. Table 10.4 shows the letters that the ethiop package recognizes. To enter any of the letters, we type the consonant on the left and the vowel at the top. For example, the letter ቂ is pronounced *qi* and is entered as qi. Accented consonants are entered by prefixing the consonant with a symbol that denotes the accent (e.g., .p). In addition, capital vowels denote long vowels.

Although the punctuation symbols look different from the punctuation symbols used in other scripts (e.g., the Latin, the Greek, or the Cyrillic) they still have essentially the same meaning. The upper row of the following table presents the input and the lower row the output that we get:

:=	:-	::	,	;			:|:	?	'?	!	'!	...
፥	፦	።	፣	፤	፡	፨	?	፧	!	፤	...

Table 10.4: The Ethiopian characters.

		a	u	i	ā	ē	e	o	wa	wi	wā	wē	we
		a	u	i	A	E	e	o	ua	ui	uA	uE	ue
			U	I				O		uI			if preferred
h	h	ሀ	ሁ	ሂ	ሃ	ሄ	ህ	ሆ			ሗ		
l	l	ለ	ሉ	ሊ	ላ	ሌ	ል	ሎ			ሏ		
ḥ	.h	ሐ	ሑ	ሒ	ሓ	ሔ	ሕ	ሖ			ሗ		
m	m	መ	ሙ	ሚ	ማ	ሜ	ም	ሞ			ሟ	ሟ	ሟ
ś	's	ሠ	ሡ	ሢ	ሣ	ሤ	ሥ	ሦ			ሧ		
r	r	ረ	ሩ	ሪ	ራ	ሬ	ር	ሮ			ሯ		
s	s	ሰ	ሱ	ሲ	ሳ	ሴ	ስ	ሶ			ሷ		
š	^s	ሸ	ሹ	ሺ	ሻ	ሼ	ሽ	ሾ			ሿ		
q	q	ቀ	ቁ	ቂ	ቃ	ቄ	ቅ	ቆ	ቈ	ቊ	ቋ	ቌ	ቍ
q̇	.q	ቐ	ቑ	ቒ	ቓ	ቔ	ቕ	ቖ		ቍ			
b	b	በ	ቡ	ቢ	ባ	ቤ	ብ	ቦ	ቧ	ቧ	ቧ	ቧ	ቧ
v	v	ቨ	ቩ	ቪ	ቫ	ቬ	ቭ	ቮ			ቯ		
t	t	ተ	ቱ	ቲ	ታ	ቴ	ት	ቶ			ቷ		
č	^c	ቸ	ቹ	ቺ	ቻ	ቼ	ች	ቾ			ቿ		
ḫ	_h	ኀ	ኁ	ኂ	ኃ	ኄ	ኅ	ኆ	ኈ	ኊ	ኋ	ኌ	ኍ
n	n	ነ	ኑ	ኒ	ና	ኔ	ን	ኖ			ኗ		
ñ	~n	ኘ	ኙ	ኚ	ኛ	ኜ	ኝ	ኞ			ኟ		
'	'	አ	ኡ	ኢ	ኣ	ኤ	እ	ኦ	ኧ				
k	k	ከ	ኩ	ኪ	ካ	ኬ	ክ	ኮ	ኰ	ኲ	ኳ	ኴ	ኵ
ḵ	_k	ኸ	ኹ	ኺ	ኻ	ኼ	ኽ	ኾ	ዀ	ዂ	ዃ	ዄ	ዅ
w	w	ወ	ዉ	ዊ	ዋ	ዌ	ው	ዎ					
‘	‘	ዐ	ዑ	ዒ	ዓ	ዔ	ዕ	ዖ					
z	z	ዘ	ዙ	ዚ	ዛ	ዜ	ዝ	ዞ			ዟ		
ž	^z	ዠ	ዡ	ዢ	ዣ	ዤ	ዥ	ዦ			ዧ		
y	y	የ	ዩ	ዪ	ያ	ዬ	ይ	ዮ	ዮ				
d	d	ደ	ዱ	ዲ	ዳ	ዴ	ድ	ዶ			ዷ		
ḍ	.d	ዸ	ዹ	ዺ	ዻ	ዼ	ዽ	ዾ			ዿ		
ǧ	^g	ጀ	ጁ	ጂ	ጃ	ጄ	ጅ	ጆ			ጇ		
g	g	ገ	ጉ	ጊ	ጋ	ጌ	ግ	ጎ	ጐ	ጒ	ጓ	ጔ	ጕ
ġ	.g	ጘ	ጙ	ጚ	ጛ	ጜ	ጝ	ጞ	ጐ	ጒ	ጓ	ጔ	ጕ
ṭ	.t	ጠ	ጡ	ጢ	ጣ	ጤ	ጥ	ጦ			ጧ		
č̣	^C	ጨ	ጩ	ጪ	ጫ	ጬ	ጭ	ጮ			ጯ		
p̣	.p	ጰ	ጱ	ጲ	ጳ	ጴ	ጵ	ጶ			ጷ		
ṣ	.s	ጸ	ጹ	ጺ	ጻ	ጼ	ጽ	ጾ			ጿ		
ç	.c	ፀ	ፁ	ፂ	ፃ	ፄ	ፅ	ፆ					
f	f	ፈ	ፉ	ፊ	ፋ	ፌ	ፍ	ፎ	ፈ	ፈ	ፏ	ፏ	ፏ
p	p	ፐ	ፑ	ፒ	ፓ	ፔ	ፕ	ፖ	ፐ	ፐ	ፗ	ፗ	ፗ
q́	'q												
k̂	'k												
h́	'h												
ǵ	'g												

~mA	(glyph)
~ri	(glyph)
~fi	(glyph)

Arabic numerals are gaining widespread acceptance in the Ethiopian script, but today the Ethiopian numerals are still used in many situations. When we type in Arabic numerals, they are printed as they are. To get Ethiopian numerals, we have to use the command \ethnum. The argument of this command can be either a number or a counter, and the value of its argument cannot be greater than 999,999. The \today command does not print the current Gregorian date where the name of the month is printed using the Ethiopian script, but prints the date according to the Ethiopian calendar. In the following table, we demonstrate the use of the commands presented in this paragraph:

Command	Output
\ethnum{2002}	፳፻፪
\ethnum{999999}	፱፻፺፱፻፺፱፻፺፱
\today	አኅጉን 4 1993 (September 9, 2001)

To have Ethiopian letters in a mathematical formula, use the \ethmath command.

➤ **Exercise 10.9** Write the code that produces the following formula:

$$\sin ሐ^2 + \cos ሐ^2 = 1$$

☐

The ethiop package provides rudimentary Λ support: users are expected to type in text using a Unicode editor. Of course, this makes it easy to type in Ethiopian text, but it is our firm belief that the TEX/Ω community must develop a new package that will incorporate Ethiopian and all of the languages covered in this chapter in a uniform way.

10.32 The Serbian Language

The serbian option of the babel package allows users to typeset Serbian documents written with the Latin script. However, there are plans to provide the necessary facilities so that the typesetting of Serbian documents written with the Cyrillic script would also be possible. Naturally, the serbian option defines a number of shorthands, which we are going to describe now. The character sequence "ℓ, where ℓ can be one of the letters c, C, s, S, z, or Z, prints the "letter" $\check{\ell}$. The shorthands "d and "D print the letters đ and Đ, respectively. In addition, the shorthands " ', " ', "<, and "> are used to print the symbols „ and " and the French guillemets.

10.33 The Sorbian Languages

The Sorbian language belongs to the Slavic family of languages and is closely related to Polish, Kashubian, Czech, and Slovak. These languages, along with the extinct Pomeranian (for example, Slovincian) and Polabian (for example, Draveno-Polabian in the Hanoverian Wendland), compose the West Slavic language group. Sorbian is still used in Upper and Lower Lusatia, where the Old Sorbian tribes of the Milceni and Luzici settled. The babel package provides the usorbian and the lsorbian options (both by Eduard Werner) that provide support for the Sorbian language spoken in Upper and Lower Lusatia, respectively. As far as it regards babel, Lower Sorbian is a "dialect" of Upper Sorbian. This simply means that one can use the lsorbian option only if the usorbian option is loaded.

The command \olddatelsorbian should be used when we want the \today command to print the date in an old-fashioned way. The command \newdatelsorbian causes \today to print the date in a "modern" way. The commands \olddateusorbian and \newdateusorbian have similar effects but for the Upper Sorbian language. The lsorbian option does not define any shorthands, but the usorbian option defines a number of shorthands. First of all, the shorthand "ℓ, where ℓ is one of the letters a, A, o, O, u, U, e, E, i, and I, prints an umlaut above ℓ. The shorthands "s and "S produce the letters ß and SS. As in the case of the ngerman option, "c, "f, and so on, are used to hyphenate cc as cc-c. In addition, the shorthands "', "‘, "<, and "> produce the German quotation marks, „ and ", and the opening and closing guillemets, « and », respectively.

10.34 The Croatian Language

The current version of the croatian option of the babel package supports only the Latin script that is used for Croatian. However, Darco Žubrinić has created a set of fonts suitable for the typesetting of various Glagolitic and Croatian Cyrillic scripts. The following is a specimen of the various forms of the Croatian scripts:

Script Type	Example
"Round type" Glagolitic	✝ⰉⰒⰔⰟⰟⰉⰞⰗⰉⰒ
Baška Glagolitic	✝ⰍⰘⰌⰟⰟⰌⰘⰈⰌⰞ
Angular Glagolitic	ⱈⰍⰆⰒⰍⰆⰘⰈⰒ
Croatian Cyrillic	ḋпостолос

Unfortunately, these fonts have not found their place in the LaTeX world—one can use them with the \font command. Of course, it is a straightforward exercise to write down the necessary font definition files and a supporting package that will provide commands such as \textangglag and so on. Such a command should simply select the appropriate font and typeset its argument using this font. For more information on these fonts, see [31, 32].

10.35 The Perso-Arabic Languages[7]

The ArabTEX system (by Klaus Lagally) is a bundle of LATEX packages that allows people to typeset virtually any kind of Arabic text. In addition, the system provides support for Hebrew document typesetting. Another option is to use the omega package that is part of the Ω distribution. However, ArabTEX is rather slow and, in our humble opinion, it is not a system suitable for the typesetting of really long documents. Nevertheless, since ArabTEX is an extremely intelligent system (e.g., it can typeset in a right-to-left direction without using any of the advanced features of Ω or ε-TEX), we will briefly describe the system and then describe the omega package.

We begin our brief presentation of ArabTEX by giving the contents of a sample input file and the typeset output (see Figure 10.1).

```
\documentclass[a4paper]{article}
\usepackage{arabtex}
\sloppy \frenchspacing
\begin{document}

\setarab \transfalse
\setnashbf \Large
\centerline {<nawAdiru>}
\normalsize
\centerline {<^gu.hA wa-.hamIruhu al-'a^saraTu>}

\setnash
\begin{arabtext}
i^starY ^gu.hA 'a^saraTa .hamIriN.
fari.ha bihA wa-sAqahA 'amAmahu,
    . . . . . . . . . . . . . . . . . . . . . . . . .
\end{arabtext}
\begin{center}
This is not Arabic text! %%%%%% ''ordinary'' text
\end{center}
\setnashbf
\transtrue
\centerline {<al-waladu wa-al-.t.tablu>}
```

7. The term refers to languages that use the Perso-Arabic script.

```
\setnash
\begin{arabtext}
.talaba waladuN min 'abIhi 'an ya^stariya lahu .tablaN .sa.gIraN.
........................
\end{arabtext}
\end{document}
```

نَوَادِرُ

بُخَا وَحَيِرُهُ العَشَرَة

إِشتَرَى بُخَا عَشَرَة حَمِيرٍ. فَرِحَ بِهَا وَسَاقَهَا أَمَامَهُ، ثُمَّ رَكِبَ وَاحِدًا مِنهَا. وَفِي الطَّرِيقِ عَدَّ حَمِيرَهُ وَهُوَ رَاكِبٌ، فَوَجَدَهَا تِسعَةً. ثُمَّ نَزَلَ وَعَدَّهَا فَرَآهَا عَشَرَة فَقَالَ: أَمشِي وَأَكسِبُ حِمَارًا، أَفضَلُ مِن أَن أَركَبَ وَأخسَرَ حِمَارًا.

This is not Arabic text!

الوَلَدُ وَالطَّبلُ *al-waladu wa-'t-tablu*

ṭalaba waladun min ʾabīhi ʾan yaštariya lahu ṭablan ṣaġīran. fa-rafaḍa 'l-wā-lidu, wa-qāla lahu: yā bunayya, law-i

طَلَبَ وَلَدٌ مِن أَبِيهِ أَن يَشتَرِي لَهُ طَبلًا صَغِيرًا. فَرَفَضَ الوَالِدُ، وَقَالَ لَهُ: يَا بُنَيَّ، لَو

'štaraytu laka ṭablan fa-sawfa tuzaʿiġunā bi-ṣawtihi.

اشتَرَيتُ لَكَ طَبلًا فَسَوفَ تُزَعِجنَا بِصَوتِهِ.

qāla 'l-waladu: lā taġḍab yā ʾabī. lā ʾuṭabbilu bihi, ʾillā wa-ʾanta nāʾimun.

قَالَ الوَلَدُ: لَا تَغضَب يَا أَبِي. لَا أُطَبِّلُ بِهِ، إِلَّا وَأَنتَ نَائِمٌ.

Figure 10.1: Output of a sample ArabTₑX input file.

To typeset documents written in the Arabic script, we must load the arabtex package. Then, we must choose a language; in this case we have chosen the Arabic language with the command \setarab. Other possible choices include the modern Iranian language called Farsi (\setfarsi), Urdu (\seturdu), the official literary language of Pakistan, Pashto (\setpashto), the principal vernacular language of Afghanistan and parts of western Pakistan, the Arab dialects spoken in Maghreb (\setmaghrib), Uigur (\set-uigur), the language of the Turkic people inhabiting the Xinjiang region in China,

Kashmiri (\setkashmiri), the Dardic language of Jammu and Kashmir, Sindhi (\set-sindhi), the Indic language of Sind, Old Malay (\setmalay), Kurdish (\setkurdish), the language of the Kurds, and Ottoman Turkish (\setturk).

The command \transfalse turns off the automatic generation of a transliteration of Arabic text. The command \transtrue turns on this feature. Just in case we only want the transliteration, we have to use the command \arabfalse. Since ArabTEX supports so many different languages, it makes sense to provide support for the various transliterations available. We can select a particular transliteration with the command \settrans, which has one argument—the name of a transliteration. The supported transliterations are: zdmg (the default), english (used in the Encyclopædia of Islam), iranica (used in the Encyclopædia Iranica), lazard (conventions set by Gilbert Lazard), urdu (follows the ALA-LC romanization tables—transliteration schemes for non-Roman scripts), kashmiri (ALA-LC conventions), and turk (similar to modern Turkish). Another option is to select the font that is used for the transliteration with the command \set-transfont{font}.

The commands \setnash and \setnashnf are used to select the medium and bold-face series of the default font. The commands \LR and \RL are used to produce non-Arabic text in an Arabic context and vice versa. An Arabic context can be created with the environments RLtext and arabtext. Both environments behave identically. The last aspect of Arabic document typesetting with ArabTEX is the way we prepare the input file. The "standard" input encoding is shown in the following table:

a	١	b	ب	p	پ	t	ت	_t	ث
^g	ج	.h	ح	_h	خ	d	د	_d	ذ
r	ر	z	ز	s	س	^s	ش	.s	ص
.d	ض	.t	ط	.z	ظ	'	ع	.g	غ
f	ف	q	ق	v	ڤ	k	ك	g	گ
l	ل	m	م	n	ن	h	ه	w	و
y	ي	_A	ى	T	ة	c	خ	^c	چ
,c	خ	^z	ژ	^n	ڭ	^l	ڶ	.r	ڔ
A	آ	I	إي	U	أو	_a	ٱ	_i	أ
_A	ئ	a	آ	i	إ	u	أ	aN	أ
iN	إ	uN	أ	^A	آ	^I	إي	^U	أو

Other input encodings are also supported by loading the appropriate package: ASMO 449 (iso9036), ISO-8859-6 (iso88596), CP-1256 (arabwin), ISIRI 3342 Persian Standard Code (isiri), and UTF-8 (utf8).

The omega package provides a number of environments that can be used to typeset Perso-Arabic text. All of the environments assume that text is entered using some ASCII transliteration. To typeset Arabic, we need to use the environments arab and smallarab. The second environment should be used only for short passages. The environments latberber and berber should be used to get Berber written with Latin or Arabic letters, respectively. The environments tifinagh, urdu, pashto, pashtop, and sindhi should be used to typeset text in the corresponding languages. Note that pashtop should be used for Pakistani Pashto. The standard ASCII transliteration is shown in the following table:

A	ا	p	پ	z	ز	`		I	ى
'a	أ	j	ج	zh	ژ	gh	غ	n	ن
y	ي	'i	إ	H	ح	s	س	f	ف
'n	ن	'y	ئ	'A	آ	kh	خ	sh	ش
q	ق	-h	ه	||	ء	"A	آ	ch	چ
S	ص	v	ڤ	"h	ة	E	ع	b	ب
b	ب	d	د	D	ض	k	ك	e	ة
dh	ذ	T	ط	g	گ	U	و	Llah	لله
th	ث	r	ر	Z	ظ	l	ل	'u	ؤ
SLh	ﷺ	Ta	ط	Ti	ط	Tu	ط	T<>	ط
Ta|	ط	TaN	ط	TiN	ط	TuN	ط		

10.36 India's Languages

India's languages are numerous and difficult to deal with in a system such as LaTeX. The main reason is the number of characters needed due to the large number of ligatures required for most of them.

Unfortunately, it is not a straightforward exercise to typeset a document written in one of India's numerous languages and dialects. The main reason is that the "letters" of these languages in many cases consist of two or more characters that may be placed before or after a main, or central, character, so it is absolutely necessary to have some preprocessor that will modify the input file (think of the preprocessor needed for Thai documents). Obviously, this is a task ideally suited for Ω, but, unfortunately, no one has really done anything in this direction. Meanwhile, the best solution seems to be the ITRANS program (by Avinash Chopde) available from http://www.aczone.com/itrans.

ITRANS is a transliteration program—the user types in the text with Latin characters using some (predefined) conventions in a file (for an example, see Table 10.5). Once we have prepared our input file, we process this file with ITRANS to get another file, which, in turn, will be processed by LaTeX. The file produced by ITRANS is far too complex and is not intended for human inspection.

ITRANS supports the languages Devanagari (Sanskrit, Hindi, and Marathi), Tamil, Bengali, Telugu, Gujarati, Kannada, Punjabi (Gurmukhi), and Romanized Sanskrit. Malayalam and Oriya are not supported at the moment of this writing, but future versions of ITRANS will probably support them.

There are some special arrangements for ITRANS to work properly. First of all, we must load the package itrans. Then, we have to inform ITRANS what transliteration tables (i.e., which languages) we will need. The transliteration tables are contained in files having the .ifm filename extension and are part of the distribution of the ITRANS program. Table 10.6 shows the available .ifm files and their corresponding language.

These files are loaded in the preamble of our input file by using the "command"

```
#bengaliifm=itxbeng.ifm
```

for Bengali and similarly for the other languages. We also have to load the proper font. For Bengali, the proper declarations are

```
\newfont{\itxbeng}{itxbeng at 14pt}
#bengalifont=\itxbeng
```

Notice that we asked here for the 14 pt size. Finally, any text to be transliterated must be inside the environment

```
{#language text #endlanguage}
```

Here is an example (from the documentation of ITRANS):

কে লইবে মোর কার্য, কহে সন্ধ্যা রবি
শুনিয়া জগৎ রহে নিরুত্তর ছবি ।
মাটির প্রদীপ ছিল, সে কহিল, স্বামী
আমার যেটুকু সাধ্য করিব তা আমি ।
--- রবীন্দ্রনাথ ঠাকুর

```
\documentclass{article}
\usepackage{itrans}
\newfont{\itxbeng}{%
    itxbeng at 14pt}
#bengaliifm=itxbeng.ifm
#bengalifont=\itxbeng
\begin{document}
{#bengali
ke la_ibe mor kaaJa^r, kahe
sandhyaa rabi \\
shuniYaa jagat.h rahe
niruttar chhabi | \\
maaTir pradiip chhila, se
kahila, sbaamii \\
aamaar JeTuku saadhya
kariba taa aami | \\
--- rabiindranaath Thaakuur
 #endbengali}
\end{document}
```

Vowels

अ	a
आ	aa *or* A
इ	i
ई	ii *or* I
उ	u
ऊ	uu *or* U
ऋ	RRi *or* R^i
ॠ	RRI *or* R^I
ऌ	LLi *or* L^i
ॡ	LLI *or* L^I
ए	e
ऐ	ai
ओ	o
औ	au
अं	aM
अः	aH

Digits

०	0
१	1
२	2
३	3
४	4
५	5
६	6
७	7
८	8
९	9

Consonants

क	ka
ख	kha
ग	ga
घ	gha
ङ	~Na *or* N^a
च	cha
छ	Cha *or* chha
ज	ja
झ	jha
ञ	~na *or* JNa
ट	Ta
ठ	Tha
ड	Da
ढ	Dha
ण	Na
त	ta
थ	tha
द	da
ध	dha
न	na
प	pa
फ	pha
ब	ba
भ	bha
म	ma
य	ya
र	ra
ल	la
व	va *or* wa

Consonants

श	sha
ष	Sha *or* shha
स	sa
ह	ha
ळ	lda *or* La
क्ष	kSha *or* xa
ज्ञ	j~na *or* GYa

Specials/Accents

क़	qa
ख़	Ka
ग़	Ga
ज़	Ja *or* za
फ़	fa
ड़	.Da
ढ़	.Dha
ॐ	AUM or OM
ऱ्ग	Rga
र्ग	rga *or* ga^r
गं	ga.n
ऑ	aa.c
डँ	Da.N
इ	D.h
दः	duH
ऽ	.a

Table 10.5: The transliteration tables for Devanagari (Hindi, Sanskrit, and Marathi).

Table 10.6: The transliteration table files for the Indian languages.

Devanagari (Sanskrit, Hindi, Marathi)	`dvnc.ifm` (uses PostScript fonts) `dvngfull.imf` (fully ligaturized METAFONT fonts) `dvng.ifm` (simplified form of `dvngfull.ifm`; it eliminates many ligatures; looks simpler/better) `xdvng.ifm` [modified form of `dvng.ifm`; some characters are different (adds "ha-ri", deletes "ja-jnh")]
Gujarati	`itxguj.ifm`
Bengali	`bnbeng.ifm` (METAFONT fonts) `itxbeng.ifm` (PostScript fonts)
Tamil	`wntml.ifm` (METAFONT fonts)
Telugu	`tlgutx.ifm` (METAFONT fonts)
Kannada	`kantex.ifm` (METAFONT fonts)
Gurmukhi	`pun.ifm` (PostScript fonts)
Romanized Sanskrit	`romancsx.ifm` (PostScript fonts)

Table 10.7 shows example commands for the fonts that need to be defined for each script.

Table 10.7: Transliteration tables and sample fonts.

`dvnc.ifm`	dnh
`dvngfull.ifm`	dvng10
`dvng.ifm`	dvng10
`xdvng.ifm`	xdvng
`itxguj.ifm`	itxguj
`itxbeng.ifm`	itxbeng

`wntml.ifm`	wntml10
`tlgutx.ifm`	tel10
`kantex.ifm`	kan18
`pun.ifm`	pun
`romancsx.ifm`	ncprcsxp

Alternative fonts can be found in the font declarations of the samples of the ITRANS distribution (check the `itx` files). For example, the font for Romanized Sanskrit comes with an italic version under the name `ncpicsxp`.

Romanized Sanskrit (ncprcsxp font)	Romanized Sanskrit (ncpicsxp font)
yogasthaḥ kuru karmāṇi saṅgaṃ tyaktvā dhanaṃjaya ǀ siddhyasid-dhyoḥ samo	*yogasthaḥ kuru karmāṇi saṅgaṃ tyaktvā dhanaṃjaya ǀ siddhyasid-dhyoḥ samo*

A nice feature of ɪᴛʀᴀɴꜱ is that it can input a file into your document. This is very useful when dealing with big documents. For this, it provides the command #input=*file-to-input*. It also supports short forms for the language markers. If the last language used was, say, Marathi, then more Marathi text follows, one can use ## instead of #marathi or #endmarathi, and ɪᴛʀᴀɴꜱ will remember to use the last language used.

The itrans package has several options, such as devanagari or talugu (check out the file itrans.sty), that help produce special characters for these languages.

There are cases when one needs to break the lexical scan of the language; that is, avoid the association (and the resulting ligatures) of some consecutive characters. Breaking of the lexical scan is done with the character _. Thus, in Marathi ai produces ऐ but a_i produces अइ.

It is very important to note that since the preprocessor does not change anything unless it is inside a group

#language text #endlanguage

it is very easy to use other languages in the same document (e.g., the languages supported by the babel package).

More information for ɪᴛʀᴀɴꜱ (for example, the transliteration tables for all languages) can be found in its documentation.

10.37 The Cherokee Language

Cherokee—more properly spelled Tsalagi—is an Iroquoian language with an innovative written syllabary invented by the Cherokee scholar Sequoya. Currently, the Cherokee syllabary consists of 84 "letters," each of them denoting a particular syllable. The cherokee package (by Alan M. Stanier) allows one to write simple phrases or words in Cherokee. The package defines a command for each syllable. The general form of these commands is \Cxxx, where xxx is any of the Cherokee syllables. For example, to typeset the Cherokee word for bread (gadu), we have to enter the commands \Cga\Cdu! However, care must be taken since some commands do not use the standard names of the Cherokee syllables.

Ꭰ a	Ꭱ e	Ꭲ i	Ꭳ o	Ꭴ u	Ꭵ v
Ꭶ ga	Ꭷ ka	Ꭸ ge	Ꭹ gi	Ꭺ go	Ꭻ gu
Ꭼ gv	Ꭽ ha	Ꭾ he	Ꭿ hi	Ꮀ ho	Ꮁ hu
Ꮂ hv	Ꮃ la	Ꮄ le	Ꮅ li	Ꮆ lo	Ꮇ lu
Ꮈ lv	Ꮉ ma	Ꮊ me	Ꮋ mi	Ꮌ mo	Ꮍ mu
Ꮎ na	Ꮏ hna	Ꮒ ne	Ꮓ ni	Ꮔ no	Ꮕ nu
Ꮖ nv	Ꮗ qua	Ꮘ que	Ꮙ qui	Ꮚ quo	Ꮛ quu
Ꮜ quv	Ꮝ sa	Ꮞ s	Ꮞ se	Ꮟ si	Ꮠ so
Ꮡ su	Ꮢ sv	Ꮣ da	Ꮤ ta	Ꮥ de	Ꮦ te
Ꮧ di	Ꮨ ti	Ꮩ do	Ꮪ du	Ꮫ dv	Ꮬ dla
Ꮭ tla	Ꮮ tle	Ꮯ tli	Ꮰ tlo	Ꮱ tlu	Ꮲ tlv
Ꮳ tsa	Ꮴ tse	Ꮵ tsi	Ꮶ tso	Ꮷ tsu	Ꮸ tsv
Ꮹ wa	Ꮺ we	Ꮻ wi	Ꮼ wo	Ꮽ wu	Ꮾ wv
Ꮿ ya	Ᏸ ye	Ᏹ yi	Ᏺ yo	Ᏻ yu	Ᏼ yv

Table 10.8: The standard Cherokee syllabary and its Latin transcription.

The ocherokee[8] package (by Apostolos Syropoulos with assistance from Antonis Tsolomitis) is an experimental package to typeset Cherokee text with Λ. The cherokee environment can be used to typeset Cherokee text using the standard Latin transcription (see Table 10.8 on page 356). Here is an example:

ᏍᏏᏆᏯ ᎠᎴ ᎠᏲᎧ.

```
\begin{cherokee}
ssiquaya ale ayoka.
\end{cherokee}
```

If we have a Unicode editor at our disposal, we should use the ucherokee environment. This way, we can enter Cherokee syllables directly. The commands \textcher and \textucher are actually shorthands of the two environments presented above.

Although the syllable nah is part of Unicode, the ocherokee package does not support this syllable, as it is not used anymore by the Cherokee people. We conclude this section with the following Cherokee blessing:

ᏅᏓ ᎤᎴᎯᏟ ᎠᎳᎣᏍ ᎠᎳᏫᏯ ᎭᏗᏒᏗ ᎾᏍᎬ ᎣᏄᏩᏙᏗᎭ.

As of this writing, there are no other packages that support the typesetting of any other Native American language. It seems that the main reason for this lack of typesetting tools is the fact that all other Native American languages use the Latin alphabet. For example, the Apache language uses all of the Latin letters and the letters a, e, i, o,

---

8. The package is accompanied by a PostScript version of the Official Cherokee font created by Tonia Williams. The author of the package wishes to thank Tonia Williams for allowing him to redistribute the PostScript version of the font.

and u with an acute accent [e.g., isdzán (woman), gídí (cat)], an ogonek [e.g., nkęęz (time)], or both [e.g., nadą́' (corn)], and the letter ł[e.g., łog (fish)]. However, there are some languages of the Americas, such as Cree or Inuktitut (the Amerindian Eskimo language), that have their own writting systems, but their support is under development (see Section 10.43).

10.38   The Hungarian Language

Currently, only the magyar option (by József Bérces) of the babel package provides the facilities to typeset Hungarian LaTeX documents. This option redefines the \caption command so that the words *táblázat* (table) and *ábra* (figure) come *after* the number and the colon is replaced by a dot (e.g., 2.1. táblázat). Since the Hungarian definite article is heavily used in the referencing mechanism, the magyar option provides a number of commands that allow LaTeX to produce the correct strings for references. But let us now present these commands.

The command \ondatemagyar works like the \today commands but produces a slightly different date format, which is used in expressions such as "... on September the 6th ...". Here is an example that shows the difference:

2004. szeptember 6.	\today
2004. szeptember 6-án	\ondatemagyar

The commands \Az and \az have a Hungarian word as argument and both print the correct form of the definite article, a nonbreakable space, and the word. The first command should be used at the beginning of a sentence. The commands \Azr and \azr have a label as argument and are used to print references with the correct definite article. In particular, when we want to reference mathematical equations, we can use the form \Azr(label) (or \azr(label)), but the *label* must be surrounded by parentheses. The commands \Azp and \azp should be used for page referencing. Both commands have one argument, which is a label. Similarly, the commands \Azc and \azp should be used for referencing bibliographical items. The magyar option also defines a number of shorthands, which are shown in the following table:

Shorthand	Explanation	Example
'c, 'C	ccs is hyphenated as cs-cs	lo'ccsan →locs-csan
'd, 'D	ddz is hyphenated as dz-dz	e'ddz\"unk →edz-dzünk
'g, 'G	ggy is hyphenated as gy-gy	po'ggy\'asz →pogy-gyász
'l, 'L	lly is hyphenated as ly-ly	Kod\'a'llyal →Kodály-lyal
'n, 'N	nny is hyphenated as ny-ny	me'nnyei →meny-nyei
's, 'S	ssz is hyphenated as sz-sz	vi'ssza →visz-sza
't, 'T	tty is hyphenated as ty-ty	po'ttyan →poty-tyan
'z, 'Z	zzs is hyphenated as zs-zs	ri'zzsel →rizs-zsel

In addition, the shorthands ' ' and ' ' produce the opening (") and closing (") quotation marks. Usually, Hungarian input files should be prepared in the Latin1 character set, so one has to use the corresponding encoding for the inputenc package.

10.39 The Turkish Language

Currently, only the turkish option (by Mustafa Burc) provides a complete solution to the problem of typesetting Turkish documents with LATEX. This option implements the typographic rule that dictates that some space must be added before the characters :, !, and =.

10.40 The Mongolian Language[9]

Mongolian writing is a fairly complex topic. In the history of the written language, numerous scripts were either accepted from other cultures or domestically designed. Important scripts with a practical significance today are Uighur and Cyrillic. Other scripts were also employed at given times in history (e.g., Chinese, Phagsba, Soyombo, and Latin, which had been used during the 1930s).

The traditional Mongolian script is called Uighur and is written in vertical lines from left to right (i.e., LTL in Ω's parlance). Now, the Uighur script is again, in legal though not in practical terms, the official script of Mongolia. Despite the legal status, the *de facto* writing system in Mongolia is Cyrillic; however, Uighur is the standard script used in Inner Mongolia, China.

The Chinese script was used for a short time during the 13th and the beginning of the 14th century, during the Yuan dynasty, in cultural and linguistic applications such as dictionaries and so on.

The Phagsba or Square Writing was developed in the 13th century by a famous Tibetan monk and scholar, Phagsba. This script incorporated features from the Tibetan script and inherited the writing direction from Chinese. In the 17th century, a second Square Writing, called Horizontal Square Writing, was developed. This script was also heavily influenced by the Tibetan script.

Another script, the conception of it politically motivated, was the Soyombo script designed by the Mongolian monk and scholar Zanabazar in 1686. This script has never managed to become a script for daily usage, although it survives in religious inscriptions. The only symbol of that script that can be seen literally everywhere in Mongolia is the Soyombo symbol (see Figure 10.2 on page 365). It is even formally described in Mongolia's constitution as a national symbol and decorates flags, money, official buildings, official documents, and seals, to name just a few examples.

---

9. The information regarding the writing systems of the Mongolian language is based on the Mongolian FAQ by Oliver Corff (see http://userpage.fu-berlin.de/~corff/mf.html).

In 1940, the then Mongolian People's Republic started using a modified Cyrillic alphabet after a short period of latinization experiments. Despite a few orthographic instabilities, the Cyrillic system is the major vehicle of written communication today in Mongolia; virtually all newspapers, books, etc., are printed in the Cyrillic alphabet.

10.40.1 Modern Mongolian — Cyrillic

The MonTEX system (by Oliver Corff with assistance from Dorjpalam Dorj) provides the necessary tools to typeset Mongolian language documents transcribed in the Cyrillic and the Uighur scripts. In addition, the system provides the necessary tools for the typesetting of other languages such as Buryat and Manchu. All of these features are supported by the mls package of the MonTEX system. The package provides a number of options that correspond to the main language of the document and the input encoding to be used. The available languages are xalx (modern Mongolian as spoken in Mongolia today), buryat (Buryat is a Mongolian language that is spoken by the Buryat people living north of the Russian-Mongolian border in the Buryat Autonomous Republic near Lake Baikal), kazakh (Kazakh is a Turkic language spoken in Kazakhstan and by the Kazakh minority in Mongolia), bicig (Uighur Mongolian), bithe (Manchu), english, and russian.

There are two methods to prepare a Mongolian document transcribed in the Cyrillic script: either we use a Cyrillic character set or a Latin transliteration. If we choose to enter text using a Latin transliteration, then we must follow the conventions presented in the following table:

А а	A a	Б б	B b	В в	W w
Г г	G g	Д д	D d	Е е	E e
Ё ё	"E/Ë "e/ë	Ж ж	J j	З з	Z z
И и	I i	Й й	"I/Ï "i/ï	К к	K k
Л л	L l	М м	M m	Н н	N n
О о	O o	Ө ө	"O/Ö "o/ö	П п	P p
Р р	R r	С с	S s	Т т	T t
У у	U u	Ү ү	"U/Ü "u/ü	Ф ф	F f
Х х	X x	һ һ	H h	Ц ц	C c
Ч ч	q q	Ш ш	\Sh sh	Щ щ	\Sc \sc
Ъ ъ	\Y \y	Ы ы	Y y	Ь ь	\I \i
Э э	"A/Ä "a/ä	Ю ю	Yu yu	Я я	Ya ya

On the other hand, if we have at our disposal an environment with a Cyrillic character set, accompanied by a suitable keyboard (or a multilingual editor such as EMACS), then we can use any of the available predefined input encodings—MLS (based on the original Mongolian Language Support for IBM computers), NCC (a popular encoding with only Cyrillic characters), MOS, MNK, DBK, CTT (four Cyrillic-only encodings),

IBMRUS (not suitable for Mongolian), KOI (a Russian-only Cyrillic encoding), 850, 852, MAC, ATARI, or ROMAN8. To switch to the first input method, use the \SetDocumentEncodingLMC command. The command \SetDocumentEncodingNeutral switches to the Cyrillic input method and requires the selection of one of the available input encodings. In addition, the commands \SetDocumentEncodingBicig and \SetDocumentEncodingBithe are used to switch to the Simplified Classical Mongolian and to transliterated Manchu input methods, respectively. Moreover, the declarations \mnr and \rnm can be used to switch to from ordinary text to transliterated Cyrillic text. The commands \xalx and \lat have one argument and can be used to temporarily switch to Mongolian or Latin text mode, respectively.

Depending on the settings of the document language, or main language, the command \today is redefined to match Buryat, Xalx, Russian, or Bicig conventions. Regardless of the main language, the internal commands for producing the proper date are nonetheless directly accessible by the user:

2001 оной ноябриин 27-ной үдэр	\BuryatToday
2001 оны арван нэгдүгээр сарын 27	\XalxToday
27 ноября 2001	\RussianToday
November 27, 2001	\today

The command \BicigToday prints the date using the Uighur script (see margin). Similarly, the commands \KazakhToday and \BitheToday produce the current date in Kazakh and Manchu, respectively.

The commands \Togrog and \togrog produce the symbol for the national currency of Mongolia—Togrog or Tugrik. These commands produce sans serif versions of the currency symbol, which is considered standard. It is possible to produce any form of the currency symbol by using the commands \MyTogrog and \mytogrog. These commands pick the current font style to render the currency symbol:

₮	\Togrog
₮	\togrog
₮	\textrm{\MyTogrog}
₮	\textbf{\mytogrog}

10.40.2 Classical Mongolian — Uighur

To prepare an Uighur Mongolian document, one can use the MLS code page and character set or, similar in structure to the Cyrillic approach described above, use one of two Latin transliteration schemes. The first MonTₑX transliteration system is identical to the broad romanization first published in the 1990s, whereas the Simplified Transliteration uses a smaller set of vowels and actually imitates some of the inherent ambiguities of the Uighur Mongolian script. The MonTEX system supports both transliteration schemes in different environments; while the broad system is to be used with small portions of text, like one-word or one-phrase insertions in scientific papers, for example, the Simplified Transliteration can be used for complete documents as the main language of the body text.

To typeset complete Uighur documents vertically, it is absolutely necessary to process your input file with ε-LATEX. If one wants to typeset individual Uighur Mongolian words (e.g., dictionary entries, and so on), then standard LATEX can be used instead. The same principle applies to Manchu (see Section 10.42).

In each column of the following table, we present the Uighur letter, the MLS, and the MonTEX transliteration employed.

letter	MLS	MonTEX	letter	MLS	MonTEX	letter	MLS	MonTEX
᠊	a	a	᠊	ä	ä, E	᠊	i	i
᠊	o	o	᠊	u	u	᠊	ö	ö, O
᠊	ü	ü, U	᠊	n	n	᠊	*ng	ng
᠊	x	x	᠊	γ	G	᠊	k	k
᠊	g	g	᠊	b	b	᠊	p	p
᠊	f	f	᠊	s	s	᠊	š	S
᠊	t	t	᠊	d	d	᠊	l	l
᠊	m	m	᠊	c	c	᠊	z	z
᠊	y	y	᠊	r	r	᠊	v	v
᠊	h	h	᠊	j	j	᠊	K	K
᠊	[-]	Q	᠊	C	C	᠊	Z	Z

To enter Uighur Mongolian, we can use the command \bicig, which is similar to \xalx, mentioned above, or the environments bicigtext and bicigpage. The environment bicigpage should be used when we want to prepare whole pages of Uighur Mongolian text. The command \bicig and the environment bicigtext should be used to typeset Uighur in horizontal mode. The command \bosoo typesets its argument vertically, so it can be used in conjunction with \bicig to typeset small passages in vertical mode. To make things easy, MonTEX provides the commands \mbosoo and \mobosoo, which produce Uighur text typeset in vertical mode. \mbosoo expects input in broad transliteration, and \mobosoo expects its input to be in Simplified style, as the following example shows:

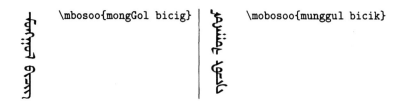

```
\mbosoo{mongGol bicig}        \mobosoo{munggul bicik}
```

Here is how we can enter the special symbols and the punctuation marks of the Uighur script:

	!		?	!?		!?		?!		?!		✱		*	MSP		-
⌢	(	⌣	)	⌃	<	⌄	>	⩘	‹	FVS1	'
FVS2	"	MVS	=	'		⋮	.	⁓	,	‥	:
❖	;	⋮	‥	𝘰	0	𝘰	1	𝘯	2	𝘯	3
𝘶	4	𝘫	5	𝘨	6	𝘦	7	𝘭	8	𝘤	9

In the preceding table, certain characters such as the Mongolian space, are intentionally presented visually. The glyph input style is shown in the following table:

	l		@		a		A		i		o		0
	n		l		L		Q		m		M
	x		X		g		I		B		b
	t		d		r		R		z		y
	s		S		q		c		v		h
	K		k		P		p		f		Z
	C		j		e		E		Y		G
	-		=		,		;		V		u
	T		U		W		ml		11

10.40.3 Classical Mongolian — Horizontal Square Writing

The Horizontal Square Writing method is known as Xäwtää Dörböljin, or Horizontal Square Alphabet (although it is called an "alphabet," it is actually a syllabary). The mxd package (by Oliver Corff) is a first attempt to allow the typesetting of documents transcribed in Xäwtää Dörböljin with LaTeX. Like the Soyombo script, the Xäwtää Dörböljin script is two-dimensional and in practice is heavily influenced by the Tibetan script.

The major area of each syllable is occupied by a main consonant. If no vowel is added, the basic vowel a is assumed. If one of the vowels i, ä, o, or ö is added, it is placed on top of the syllable; otherwise, if one of the vowels ü or u is added, then it is placed at the bottom of the character box. Long vowels are marked by a protruding tip of the right-side beam. As is obvious, Xäwtää Dörböljin syllables are constructed in a manner similar to Soyombo and Tibetan syllables.

The mxd package provides the \mxd command, which should be used to switch to the Xäwtää Dörböljin mode. Since there is at the moment no way to switch back to the previous mode, the user should use this command in a local scope. The vowels *o*, *ö*, *u*, *ü*, and *ä* are entered as o, O, u, U, and e, respectively. In addition, the package supports four different script styles, which are shown below:

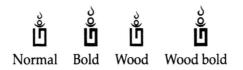

Normal   Bold   Wood   Wood bold

The wood style is actually the italic shape of the font used. Table 10.9 shows the characters that are currently supported.

Table 10.9: Xäwtää Dörböljin script input.

glyph	code	glyph	code	glyph	code	glyph	code	glyph	code
	‘		\sA	|	.	||	..		’
	‘-		‘i		‘i-		‘e		‘e-
	‘U		‘U-		‘u		‘u-		‘o
	‘o-		‘O		‘O-		‘W		‘I
	g		k		x		z		c
	=		d		t		n		b
	p		m		y		r		w
	l		\$		s		h		\sks
	‘G		‘K		‘*		‘D		‘N
	‘B		‘M		‘R		‘L		‘Q
	‘S		‘-a		\sri		\sri-		\sli
	\sli-		‘o		‘H		\sg		\sgh
	j		\sjh		T		\sth		\sdd
	\sdh		\sdn		\sD		\sDH		\sB
	\sBH		\sds		\sky		\skr		\skl
	\slk		ssk		\srk		X		q
	@								

10.40.4 Classical Mongolian – Soyombo

The soyombo package (by Oliver Corff) provides support for the typesetting of Mongolian text using the Soyombo script. The \Soyombo command produces the Soyombo symbol (see Figure 10.2). To switch to the Soyombo font, use the \soyombo command. It makes sense to briefly describe the way Soyombo syllables are constructed. There is always a basic consonant attached to the top left corner of a vertical beam. If no vowel sign is added, the basic vowel *a* is assumed. If one of the vowels *i, ä, o,* or *ö* is added, then it is placed on top of the syllable; otherwise, if one of the vowels *ü* or *u* is added, then it is positioned in the lower third of the syllable. Long vowels are indicated by placing a tip under the beam. Closing consonants are placed in the right third of the lower third of the syllable. Figure 10.2 presents the structural layout of the Soyombo syllables. But how do we enter Soyombo syllables? Initial consonants are entered as such (see Table 10.10). The vowel *a* is not marked; other vowels are entered by using lowercase letters for short vowels and uppercase letters for long vowels. In particular, the vowels *o, ö, u, ü,* and *ä* are entered as o, O, u, U, and e, respectively. The "letters" u and U have a shorter form accessed with the "letters" v and V, respectively. These shorter forms are used for combinations with a final consonant. Syllable final consonants are always entered in uppercase. Commands that typeset special symbols for writing Sanskrit and Tibetan are also provided and are shown in Table 10.10.

	‘		.		..		\s0		-		i
	i-		e		e-		U		U-		u
	u-		o		o-		O		O-		W
	I		g		k		x		z		c
	=		d		t		n		b		p
	m		y		r		w		l		\$
	s		h		\sks		G		K		*
	D		N		B		M		R		L
	Q		S		-a		\sri		\sli		\sli-
	O		H		\sg		\sgh		j		\sjh
	T		\sth		\sdd		\sdh		\sdn		\sD
	\sDH		\sB		\sBH		\sds		\sky		\skr
	\skl		\skm		\skk		\snk		\snc		\snt
	\snd		P		\slk		\sSk		\ssk		\srk
	z		c		J		X		q		Q

Table 10.10: Soyombo character input method.

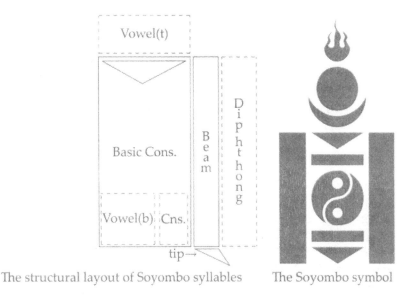

The structural layout of Soyombo syllables   The Soyombo symbol

Figure 10.2: Soyombo letters and symbol.

10.41 The Vietnamese Language

Werner Lemberg has produced the necessary files that allow people to typeset Vietnamese documents that have been prepared using one of the following input encodings:
VISCII, VPS, and TCVN. The vietnam package by Werner Lemberg and Hàn Thế Thành
has three options, which correspond to the input encoding method employed to prepare
the input file. The default option is visci. The vietnam package redefines the commands
that hold the predefined names so that they print Vietnamese names. Vietnamese is
written in the Latin script augmented with a number of letters with various accents.
This means that one cannot prepare a Vietnamese input file with an ordinary editor. A
suitable solution is to use the EMACS mule mode. For example, the following text has
been prepared with this editor:

> Xin chào Apostolos Hôm qua tôi đã biết cách dùng Việt trong LATEX.
> Tuy nhiên chất lượng font không được tốt, đặc biệt là khi hoán
> chuyển sang PDF format.

Of course, our Vietnamese friend complained that he could not create quality output
from pdfLATEX. But this just happened because he was not aware of software that can
create PostScript Type 1 fonts from METAFONT sources. For example, he could use
TEXTRACE (by Péter Szabó). To typeset the same document with Λ, we must put the
following in our documents preamble:

```
\ocp\viscii=inviscii
\ocplist\InVISCII=
   \addbeforeocplist 1 \viscii
   \nullocplist
\pushocplist \InVISCII
```

Of course, we must also put the following command in the body of our input file:

$$\texttt{\textbackslash fontfamily\{omlgc\}\textbackslash selectfont}$$

to make Λ use the Ω fonts. Here is how Λ will typeset the text above:

> Xin chào Apostolos Hôm qua tôi đã biết cách dùng Việt trong LaTeX. Tuy nhiên chất lượng font không được tốt, đặc biệt là khi hoán chuyển sang PDF format.

If we have prepared our file using a Unicode editor that produces UCS-2, we do not need to use any ΩCP. If our Unicode editor produces UTF-8 files, then we must use the inutf8 ΩCP.

10.42 The Manchu Language

The Manchu language is part of the Tungusic language family, a subdivision of the Altaic language family, which includes Mongolian, Turkic, and according to some scholars, Japanese and Korean (under the title "macrotungusic"). Manchu is written using a modified form of the Uighur script. This form includes dots and circles and is called *tongki fuka sindaha hergen* (script with dots and circles). The MonTeX system provides support for the Manchu language. To enter Manchu text, we must use the following transliteration:

	a		e		i		o		u
	v		n		k		g		h
	b		p		s		s'		t
	d		l		m		c		j
	y		k'		g'		h'		r
	f		w		sy *		cy *		j' *
	dz *		tsh *		tshy *		zr *		z
	zh †		ts †		ng' †		1' †		p' †
	t' †		aii						

Manchu characters marked with an asterisk are special characters listed in major dictionaries, whereas the characters marked with a dagger are used to transcribe the Tibetan alphabet in the Manchu script. Note that the diphthong *ai* is transliterated as aii. The environments bithetext and bithepage are the Manchu equivalents of the environments bicigtext and bicigpage. The \mabosoo command is the Manchu equivalent of the \mobosoo command.

10.43 The Inuktitut Language

Inuktitut is the language of the Inuit (also known as Eskimos, but the term is considered offensive by Inuit who live in Canada and Greenland). The language is spoken in Greenland, Canada, Alaska, and the Chukotka Autonomous Okrug, which is located in the far northeast region of the Russian Federation, by approximately 152,000 people. The Inuktitut syllabics are used by Inuit who live in Canada, especially in the new Canadian territory of Nunavut. This writing system was invented by Reverend James Evans, a Wesleyan missionary. This system was based on earlier work on the Cree language, which, in turn, was based on work on the Ojibway language. The following table shows the Inuktitut syllabics and the Latin transcription of the Inuktitut symbols:

ᐃ	i	ᐅ	u	ᐊ	a	H	h
ᐱ	pi	ᐳ	pu	ᐸ	pa	‹	p
ᑎ	ti	ᑐ	tu	ᑕ	ta	⸀	t
ᑭ	ki	ᑯ	ku	ᑲ	ka	ᑊ	k
ᒋ	gi	ᒍ	gu	ᒐ	ga	ᒡ	g
ᒥ	mi	ᒧ	mu	ᒪ	ma	ᒻ	m
ᓂ	ni	ᓄ	nu	ᓇ	na	ᓐ	n
ᓕ	li	ᓗ	lu	ᓚ	la	ᓪ	l
ᓯ	si	ᓱ	su	ᓴ	sa	ᔅ	s
ᔨ	ji	ᔪ	ju	ᔭ	ja	ᕝ	j
ᕆ	ri	ᕈ	ru	ᕋ	ra	ᕐ	r
ᕕ	vi	ᕗ	vu	ᕙ	va	ᕝ	v
ᕿ	qi	ᖁ	qu	ᖃ	qa	ᖅ	q
ᖏ	ngi	ᖑ	ngu	ᖓ	nga	ᖕ	ng
ᖠ	lhi	ᖢ	lhu	ᖤ	lha	ᖦ	lh
ᙱ	nngi	ᙳ	nngu	ᙵ	nnga	ᖖ	nng

The oinuit package (by the first author of this book) is an experimental package that tries to solve the problem of typesetting Inuktitut text with ∆. The package provides the options nunavut, quebec, iscii, utf8, and ucs2. The first two options should be used to enter Inuktitut text using the Latin transliteration presented above. The text is typeset using the Anglican or the Catholic syllabic orthography, respectively. The inscii option should be used when preparing a file using a particular extended ASCII that includes the Inuktitut syllabics. The declaration \inuittext changes permanently the font encoding, the input method, and the hyphenation rules in effect. If we do not wish to globally alter these parameters, then we should use the command \textinuit

Figure 10.3: A bilingual text typeset with Λ and the oinuit package.

ᐃᒪᕆᐊᑐᖅ ᑕᖅᑐ	Imaruituq Taqtu
ᓱᕈᓯᐅᑎᓪᓕᒃ ᐊᑐᓚᐅᕐᒋᒃ ᐱᑦ ᒍᑕᓂᒥ ᐊᖅᕐᐋᓗ ᑯᐊᐊᕐᑲᓚᐅᕐᒋᒃ ᐱᓕᖅᕐᓂᕐᑎᐊᖅᓲᒃ. ᑕᐸᐸ ᐱᑦ ᐊᕐᐊᕐᐸᕐᐋᓄᖅᒃ ᑕᒪ ᓚᓪᑲᐅᐳᓄ ᐊᐊᓂᓗ ᐱᕈᓐᑎᑕᐅᐸᓚᐅᕐᒋᒃ. ᐊᖅᓗᓄᒃ ᐊᓗ ᐱᐊᕐᓗᐊᖅ ᓇᑐᕐᐸᕐᐸᕐᒐ. ᓚᑦ ᐊᐊᐊᓚ ᐅᑲᐊᑎᐅᐳᓚ ᑲᐊᓚᖅᑎᐊᐊᓄᓗ ᐱᐊᕐᒐᕐᐊᖅ ᐅᐸᓚᑦ ᑲᐊᓚᖅᑎᐊᐊ ᓲᐊᖅᓄᓚ ᐊᐅᑲᐊᑐᐳᑎᐅᐊᕐᒍᑕᑎᑦᑎ ᐊᐅᒍ ᑕᒪ ᒪᓂᐅᐸᓗᐊᕐᒍᕐᕐᓗᒍ ᓗᐊᐳᒃᓂ ᐅᑲᐱᖅᓄᐊᖅᕐᓚᕐᒐᖅ ᕐᓗᓂ.	When we were children we never had anything to worry about, all we had to do was play. It was all there was and we were very happy. But as we grew older, our parents, especially our mothers, started to teach us the things we had to know, such as how to look after a house. My mother told me that she wanted me to learn these things because I would have a house of my own when I grew up, but I didn't believe it.

or the environment inuit. Figure 10.3 shows some bilingual text typeset with Λ and the oinuitn package. The package uses a PostScript version of the Nunacom TrueType font developed by Nortext (http://www.nortext.com), which is redistributed with permission from Nortext.

10.44 Archaic Writing Systems

Peter R. Wilson has developed a number of packages that can be used to typeset text written in certain archaic writing systems, which include:

Package	Script	Package	Script
coptic	Coptic	cypriot	Cypriotic Greek
etruscan	Etruscan	greek4cbc	Greek of the 4th century B.C.
hieroglf	Hieroglyphic	greek6cbc	Greek of the 6th century B.C.
ugarite	Ugaritic	linearb	Linear B (preclassical Greek script)
phoenician	Phoenician	protosem	Protosemitic
runic	Runic	oldprsn	Old Persian (cuneiform writing)

In addition to the coptic package, one can also use the copte package (by Serge Rosmorduc) to typeset Coptic text.

It should be noted that as far as the hieroglyphs are concerned, the package provides access to a small subset of the glyphs of the Sesh Nesout system created by Serge Rosmorduc. Wilson chose about 70 of the most common glyphs from this package to create the hieroglyph package.

Table 10.11 shows the new font selection commands provided by the packages described in this section. The short passage commands take as argument a piece of text

Table 10.11: New commands provided by the packages that provide support for archaic writing systems.

Package	Font Family Selection	Short Passage Command
copte	—	\textcopte
cypriot	\cyprfamily	\textcypr
etruscan	\etrfamily	\textetr
greek4cbc	\givbcfamily	\textgivbc
greek6cbc	\gvibcfamily	\textgvibc
hieroglf	\pmhgfamily	\textpmhg
linearb	\linbfamily	\textlinb
oldprsn	\copsnfamily	\textcopsn
phoenician	\phncfamily	\textphnc
protosem	\protofamily	\textproto
runic	\futfamily	\textfut
ugarite	\cugarfamily	\textcugar

in the corresponding language. In addition, to typeset Coptic language texts, we need to load the font encodings COP and T1. In other words, we must include the following command in the preamble of our file:

\usepackage[COP,T1]{fontenc}

In what follows, we present the transliterations defined by the various packages that can be used to access individual letters or symbols, in general.

———————————— 6th century Greek ————————————

ΑΒΓΔ ΕΙΘ⊕ ΙΚΓⲘ ΝΞΟΓ ϷΣΤΥ ΧΦΥΩ

\textgvibc{ABGD EZH\TTheta\ IKLM N\TXi OP RSTU X\TPhi\TPsi\TOmega}

──────────── 4th century Greek (smooth) ────────────

ΑΒΓΔ ΕΖΗΘ ΙΚΛΜ ΝΞΟΠ ΡΣΤΥ ΧΦΨΩ

`\textgivbc{ABGD EZH\TTheta\ IKLM N\TXi OP RSTU X\TPhi\TPsi\TOmega}`

──────────── 4th century Greek (rough) ────────────

ΑΒΓΔ ΕΖΗΘ ΙΚΛΜ ΝΞΟΠ ΡΣΤΥ ΧΦΨΩ

`\textgivbc{abgd ezh\tTheta\ iklm n\tXi op rstu x\tPhi\tPsi\tOmega}`

──────────── Etruscan ────────────

ABΓD FFI⊟ ⊗IKL ᛗᚱ⊟O ᚱᛗΦᛉ ᛟTYX ΦΥ8   A⊟ᛌD ᛺ᚱI⊟
⊗IᚷJ ᛗᚤ⊟O ᛌᛝᚤ ᛟTYX ΦΥ8

`\textetr{ABGD EFZH \TTheta IKL MN\TXi OP\Tsade Q RSTU X\TPhi\TPsi`
`8 abgd efzh \tTheta ikl mn\tXi op\tsade q rstu x\tPhi\tPsi 8}`

──────────── Phoenician ────────────

𐤃ᚱTΔ FFI⊟ ⊗IᛟL ᛗᚱ≢O ᚱᛝΦᛉ Wᛏ   ᛚᛝTΔ ᛺ᚱI⊟ ⊗Iᛘ
ᛗᚤ≢O ᛝᚱΦᛝ Wᛏ

`\textphnc{ABGD EFZH \TTheta IKL MN\TXi O P\Tsade QR ST`
` abgd efzh \tTheta ikl mn\tXi o p\tsade qr st}`

──────────── Runic ────────────

ᚠᚢᚦᛘ ᚱᛚᚷᛈ ᚺᛏIᛈ ᛌᚲᛃᚺ ᛏᛒᛗᛗ ᚱᛪᚺᚱ :

`\textfut{FU\Fthorn A RKGW HNIJ YPXS TBEM L\Fng DO :}`

──────────── Ugaritic ────────────

`\textcugar{abgH dhwz IJyk SlmD nZs' pxqr TGti uX:}`

abġ ḥ dhwz ḫṭyk ślmḏ nẓṣ' pṣqr ṭġti uṣ:

`\translitcugar{\Ua\Ub\Ug\Uhu \Ud\Uh\Uw\Uz  \Uhd\Utd\Uy\Uk`
`\Usa\Ul\Um\Udb \Un\Uzd\Us\Ulq \Up\Usd\Uq\Ur \Utb\Ugd\Ut\Ui`
`\Uu\Usg\Uwd}`

—————————————————— Linear B ——————————————————

⊓├⊟⊕ ⌿⊽≠⌽ ⌞⌃⌅⍁⊓ ⌽ ⍍⍓⍍⍓ ⊓⍦⊝ ⌥⍦≠⎓ ⍔
⌣⊤⍦⎇ ⌥⍍⊤⍌ ⍀⋂⍟ ⊟⍦⍓⍦ ⌃⏚⌅⌥ ┼⍀⊤⍓ ⌥ ⊬⍀⍀⋔⌃
⌥⍀⍀⌥ ⊟⏀

\textlinb{adjk mnpq rstw z eDJK MNPQ RSTW Z ifcy CGXO
Y36 ogbh AEH8 U147 9 uxLv BFIV 25}

Or

\textlinb{\Ba\Bda\Bja\Bka \Bma\Bna\Bpa\Bqa \Bra\Bsa\Bta\Bwa \Bza
\Be\Bde\Bje\Bke \Bme\Bne\Bpe\Bqe \Bre\Bse\Bte\Bwe \Bze
\Bi\Bdi\Bki\Bmi \Bni\Bpi\Bqi\Bri \Bsi\Bti\Bwi \Bo\Bdo\Bjo\Bko
\Bmo\Bno\Bpo\Bqo \Bro\Bso\Bto\Bwo \Bzo \Bu\Bdu\Bju\Bku
\Bmu\Bnu\Bpu\Bru \Bsu\Btu}

—————————————————— Hieroglyphs ——————————————————

\textpmhg{ABCD EFGH IJKL MNOP QRST UVWX YZ abcd efgh ijkl mnop
qrst uvwx yz +?/|}
ꜥbḥd šꜣibḥ ḥmkiw mḥtꜣst tpwrsṭ wḏwḥ ꜣw ꜣbḏꜥmd irfgh iprkl
mnwꜣp ḥrrst ḥdnbww yš imyawtkmꜣwꜥ msḏḥwtybꜣ gm
\translitpmhg{\HA\HB\HC\HD \HE\HF\HG\HH \HI\HJ\HK\HL \HM\HN\HO\HP
\HQ\HR\HS\HT \HU\HV\HW\HX \HY\HZ \Ha\Hb\Hc\Hd \He\Hf\Hg\Hh
\Hi\Hj\Hk\Hl \Hm\Hn\Ho\Hp \Hq\Hr\Hs\Ht \Hu\Hv\Hw\Hx \Hy\Hz
\Hplus\Hquery\Hslash\Hvbar \Hms\Hibp\Hibw\Hibs \Hibl\Hsv}

It is important to stress that the hieroglf package gives access to a small subset of the most common glyphs chosen among the many more provided by the Sesh Nesout system (by Serge Rosmorduc), which is available from http://khety.iut.univ-paris8.fr/~rosmord. This is, however, a much more difficult system to use, as it requires a (provided) preprocessor (called Sesh Nesout) that must be used to preprocess the input file. Obviously, one can work on the creation of an ΩTP that will eventually substitute the preprocessor.

─────────────────────── Protosemitic ───────────────────────

```
\textproto{abgd zewh iykl mnop uvqr sxt ABGD ZEWH IYKL MNOP UVQR
SXT}
```

─────────────────────── Old Persian ───────────────────────

```
\textcopsn{aiuk KxgG cjJt ToCd PDnN pfbm wMyr RlvV sSzh XqQL
BeEF :}
```

```
\textcopsn{\Oa\Oi\Ou\Oka \Oku\Oxa\Oga\Ogu \Oca\Oja\Oji\Ota
\Otu\Otha\Occa\Oda \Odi\Odu\Ona\Onu \Opa\Ofa\Oba\Oma
\Omi\Omu\Oya\Ora \Oru\Ola\Ova\Ovi \Osa\Osva\Oza\Oha}
```

─────────────────────── Cypriot ───────────────────────

```
\textcypr{agjk lmnp rstw eKLM NPRS TW icdy CGOY 36 obhf AEHU 1479
uvqB FIV2 5}
```

Or

```
\textcypr{\Ca\Cga\Cja\Cka \Cla\Cma\Cna\Cpa \Cra\Csa\Cta\Cwa
\Ce\Cke\Cle\Cme \Cne\Cpe\Cre\Cse \Cte\Cwe
\Ci\Cki\Cli\Cmi \Cni\Cpi\Cri\Csi \Cti\Cwi
\Co\Cjo\Cko\Clo \Cmo\Cno\Cpo\Cro \Cso\Cto\Cwo\Czo
\Cu\Cku\Clu\Cmu \Cnu\Cpu\Cru\Csu \Ctu}
```

─────────────────────── Coptic ───────────────────────

```
\textcopte{8ABC DEFG HIJK LMNO PQRT UWXY ZÉËÍ ÏÑÓ
abcd efgh ijkl mnop qrtu wxyz ÈÊÌÎ ÐÒÔ}
```

In addition to the above, the hieroglf package allows for the creation of cartouches with the commands \cartouche and \Cartouche (the second command gives wider lines and curves for the cartouche). Here is an example:

`\Cartouche{\textpmhg{\Hp:\Ht-\Ho-\Ho:`
`\HM-\Hy-\Hs}}`

Vertical cartouches can be created using the \vartouche and \Vartouche commands.

11

To Err Is Human

To recapitulate on the use of LATEX, the reader may recall that we type our text containing the formatting commands in an input file using our favorite editor. We then invoke the TEX typesetting program, which processes the input file using the LATEX format, outputting a minimum of three files: a DVI file, an auxiliary file used by the program to generate things such as cross-references, and a log file that contains information about what TEX encountered, including details of any warnings or errors. While it is typesetting a file, TEX tells you about its progress either in a window or, on older systems, on the full screen of your computer. Any warnings are described without the program halting, and on faster systems some of this can be rather fleeting, but since they are reproduced in the log file, this can be examined afterwards. When an error is encountered, the program actually halts, giving an indication of the nature of the error, on which line of your input file the error might be found, and a ? prompt. Warning messages indicate problems that are not serious but that are likely to affect the output (e.g., problems with hyphenation, line-breaking, cross-references and labels, finding a particular font, etc.), while errors are more serious, causing the TEX program to stop (e.g., the messages environment undefined, an omitted item in a list making environment, and misplaced alignment tab character &, among others).

Minor errors can be skipped over by just pressing the Enter key. This allows you to check for more than one error at a time. However, sometimes a single mistake can lead to multiple errors and you can *exit* the program by typing the letter x followed by the Enter key. An alternative is to ask for assistance with editing by typing in the letter e followed by the Enter key. This will print out a message stating on the line on which the error may be found, and on some systems it will start up your editor and then place your prompt by the offending text. Sometimes, instead of a ? prompt you will see a * with no error message. This usually indicates that the \end{document} command has been forgotten and you can exit the program by typing \stop followed by the Enter key. Either method of stopping TEX will make it produce the log file. This file can be opened up in your text editor to take another look at the error and warning messages that were generated during the latest run of your input file through the typesetting engine.

Various alternative responses from the user are possible at the prompt following an error message ?; uppercase or lowercase letters may be typed. *Scroll* mode is started by typing the letter s, followed by the Enter key, and it causes TEX to proceed through the file and display any error messages on the screen without stopping unless a file mentioned in a \input or \include command cannot be found on the computer. *Run* mode occurs when the user types in a letter r followed by the Enter key and is the same as *Scroll* mode apart from the fact that it will continue even in the absence of a file named in a \input or \include command. *Quiet* mode involves the user typing a letter q followed by the Enter key and acts the same as *Run* mode, but no messages are written to the output window, although they still go to the log file. An *Insert* mode is available by typing the letter i followed by the Enter key. This allows the user to type in a line of commands/text from the keyboard to correct the error. This will only affect the current run through the typesetting engine; it will not change the user's input file. It is useful where something has been omitted (e.g., a math environment delimiter, such as $). One can obtain more information about an error through using the *help* facility, invoked by typing in the letter h followed by the Enter key. This will also produce hints about how the problem might be solved.

An alternative to working through the errors in your file interactively is to place the command \batchmode at the start of your LATEX input file. Invoking LATEX will then cause it to automatically work as far as it can through the input file, although you may end up with a long list of errors in your log file since the effects of each can be cumulative. Note that sometimes an error can lead to LATEX producing error messages that are purely artifactual: if you correct that error, the others will go away. However, after TEX outputs a page of output, the effect of the error has often passed, and the next error encountered is likely to be a genuine one worth checking out.

The way in which TEX does its typesetting has an impact upon the messages that it generates for errors and warnings. TEX switches between generating its output on an expanding "scroll" of typeset output and "cutting off" pages of output that are written out. Lamport [20] points out how this is analogous to the way in which traditional typesetters produce lengths of metal type, called galleys, that are manageable lengths from a hypothetical scroll of text that opens out vertically from the beginning of the document downwards. Because of historical limits to the amount of memory available to TEX, it does not keep much more than a page in memory at a time. After it generates each paragraph unit, it checks to see if there is enough material to typeset a page; if so, the page is added to the DVI file with any header and page number added. Following each page of output, TEX displays a message in the output window giving the page number in square brackets. This means that the offending material from our input file that generated the error probably occurred in the section of input text that corresponds to the typeset page. However, it is possible that the error actually occurred in the text that had been typeset since the last page of output (i.e., those paragraphs placed on the "scroll" that did not yet make up enough material for a complete page output). Thus, the error might lie on the last generated page of output, or it might be in the next couple of paragraphs awaiting output.

11.1 LATEX's Error Locator

We have described, in outline, how, when TEX discovers a problem with the input file, an indicator of the error is produced in the output window and the log file. In addition, a locator for the error informs you how far through your input file the typesetting engine had gone, before discovering the error. Often the line shown by the error indicator will indicate the error, although due to the way that TEX only feeds whole pages to the DVI file, it is possible that the error may not be shown in the typeset output that has been generated thus far. Running LATEX on a simple input file with the contents:

```
\documentclass[12pt]{article}
\begun{document}
 My sample text.
\end{document}
```

generates this error message on the screen and in the log file

```
! Undefined control sequence.
l.2 \begun
            {document}
    ?
```

which informs us that a TEX error occurred at line 2 of the document input file. In this case, it is the misspelled command name \begun, which should be corrected to \begin for the example to typeset correctly.

Occasionally, the error is not detected when TEX is generating the scroll; instead, it happens when it is cutting off the page for output to the DVI file. These are called *outputting* errors and are indicated by the text <output> appearing above the error locator, the latter suggesting where TEX got to while typesetting the scroll, the error itself having occurred there or at some point since the last page was cut off the scroll.

When LATEX processes the \begin{document} command, the auxiliary file is read in, and at this point errors can be detected. In the case of an error generated by the \begin{document}, it is likely that the error was produced the previous time that you ran LATEX on the file. However, if the error was produced by the \end{document}, this suggests that the problem arose from moving an argument that contains a fragile command. Similarly, some of the additional files that LATEX generates can generate errors [e.g., tables of contents (.toc files), lists of figures (.lof files), and lists of tables (.lot files)]. The appearance of the error means that it was the previous run of the input file through LATEX that produced the error that is now detected. Often, this is a problem with a captioning or sectioning command.

If you cannot detect the nature of the error from the error messages that LaTeX displays in the output window and prints into the log file, then the next step is to use the program for displaying your DVI files to examine the typeset output. Entering an exit command x at the error prompt ? followed by the Enter key can leave the portion of the typeset text with the error on the scroll, for if TeX has not generated the error in a complete page of output, then it will not have been sent to the dvi. You can get around this by using one of the alternative mode commands at the error prompt (e.g., s, r, and q will continue typesetting through the remainder of the document). An easier alternative, when many pages follow the error, is just to keep pressing the Enter key until TeX indicates that it has produced one more page of output. Failing this approach, you will have to try, by trial and error, to locate the smallest fragment of your text that generates the error. The use of the % symbol can be useful since you can try "commenting out" suspect pieces of text to see if the error then disappears. If your file with the commented out text now typesets correctly through the area where the error was, then we can be confident that the error lies in the isolated piece of text.

As a learning aid, we shall be writing some simple files that illustrate the kind of errors that one can encounter during the course of preparing a LaTeX source file. After carefully typing in the example, you will need to take a look at the messages that appear in the output window of your typesetting engine, and you may also need to examine the log file with your editor. At an error prompt, you will be guided as to which command mode to ask for and shown how to correct the error.

11.2   Error Messages

One can determine whether an error was from LaTeX or actually from TeX by examining the error message (see Table 11.1). A message from a LaTeX error will begin with the text

```
! LaTeX Error:
```

For example,

```
! LaTeX Error: Lonely \item--perhaps a missing list environment.
```

is produced if you placed a list item outside of its list environment, whereas a TeX error simply begins with a ! followed by the message; for example,

```
! Extra alignment tab has been changed to \cr.
```

occurs if you put too many entries in a single row of a tabular or array environment. This information can then guide you to the appropriate section to look at since we have divided errors and warnings into their LaTeX and TeX counterparts.

Table 11.1: LaTeX error messages and their probable causes.

LaTeX Error Message	Probable Cause
`Bad \line or \vector argument.`	A negative length or invalid slope was given as an argument to a `\line` or `\vector` command.
`Bad math environment delimiter.`	Unmatched delimiters for math mode, or braces; for example, `\(` or `\[` used in math mode, `\]` or `\)` used in paragraph or LR mode.
`\begin{...}on input line ... ended by \end{...}.`	An `\end` command that does not match the associated `\begin` command.
`Can be used only in preamble.`	A command was used that must only occur in the preamble, `\includeonly`, `\makeindex`, `\nofiles`, or `\usepackage` should go before `\begin{document}`.
`Cannot determine size of graphic ... (no BoundingBox).`	LaTeX was unable to find the bounding box comment in an included graphics file.
`Command ... already defined.`	One of `\newcommand`, `\newenvironment`, `\newlength`, `\newsavebox`, or `\newtheorem` was used to define an existing name. Try a different name or use `\renewsomething`.
`Command ... invalid in math mode.`	The named command cannot be used in math mode.
`Counter too large.`	A counter for a numbered entry was set with a number that was too big, or an enumerated list is too long.
`Environment ... undefined.`	A `\begin` command occurred for an environment that does not exist.
`File ... not found.`	The named file with extension `tex`, document class with extension `cls`, or package with extension `sty` does not exist.
`Illegal character in array arg.`	An `array` or `tabular` environment, or the second argument of a `\multicolumn` command, contained an illegal character.
`\include cannot be nested.`	A `\include` command was used to insert a file that also contained a `\include` command.
`Lonely \item--perhaps a missing list environment.`	An `\item` occurred outside of a list environment.
`Missing \begin{document}.`	LaTeX found something that caused it to start typesetting before actually encountering the `\begin{document}` command.
`Missing p-arg in array arg.`	An `array` or `tabular` environment, or the second argument of a `\multicolumn` command, contained a p not followed by an expression in braces.

Table 11.1: Continued.

LaTeX Error Message	Probable Cause
`Missing @-exp in array arg.`	An `array` or `tabular` environment, or the second argument of a `\multicolumn` command, contained an @ not followed by an @-expression.
`No counter '...' defined.`	A `\addtocounter` or `\setcounter` command, or an optional argument to a `\newcounter` or `\newtheorem`, was requested with a counter that does not exist.
`No \title given.`	No `\title` command appeared before the use of `\maketitle`.
`Not in outer par mode.`	A `\marginpar` command, or a `figure` or `table` environment, occurred inside a parbox, a minipage, or in math mode, or a floating object occurred within another floating object.
`Option clash for package ...`	Different options were used for the same package, which was loaded twice (possibly by another package).
`\pushtabs and \poptabs don't match.`	Either a `\end{tabbing}` command appeared with an unmatched `\pushtabs` command(s), or a `\poptabs` command had no matching `\pushtabs`.
`Something's wrong --perhaps a missing \item.`	Possible causes include an omitted `\item` from a list environment, or an argument to a `thebibliography` environment is missing.
`Tab overflow.`	The maximum number of tab stops has been exceeded by a `\=` command.
`There's no line here to end.`	A `\\` or `\newline` command occurs incorrectly between paragraphs. Try a `\vspace` command instead.
`This file needs format ... but this is ...`	A document class or package was used that is not compatible with this version of LaTeX, or you have a LaTeX installation problem.
`This may be a LaTeX bug.`	This is unlikely to be an actual LaTeX bug. Probably, a previously announced error has confused LaTeX.
`Too deeply nested.`	You have too many list-making environments nested within each other.
`Too many columns in eqnarray environment.`	Three & column separators are used in an `eqnarray` environment without a `\\` command between them.
`Too many unprocessed floats.`	Too many figures and tables have been saved by LaTeX, or one of your pages has too many `\marginpar` commands on it.

Table 11.1: Continued.

LaTeX Error Message	Probable Cause
`Undefined color '...'.`	The named color was not defined with \definecolor.
`Undefined color model '...'.`	The color model requested in \definecolor is unknown.
`Undefined tab position.`	A \= command has not been used to define the tab position sought by one of \<, \>, \+, or \-.
`Unknown graphics extension ...`	An unknown file extension was found when the \includegraphics command tried to determine the file type of the graphic.
`Unknown option ... for ...`	An unavailable option was specified in a \documentclass or \usepackage command.
`\verb ended by end of line.`	The text following a \verb command goes beyond the present line. You may have omitted an end character.
`\verb illegal in command argument.`	The argument to a command contains a \verb command.
`\< in mid line.`	A tabbing environment contains a \< in the middle of a line rather than at the beginning of the line.

11.2.1 Errors found by LaTeX

As we mentioned earlier, sometimes a single error can generate others in a knock-on effect. The most common example is an input file with a problematic list environment. In our version, there are three items in an enumerated list that has an error in that the environment starts with \begin{numbering} instead of \begin{enumerate}. In addition, a simple error is employed to demonstrate the error recovery facilities TeX offers.

The following text can be typed in as a complete example of error propagation:

```
\documentclass{article}
\begin{document}
    $$\sun x^2 + \cos x^2 = 1$$
    Computer languages are considered to be
    \emph{object-oriented} if they support the
    following properties:
    \begin{numbering}
        \item abstraction
```

```
                    \item encapsulation
                    \item inheritance
                    \item polymorphism
                \end{enumerate}
            \end{document}
```

Running LaTeX on this input file generates the first error:

```
! Undefined control sequence.
1.3            $$\sun
                    x^2 + \cos x^2 = 1$$
?
```

At the prompt, we can type the letter i and then the correct name of the command:

```
? i
insert> \sin
! LaTeX Error: Environment numbering undefined.
See the LaTeX manual or LaTeX Companion for explanation.
Type H <return> for immediate help.
...
1.7 \begin{numbering}
?
```

Unfortunately, it is not now possible to do the same and to replace the erroneous piece of code, as it consists of more than one token. So, all we can do is to press return and get a succession of error messages. Thus, we get these three \lonely \item messages caused by our failure to call up an enumerate environment so that LaTeX finds the items outside of a list environment:

```
! LaTeX Error: Lonely \item--perhaps a missing list environment.

See the LaTeX manual or LaTeX Companion for explanation.
Type  H <return>  for immediate help.
 ...

1.6 \item a
            bstraction
            ?

! LaTeX Error: Lonely \item--perhaps a missing list environment.

See the LaTeX manual or LaTeX Companion for explanation.
Type  H <return>  for immediate help.
 ...
```

```
l.7 \item e
            ncapsulation
            ?

! LaTeX Error: Lonely \item--perhaps a missing list environment.

See the LaTeX manual or LaTeX Companion for explanation.
Type  H <return>  for immediate help.
 . . .

l.8 \item i
            nheritance
            ?

! LaTeX Error: Lonely \item--perhaps a missing list environment.

See the LaTeX manual or LaTeX Companion for explanation.
Type  H <return>  for immediate help.
 . . .

l.9 \item p
            olymorphism
            ?
```

They are followed by a final error message

```
! LaTeX Error: \begin{document} ended by \end{enumerate}.

See the LaTeX manual or LaTeX Companion for explanation.
Type  H <return>  for immediate help.
 . . .

l.10 \end{enumerate}

?
```

that indicates that the program attempted to terminate an enumerate environment that was never started up. Pressing enter one more time causes the program to terminate. So, we have seen that a single error in our input file can lead to a cascade of other error messages. In such cases, one is often better off typing I\stop rather than X followed by the return key since the former will include the final material that has been processed in the output.

11.2.2 Errors in LaTeX Packages

Many classes and packages are available to extend the facilities offered by LaTeX. Usually, additional packages and classes have error and warning messages that concern their own use and are described in the documentation that accompanies them. For example, adding the command

```
\usepackage{babel}
```

to the preamble of our document produces the following error message:

```
! Package babel Error: You haven't specified a language option.
```

By adding the option german to the options list of the \usepackage{babel} command, we are enabled to typeset German language documents (i.e., \usepackage[german,english]{babel}).

11.2.3 Errors Found by TeX

One error that can cause confusion or panic when first encountered is the message

```
!  TeX capacity exceeded, sorry [...].
```

This happens when TeX halts its execution because the internal space required to process your document was used up. This usually has nothing to do with not having sufficient capacity for your document but is more likely to be a knock-on effect of another type of error in your input file. The following example attempts to define two new commands, \esmile and \efrown, that are used as shorthands to express feelings in informal messages (usually e-mails). Table 11.2 lists TeX error messages and their causes.

```
\documentclass{article}
\newcommand{\esmile}{a \esmile}
\newcommand{\efrown}{\texttt{;(}}
\begin{document}
Today is my birthday \esmile \\
Unfortunately, I have to work late \efrown
\end{document}
```

However, a careless slip has resulted in \esmile being defined in terms of itself rather than the string of characters that makes up the e-mail icon. As a result of this, the following message was generated by TeX:

```
! TeX capacity exceeded, sorry [main memory size=2000001].
\esmile ->a
            \esmile
            l.5 Today is my birthday \esmile
If you really absolutely need more capacity,
you can ask a wizard to enlarge me.
```

Table 11.2: T_EX error messages and their probable causes.

T_EX Error Message	Probable Cause
! Double subscript.	Two adjacent subscripts have occurred in a math environment. Try nesting the braces (e.g., $y_{2_{4}}$ gives y_{2_4}).
! Double superscript.	Two adjacent subscripts have occurred in a math environment. Try nesting the braces (e.g., $y^{2^{4}}$ gives y^{2^4}).
! Extra alignment tab has been changed to \cr.	Too many & column separators in one row of an array or tabular environment. Probably a forgotten \\ command.
! Extra }, or forgotten $.	Unmatched math model delimiters or braces probably caused by a missing { \[\(or $.
! I can't find file '...'.	Your named file does not exist.
! Illegal parameter number in definition of ...	Incorrect use of a # in one of the \newcommand, \renewcommand, \providecommand (the package writer's verion of \newcommand), \newenvironment, or \renewenvironment. Nesting these commands also causes this.
! Illegal unit of measure (pt inserted).	Possibly the same problem as with the message: ! missing number, treated as zero, or you forgot units of a length argument (e.g., 9 instead of 9pt).
! Misplaced alignment tab character &.	You typed the special character & in a passage of text rather than an array or tabular environment. Try a \&.
Missing control sequence inserted.	A first argument that is not a command name was given as an argument to one of the \newcommand, \newlength, \newsavebox, or \renewcommand commands.
! Missing number, treated as zero.	1) T_EX expected a number or length as the argument to a command but did not get one. 2) A square bracket in some text was mistaken for the start of an optional argument. 3) \protect was placed in front of a length or \value command.
! Missing {inserted or ! Missing }inserted	At this point T_EX is probably confused, and the error locator indicates a place too far beyond the actual error.
! Missing $ inserted or ! Missing $$ inserted.	A math mode command occurred when T_EX was not in math mode or a blank line while it was in math mode.
! Not a letter.	An inappropriate argument to a \hyphenation command was used.

Table 11.2: Continued.

TEX Error Message	Probable Cause
! Paragraph ended before ... was complete.	A command argument contained an inappropriate blank line. You may have left off the right brace to finish an argument.
! TeX capacity exceeded, sorry [...].	An error in your input file is the most likely cause, rather than TEX actually running out of space. Probably, TEX is looping endlessly because of a wrong command.
! Text line contains an invalid character.	Your input file contains a nonprinting character. Use an editor that just produces ASCII characters, or choose "save as ASCII" from your word processor.
! Undefined control sequence.	You may have misspelled or misplaced a command name. Alternatively, you have omitted a \documentclass or \usepackage command.
! Use of ... doesn't match its definition.	1) If ... is a command for LATEX, then you may have used the incorrect syntax for an argument to a picture command. 2) If ... is a \@array, there is an error in the @-expression in the argument of an array or tabular environment (try \protect with a fragile command). 3) A fragile command having an optional argument that occurs in a moving argument can also cause this.
! You can't use 'macro parameter character #' in ... mode.	You typed the special character # in a passage of normal text. Try using \#.

See if you can help our user out by rewriting the definition of \esmile in terms of a suitable e-mail icon such as :-), and then try running TEX. You should then find that TEX runs smoothly and processes the input file without any difficulty.

The present generation of computers has sufficient memory to give TEX the space that it needs for most documents, but a given installation of TEX only has a fixed amount of space set up. For this reason, the version installed on your computer may need to be run with different settings, or a bigger version may need to be obtained. Lamport (in [20] pages 142–144) gives more discussion of the types of space that may be used up and some solutions to get around the problem.

When writing mathematics, errors often arise from the omission of a closing command to return to a text environment, such as $ or a forgotten closing brace }. Continually pressing the return key will usually get TEX to finish processing the file, but a more convenient alternative sometimes is to use the scroll mode (type S and the return key at the error prompt), which will proceed all the way through the file and allows one to look at the typeset result in the output DVI file to see the nature of the error.

11.3 Warnings

11.3.1 Warnings Generated by LaTeX

You can tell whether a warning is generated by LaTeX (see Table 11.3) since it will begin with the text LaTeX Warning: . For example, if we reference an undefined label as in this passage

```
\documentclass{article}
\begin{document}
The British philosopher Gilbert Ryle introduced the term ''the Ghost
in the Machine'' to characterize the Cartesian view of the mind.
Section
\ref{volition} introduces his view of mental processes.

\section{The Myth of Volitions, According to Ryle's (1949) book The
Concept of Mind}
\end{document}
```

then the following LaTeX warning is generated:

```
LaTeX Warning: Reference 'volition' on page 1 undefined
on input line 5.
```

This can be corrected by adding the command \label{volition} immediately following the closing brace of the sectioning command. Running LaTeX twice on the file will generate the correct cross-reference and will omit the warning on the second run through. An additional warning is still generated on the screen and written to the log file, namely

```
Overfull \hbox (15.433pt too wide) in paragraph at lines 8--8
[]\OT1/cmr/bx/n/14.4 The Myth of Vo-li-tions, Ac-cord-ing to
Ryle (1949)[]
```

We can see that this is a TeX warning since it has no ? character preceding the message indicating an error and because it is not preceded by the words LaTeX Warning: . It is telling us that TeX could not find a good place to break the line containing the section heading. We leave it as an exercise for the reader to assist TeX in correctly breaking the line.

Table 11.3: LaTeX warnings and their probable causes.

LaTeX Warning Message	Probable Cause
`Citation '...' on page ...` `undefined.`	You have not defined the key in `\cite` command with a `\bibitem` command.
`Command ... invalid in math` `mode.`	You used the named command in math mode when it is not allowed there.
`Float too large for page by ...`	A table or figure is too long by the stated length in units of points. It is printed on a separate page.
`Font shape '...' in size ...` `not available`	A font was specified that is unavailable on your system, and it was replaced by the font indicated on the next line.

`h float specifier changed to`
`ht. or !h float specifier changed`
`to !ht.`
	A table or figure with an optional h or !h argument could not fit on the present page and was placed on the next page.

`Label '...' multiply defined.`	The same arguments were used for two `\bibitem` or `\label` commands. This occurred on the previous run through LaTeX.

`Label(s) may have changed.`
`Rerun to get cross-references`
`right.`
	Indicates that the values given by `\cite`, `\ref`, or `\pageref` could be wrong if the correct values have altered since the last run through LaTeX.

`Marginpar on page ... moved.`	A marginal note had to be printed lower than the text it refers to, so as not to overprint an existing marginal note.
`No \author given.`	A `\author` command did not occur before `\maketitle`.
`Optional argument of \twocolumn` `too tall on page ...`	A box too big for the page was specified by the optional argument of a `\twocolumn` command.
`Oval too small.`	A poor approximation to the requested oval occurred because the required quarter circles were not available that small.
`Reference '...' on page ...` `undefined.`	A `\label` command was not used to define the argument of a `\ref` or `\pageref` command.

`Some font shapes were`
`not available, defaults`
`substituted.`
	A font was specified that is unavailable on your system and substituted with a default alternative.

`There were multiply-defined` `labels.`	Two different `\label` commands were used in the definition of a label.
`There were undefined references` `or citations.`	A nonexistent bibliography entry or `\label` was referred to by a `\cite` or `\ref` command.
`Unused global option(s): [...].`	The `\documentclass` command, or packages that were loaded, did not recognize the indicated options.

`You have requested release`
`'...' of LaTeX, but only`
`release '...' is available.`
	Your release of LaTeX does not work with a specified document class or package. A later version of LaTeX will be required to work with that.

When using babel, a common error is a missing hyphenation package for a language that we wish to write in. If we typeset the example

```
\documentclass[a4paper,11pt]{article}
\usepackage[spanish,english]{babel}
\begin{document}
A verse from the Guantanamera song. Lyric adaption by
Julian Orbon, based on a poem by the Cuban poet
Jos\'{e} Mart\'{\i}:
\begin{verse}
\selectlanguage{spanish}
Yo soy un hombre sincero \\
De donde crece la palma, \\
Y antes de morirme quiero \\
Echar mis versos del alma.
\end{verse}
\end{document}
```

the log file informs us that there is a possible problem with the typeset output arising during the use of the add-on package babel:

```
Package babel Warning: No hyphenation patterns were loaded for
(babel)                 the language 'Spanish'
(babel)                 I will use the patterns loaded for
                        \language=0 instead.
```

This tells us that it could not find hyphenation patterns for Spanish, so it will use those for the default language, which will be the first one embedded in the format file (e.g., American English). In order to generate a format file, we have to use INITEX (or INIOMEGA, and so on). This program transforms the file latex.ltx into a fast loadable binary form and includes hyphenation patterns for various languages. Since we are typesetting poetry in Spanish with lines narrower than the width of the page, we can choose to ignore the warning since LATEX does not need to try and hyphenate the verse. However, if we were typesetting a continuous piece of Spanish prose, then we would like to load the hyphenation patterns for the Spanish language. This is done by locating the file language.dat in our TEX installation and adding the line

```
spanish           spanhyph.tex  % Spanish
```

to the file. Following this, a new LATEX format file must be generated by running INITEX on the file latex.ltx of our system. This will produce a format file for LATEX that will allow for loading the Spanish hyphenation patterns when the Spanish language is selected using the Babel package. This example illustrates the difference between TEX or LATEX errors and warnings. In the case of an error, the execution of TEX halts and some action must be taken by the user to correct the mistake, while a warning highlights a possible problem with the typeset output but program execution continues and the

DVI file is produced. Depending on the nature of the warning, this may or may not have an actual effect on the typeset output.

11.3.2 Warnings Generated by TₑX

Table 11.4 gives the subset of TₑX warnings highlighted in the LATₑX manual. These focus on places where TₑX had difficultly in breaking a line or page and can be aided in the process by the judicious use of some hinting commands from the user.

Table 11.4: TₑX warnings and their probable causes.

TₑX Warning Message	Probable Cause
Overfull \hbox ...	TₑX had difficulty finding a good place to break the line. You may need to indicate suitable places for the hyphenation of an unusual word or add a \linebreak or \newline command.
Overfull \vbox ...	TₑX had difficulty finding a good place for a page break and put too much on the page. It needs some assistance from you. Try using a \pagebreak or \enlargethispage* command.
Underfull \hbox ...	Two successive \\ or \newline commands added vertical space to your document. Alternatively, a \sloppy declaration, a sloppypar environment, or a \linebreak command may produce this warning.
Underfull \vbox ...	TₑX had difficulty finding a good place for a page break and put too little on the page. It needs some assistance from you. Try adding a \nopagebreak command to deter TₑX from breaking the page there.

11.4 The Last Straw: Strategies for Dealing with Resistant Errors

Occasionally, the situation may be reached where the TₑX program cannot be stopped following an error (e.g., when a serious error propagation causes the text processing to continue indefinitely). In this case, one may need to halt the program with the operating system interrupt, the nature of which will depend on your particular operating system, although typically simultaneously pressing the Ctrl and c keys or Ctrl and Break will do the trick.

With obscure errors, a strategy of divide and conquer can be helpful. By inserting a \end{document} command part of the way through, running LATEX, and examining the output, you can see if the first part of the file is free of errors. If it is, then you can cut and paste the \end{document} further down, typeset the altered document, and see if the error has occurred in the text between where you last ended the input and the current line where you have ended it. Of course, you will need to make sure that any active environments are also ended (e.g., placing the \end{document} command after the end of a quotation environment). This successive moving of the ending command down through the document, together with the judicious use of a comment character % to temporarily omit suspect lines, can help a great deal in tracing an erroneous piece of LATEX input. Some LATEX-oriented editing programs (such as EMACS) allow you to select pieces of input text and typeset the passages automatically without having to create a complete TEX document, which is very convenient for tracing errors (see Appendix B).

12

INSTALLING NEW TYPE

Most common font formats (PostScript Type 1, Type 3, etc., and TrueType fonts) can be used with any of the LaTeX forms (including standard LaTeX, Λ, and pdfLaTeX), and, of course, LaTeX uses by default fonts created with METAFONT. Newer formats, such as the OpenType format, can also be used since it is possible to convert them to Type 1 fonts. Direct support of OpenType may be added in the future.

The standard way to install PostScript Type 1 fonts is by using the FONTINST program by Alan Jeffrey and Rowland McDonnell. This program is actually a TeX application! Although fontinst is quite a powerful program, it cannot handle all possible cases, so we will fully describe the installation procedure of virtually any PostScript Type 1 font. It is very important for the user to understand the installation of Type 1 fonts, as it is possible to convert all common font formats to Type 1. We will, however, start with the default METAFONT fonts.

12.1 Installing METAFONT Fonts

Most of the time, METAFONT fonts come from "TeX-aware" people or the CTAN archives, and thus they come with installation instructions. In any case the installation of METAFONT fonts is simple. If the fonts come with support files, then you just place the METAFONT sources in the TeX trees (usually in `texmf/fonts/source/`) and the accompanying support files (packages and font definitions) anywhere in `texmf/tex/generic/` or `texmf/tex/latex/`. You do not really need the TFM files, as these will be generated automatically from the METAFONT sources when you use the fonts. However, if TeX font metrics are provided and you want to save computing time, you can put them anywhere in `texmf/fonts/tfm/`.

Now, you must refresh the "filename database." Unfortunately, there is no single name for this operation. Its name depends chiefly on the TeX implementation in use, so on a Unix system, the system administrator can perform the operation above by issuing

```
$ mktexlsr or $ texhash
```

On Windows installations, this operation is often linked somewhere in the start menu.

If the fonts are not accompanied by supporting package files, then you can simply use them as described in Section 3.4 using the \font command, and, of course, you may create your own package and font definition files. But, we will discuss these things later, in Section 12.4.

12.2 Installing Type 1 Text Fonts in LaTeX

Type 1 fonts come in two flavors—binary and ASCII. The binary form is actually an encrypted and compressed version of the ASCII one. The usual filename extensions are .pfa for the ASCII form and .pfb for the binary form. Both forms can be used with LaTeX by following the same procedure. Type 1 fonts are usually accompanied by the so-called "Adobe Font Metric" file which usually has the .afm filename extension. We will see the importance of the AFM file in the next paragraphs.

Suppose that we have a Type 1 font named font.pfb (the same applies to .pfa fonts). The typesetting engine needs only to know the dimensions of the glyphs, so it is not really concerned about the actual shape of the glyphs. After all, for TeX, each letter is just a box, as we have already explained, and as the reader may recall, a box has a height, a width, and possibly a depth. However, we must stress that there is a fourth parameter that we intentionally did not present until now. It is called *italic correction* and is the amount of additional white space to be added after the character to avoid the collision of the slanted character with the next one (compare '*leaf* b' with '*leaf b*'). Of course, you will need the file font.pfb (i.e., the glyphs themselves) when you want to print or preview your document. Remember: TeX typesets and it does *not* print! So, we do need the font metrics to make available for use with LaTeX. The actual glyphs will be used by the driver program.

Usually, each Type 1 font is accompanied by its font metrics file, but just in case we have a font but not the font metrics, there is a simple procedure by means of which one can get the font metrics. This procedure is described in the next subsection and can be safely skipped on a first reading.

12.2.1 Extracting Metric Information

The easiest way to get the font.afm file is by opening font.pfb in a font editor. There are several font editors, depending on the platform we work with. For Unix systems, one can use the PFAEDIT font editor by George Williams available from http://pfaedit. sourceforge.net. For other platforms, there are up to now only commercial products.

If you have such a program, open the font.pfb file and use the extract utilities provided in its menus. If you do not have such a program, another way to do the job is by using Ghostscript. Since we use Ghostscript, the font must already be known to the program with a proper declaration in the Fontmap file of your Ghostscript installation (or the Fontmap.GS file in newer releases). Of course, we can copy this file into our current working directory and make the changes temporal. Here is the recipe to get the Adobe font metric:

1. Find out the internal name of the font at hand font.pfb, convert the .pfb file to the ASCII format by issuing pfb2pfa font.pfb. This creates font.pfa, which when opened in any text editor allows us to read the font's name by looking at the line that starts with /FontName. If the name is, for example, Times-NewRoman, then this line will look like this: /FontName /Times-NewRoman.

2. Modify the Fontmap (or Fontmap.GS) file by adding to it a line like

    /Times-NewRoman (/path/to/font.pfb);

   where /path/to/ is the location of font.pfb in our system.

3. Get the afm file with the command

    $ getafm font.pfb | gsnd - >font.afm

(the command above is written for the Unix environment).

Now, we have to prepare metric files that TEX can understand since it cannot understand AFM files. TEX can deal only with TEX font metrics. Here, we have one serious restriction: each TFM file cannot contain metric data for more than 256 glyphs. This is the most serious restriction that Ω removes, and it is also the reason why when switching languages, for example from English to Greek, we have to switch fonts and, consequently, we (artificially!) need to have commands such as \textlatin, as described in Chapter 10.

But an AFM file may contain metric information for many glyphs. Thus, we have to create our TFM by selecting 256 glyphs from the many glyph metrics in the AFM file. This is done by what we call an *encoding vector*. Actually, an encoding vector is something more. It lists in a sequence the names of the glyphs for which we want to get information, and the order *is* important, as TEX really identifies each glyph by its position in this row of 256 glyphs.

12.2.2 Encoding Vectors

Using Computer Science nomenclature, we can say that an encoding vector is just an array of glyph names that describes the arrangement of glyphs in a font. Encoding vectors are used to rearrange the positions of the various glyphs in a font. Naturally, in this section we will deal only with encoding vectors that can be applied to PostScript fonts. The definition of an encoding vector consists of its name, a sequence of glyph names enclosed in brackets, and the keyword def. The encoding name and the glyph names must be prefixed with a slash (/), as each encoding vector is a valid PostScript data structure and as such it must follow the conventions of the PostScript language. Here is an example definition:

    /greek [/alpha /beta /gamma /delta ...] def

Obviously, we must know the names of the glyphs of an existing font in order to apply an encoding vector to this font. Usually, Latin fonts use standard names for each glyph, but we cannot rely on this assumption, so it is best to check the names of the glyphs of the font that we want to reencode. This can be done by inspecting the PostScript font with a font editor or a font viewer such as GFONTVIEW. This font viewer is part of the Gnome desktop environment (http://www.gnome.org). Now, the problem that we have to face is to decide how to order these glyphs in our encoding vector. The choices that we will have to make depend heavily on how we are going to use a particular font. Thus, a font with Greek letters can be used as a mathematical font or as a font that will be used to typeset ordinary Greek text. Once such issues have been decided, we define the encoding vector. For example, if one is going to use a Greek font to typeset Greek text, then this font must follow the conventions of the fonts designed by Claudio Beccari. Although this particular glyph arrangement is not universally accepted, it can be safely used to reencode fonts. The task of defining an official encoding for Greek fonts is an ongoing project. Since there is no official definition, we have to find the order of the glyphs using the default fonts by using the nfssfont.tex input file. We run LaTeX on this file and follow the instructions:

```
$ latex nfssfont


***********************************************
* NFSS font test program version <v2.0e>
*
* Follow the instructions
***********************************************

Name of the font to test = grmn1000
Now type a test command (\help for help):
*\help

\init switches to another font;
\stop or \bye finishes the run;
\table prints the font layout in tabular format;
\text prints a sample text, assuming TeX text font conventions;
\sample combines \table and \text;
\mixture mixes a background character with a series of others;
\alternation interleaves a background character with a series;
\alphabet prints all lowercase letters within a given background;
\ALPHABET prints all uppercase letters within a given background;
\series prints a series of letters within a given background;
\lowers prints a comprehensive test of lowercase;
\uppers prints a comprehensive test of uppercase;
\digits prints a comprehensive test of numerals;
\math prints a comprehensive test of TeX math italic;
```

```
\names prints a text that mixes upper and lower case;
\punct prints a punctuation test;
\bigtest combines many of the above routines;
\help repeats this message;
and you can use ordinary TeX commands (e.g., to \input a file)

* \table
* \bye

[1]
Output written on nfssfont.dvi (1 page, 10704 bytes).
Transcript written on nfssfont.log.
```

Now, we can create the encoding vector and store it in a file with a reasonable name. We first print the output of the LATEX run above and use it to define the encoding vector. Of course, this task assumes that we are familiar with the glyphs and their names.

The file where we store an encoding vector is also the right place for defining ligatures between glyphs. Ligatures are defined after the definition of the encoding vector. Each line defining a ligature is like the following one

% LIGKERN questiondown questiondown =: guillemotright ;

It starts with % LIGKERN and the actual definition of the ligature. The ligature definition must be terminated with a semicolon. Note that we can have more than one ligature definition as in the following example:

% LIGKERN hyphen hyphen =: endash ; endash hyphen -=: emdash ;

Of course, as is evident, the two definitions are related.

We now explain how to set up ligature rules. Assume that we want two glyphs to combine into a new one. Such a case is the fi ligature, where the letter f followed by an i becomes fi. The fi ligature is specified as follows:

% LIGKERN f i =: fi ;

Note that spaces are important in the above. To add the ff and ffi ligatures, we may write:

% LIGKERN f i =: fi ; f f =: ff ; ff i =: ffi ;

This is also how we have access to accented letters in languages such as Greek, Hebrew, and others. For example, we use

% LIGKERN tonos alpha =: alphatonos ;

to specify that the acute ' followed by α should become a ά.

Another type of ligature is when *one* of the letters is substituted by something else. For example, in Greek, a sigma (σ) appearing at the end of a word must change to a final sigma (ς). Thus, if a sigma is followed by, say, an exclamation mark, then the sigma must become a final sigma (which is usually called inside a font sigma1) and the exclamation mark should remain in its position. This simple rule can be expressed as follows:

```
% LIGKERN sigma exclam =:| sigma1 ;
```

The symbol =:| is used to say that *if sigma is followed by an exclamation mark replace the sigma with sigma1 but leave the exclamation mark at its position* (this is what the | character stands for). If the first character is to remain intact but the second character must change, then the =:| symbol should be replaced by the |=: symbol. If the ligature mechanism should skip one or two characters before resuming the search of ligatures, then we use =:> and =:>> or =:|>, =:|>>, |=:>, and |=:>>. Unfortunately, it is still not possible to rescan the previous characters for ligatures. Thus, symbols such as <=: are not available. This is a missing feature that would be very useful for complex typefaces.

An interesting thing is how we can incorporate special characters, such as the space character, into ligature definitions. To do this, we first define a shorthand for this character. For example, the following code defines that the symbol || will stand for a space:

```
% LIGKERN || = 39 ;
```

We always use this number and then use it for ligatures like this:

```
% LIGKERN s || =: sigma1 ;
```

Note that here we did not use =:|. With the same mechanism, we are able to use initial and final forms of letters. For example, the word "book" in Greek is βιβλίο. Note that the two betas are different. To get this effect, one can use the following code:

```
% LIGKERN || beta =: beta1 ;
```

Ready-to-use encoding vectors are provided by the distribution of the *kerkis* font family available from http://iris.math.aegean.gr/software/kerkis.

12.2.3 Creating Virtual Fonts and Metric Files

As we have already explained, TEX is a typesetting system that needs to know only the dimensions of each glyph of a given font to start typesetting using this font. Consequently, TEX does not really care where the glyphs are stored and how they are stored—this is something a driver must be aware of. This may lead someone to conclude that we can typeset a document using a *virtual* font that contains glyphs from different real fonts. Well, this is not quite true in the sense that TEX did not provide this facility originally. The designer of TEX introduced virtual fonts (an idea that was devised by David Fuchs) at a later stage to allow people to do exactly what we described—to use glyphs from different fonts in a transparent way so that it is not obvious that we are actually using different fonts. Now, we will describe how we can get a TFM file from an AFM file.

Let us assume that we have the Times-NewRoman font stored in the file `tnr.pfb`. Moreover, assume that the font metrics are stored in the file `tnr.afm` and that we want to reencode the font using an encoding vector stored in file `myenc.enc`. If we want to use the font without kerning pairs and ligatures, we can use AFM2TFM (by Tomas Rokicki) to get the font metrics:

```
$ afm2tfm tnr.afm rtnr.tfm >> tnr.map
```

If we want to reencode the font, we have to use the following command:

```
$ afm2tfm tnr.afm -T myenc.enc -v tnr8a tnr8r >> tnr.map
```

In the examples above we stored the output of AFM2TFM to file `tnr.map`. This file contains information that must be edited. The modified information should be copied to the file `psfonts.map` or else this file must be copied to the directory where `psfonts.map` resides. In the latter case, we must also add the following line in the file `config.ps`:

```
p +tnr.map
```

The command above generates two files: the "raw" TFM file `tnr8r.tfm` and the *virtual property list* file `tnr8a.vpl`. This file can be further processed before it will be transformed to a virtual font with the following command:

```
$ vptovf tnr8a tnr8a.vf tnr8a.tfm
```

We could use other names for the font files instead of the names tnr8a and tnr8r. But, due to the great amount of available fonts, there are standard rules for creating these names in order to avoid conflict. The reader is advised to consult [3] if he or she intends to share fonts with other people.

Now, open the file `tnr.map` in a text editor. It should have a line like

```
tnr8r Times-NewRoman " myenc ReEncodeFont " <myenc.enc
```

The way this line appears in the file will make DVIPS believe that Times-NewRoman is a resident font of our PostScript printer or emulator. Generally, this is not the case, so we need to modify this entry as follows:

```
tnr8r Times-NewRoman " myenc ReEncodeFont " <myenc.enc <tnr.pfb
```

With this line, we instruct DVIPS to embed the font `tnr.pfb` whenever it generates a PostScript file from a DVI file that uses the font tnr8r.

We are now ready to use our font by using the \font command to call the font tnr8r in an input file (see Section 3.4) provided that all of the font's files that we just created are in the same directory with our document (for a systemwide installation, see Section 12.5). If we have at our disposal all of the shapes and series of this font, we can create the necessary font definition files that will allow us to access shapes and series with the commands that we have learned. For more information on the construction of font definition files, see Section 12.4.

12.2.4 Creating More Fonts from a Type 1 Font

A Type 1 font can be used to create additional shapes from the glyphs of the font. For example, we can easily create slanted glyphs, although slanted glyphs may not exist in our original font, or we can use *extended* or *condensed* glyphs or even (fake) small capitals. Extended and condensed glyphs are glyphs that are scaled *only* horizontally by a factor bigger or smaller than 1, respectively. This is done when we create TFM files. Here are the necessary commands:

- Use

    ```
    $ afm2tfm tnr.afm -T myenc.enc -s 0.167 -v tnro8a tnr8r
    ```

    in order to create the (virtual) font tnro8a whose glyphs are slanted to the right at 16.7%. Use a negative number to get slant to the left!
- Use

    ```
    $ afm2tfm tnr.afm -T myenc.enc -e 1.2 -v tnre8a tnre8r
    ```

    in order to create the (virtual) font tnre8a that is extended by a factor of 1.2, or use a number less than 1 to get a condensed font.
- Use

    ```
    $ afm2tfm tnr.afm -T myenc.enc -V tnrsc8a tnrsc8r
    ```

    in order to create a fake small capital font (tnrsc8a) if your font family does not provide a real small capital font. Note that we use a capital V instead of a v). The default scaling factor is 80% but it can be changed with the -c option. For example, in order to create (fake) small capitals with a scaling factor of 75%, we should use

    ```
    afm2tfm tnr.afm -T myenc.enc -c 0.75  -V tnrsc8a tnrsc8r
    ```

12.3 Virtual Property List Files

In this section we discuss the structure of virtual property list files. This file format is human-readable and, consequently, it can be modified in order to add more features to the virtual fonts. A virtual property list file consists of three parts or lists[1]: the header list (from the start up to the LIGTABLE list), the ligature list, and the main list that describes each character that goes into the virtual font. Here is an example (of a part) of a virtual property list file:

```
(VTITLE Created by afm2tfm k.afm -T kerkisec.enc -v ek8a)
(COMMENT Please edit that VTITLE if you edit this file)
(FAMILY TeX-ek8r)
(CODINGSCHEME kerkisec)
```

---

1. The term list has its roots in the Lisp programming language, which uses a similar syntax.

```
(DESIGNSIZE R 10.0)
(DESIGNUNITS R 1000)
(COMMENT DESIGNSIZE (1 em) IS IN POINTS)
(COMMENT OTHER DIMENSIONS ARE MULTIPLES OF DESIGNSIZE/1000)
(FONTDIMEN
    (SLANT R 0)
    (SPACE D 320)
    (STRETCH D 200)
    (SHRINK D 100)
    (XHEIGHT D 485)
    (QUAD D 1000)
    (EXTRASPACE D 111)
    )
(MAPFONT D 0
    (FONTNAME ek8r)
    (FONTAT R 1200)
    (FONTSIZE R 1000)
    )
(LIGTABLE
    (LABEL C f) (COMMENT f)
    (LIG C i 0 2)
    (LIG C j 0 30)
    (STOP)
    (LABEL 0 23) (COMMENT ff)
    (LIG C l 0 25)
    (LIG C i 0 24)
    (STOP)
    (LABEL C A)
    (KRN C w R -68)
    (KRN C v R -75)
    )
(CHARACTER C V
    (CHARWD R 676)
    (CHARHT R 681)
    (CHARIC R 49)
    )
```

The header list provides some general information. In the example above we see that the family name of the font is TeX-ek8r encoded according to the kerkisec encoding scheme. The design size of the font is 10 pt, and all other sizes are given in design units, which are 1000 for this font (1000 units equals 1 em). Entering the FONTDIMEN list, we see (in the order of the example) that this font is not slanted (zero SLANT), the interword space is 320 units, and it can stretch 200 units or shrink 100 units. The XHEIGHT and QUAD lists specify the length of the dimensions 1 ex and 1 em. The EXTRASPACE list defines the space that TeX puts at the end of a sentence provided that \nofrenchspacing is enabled. This command cancels the effect of the \frenchspacing command. The sublists of the FONTDIMEN list correspond to the font dimensions described in Section 12.4. The COMMENT list is used to introduce comments in a property list.

One of the most important lists is the MAPFONT list. In the example above, it is used to say that the default external font to use for the virtual font is named ek8r. In addition, we specified the actual design size of the external font and the scaling factor. The rôle of this list is to make it possible for one virtual font to draw characters from more than one real font. This is done by assigning a number to each real font that we want to use. Thus, the lines

```
(MAPFONT D 0
   (FONTNAME Times-Roman)
   )
(MAPFONT D 1
   (FONTNAME symbol)
   )
(MAPFONT D 2
   (FONTNAME cmr10)
   (FONTAT D 20)
   )
```

load two real fonts, Times-Roman and symbol, and one more, cmr10, at the size of 20 units; each one is assigned to one of the numbers 0, 1, and 2, respectively. These numbers are used in order to select the real font to be used in the third part of the virtual property list file. In this third part, we have a sequence of statements, one for each glyph. Each statement describes the dimensions of the glyph from which the real font is to be drawn and how it is drawn. The first line identifies the glyph in the virtual font. Its code point can be expressed in four different ways: by the letter D and a decimal number, by the letter O and an octal number, by the letter H and a hexadecimal number, and by the letter C and the glyph name. Here is how one can express the same thing in all four possible ways:

```
CHARACTER C V      CHARACTER D 68
CHARACTER O 126    CHARACTER H 44
```

In the example above we see this information for the character V. Its width is 676 units, its height 681 units, and its italic correction is 49 units.

Another variable not appearing in the example of the character V is the CHARDP variable, which controls the depth of the character (i.e., how much the character extends below the baseline). For example, in the font above, the ampersand appears in the virtual property list file as

```
(CHARACTER O 46 (COMMENT ampersand)
```

```
(CHARWD R 800)
(CHARHT R 694)
(CHARDP R 13)
)
```

that is, it extends below the baseline by 13 units. Note that for the character V we had used its name in the CHARACTER statement, but for the ampersand we used its position in the font (which is the position of the ampersand in the encoding vector kerkisec.enc).

The glyph description above is the description of a real glyph. If we want to have the description of a glyph that actually belongs to some other font, we must have a glyph description like the following one:

```
(CHARACTER H AF (COMMENT code point is 175)
    (CHARWD R 0.665)
    (CHARHT R 0.799)
    (CHARIC R 0.065)
    (MAP
        (SELECTFONT D 0)
        (SETCHAR O 41) (COMMENT code point is 27)
        )
    )
```

Here, we see that the dimensions are expressed in decimal point units. This is possible when there is no DESIGNUNITS definition. Now, the description above says that we have to map the character with code point 175 to the character 27 of the real font. The following is a more complex example:

```
(CHARACTER D 197
    (MAP
        (PUSH)
        (SETCHAR C A)
        (POP)
        (MOVEUP R 0.937)
        (MOVERIGHT R 1.5)
        (SETCHAR O 312)
        )
    )
(CHARACTER O 200
    (MAP
        (MOVEDOWN R 2.1)
        (SETRULE R 1 R 8)
        )
    )
(CHARACTER O 201
    (MAP
```

```
          (SPECIAL ps: /SaveGray currentgray def .5 setgray)
          (SELECTFONT D 2)
          (SETCHAR C A)
          (SPECIAL ps: SaveGray setgray)
          )
   )
```

The first list says that the character with code point 197 is set as follows: an 'A' is typeset, and this is enclosed by PUSH and POP, which restore the original position. Then, the character with code point 130 is typeset after it is moved up by 0.937 units and to the right by 1.5 units. The last list is more complex. Before we actually explain what it does, we must warn the reader that it uses real PostScript code. Therefore, this example is useful only if the reader is accustomed to the basics of the PostScript language. Now, back to our example. The code says that in order to typeset the character with code point equal to 129 in this virtual font, we set the PostScript color to 50% gray and then typeset an 'A' from cmr10 at 20 units in this color. The SPECIAL ps: command is used to pass its argument to the PostScript driver (such as DVIPS). However, we chose to use these PostScript commands since they will be useful in one of our applications (see Section 12.3.1).

The second list of a virtual property list file is the part that holds the ligature and kerning information. In our virtual property list example above, we had

```
   (LIGTABLE
        (LABEL C f) (comment f)
        (LIG C i O 2)
        (LIG C j O 30)
        (STOP)
        (LABEL O 23) (comment ff)
        (LIG C l O 25)
        (LIG C i O 24)
        (STOP)
        (LABEL C A)
        (KRN C w R -68)
        (KRN C v R -75)
        )
```

This table describes the ligatures for 'f' and for 'ff', and two kerning pairs for 'A'. We first choose (label) the 'f' character and then state that if the next character is an 'i' we substitute both with the glyph in the (octal) position 2 (which is the 'fi' for our font) and similarly for 'fj'. Next, the 'ffl' and 'ffi' ligatures are defined similarly. Finally, we state that if the character 'w' follows the character 'A', then 'w' should be kerned to the left by 68 units and the same for the character 'v' with 75 units. More complex ligatures can be stated here using instead of the LIG function the functions

```
LIG /LIG /LIG> LIG/ LIG/> /LIG/ /LIG/> /LIG/>>
```

that correspond to the functions

$$=: \quad |=: \quad |=:> \quad =:| \quad =:|> \quad |=:| \quad |=:|> \quad |=:|>>$$

in the afm file, respectively (see Section 12.2.2).

12.3.1 Two Applications

We give here two nontrivial applications of manually editing the virtual property list file before the production of the virtual fonts. First, we give the easy one. Assume that we want to use a font that does not come with small capitals. A way to bypass this problem (although not typographically correct) is to produce fake small capitals using the -V option of the AFM2TFM program. However, there are languages, such as Greek, where a capital letter corresponds to more than one lowercase letter. For example, the capitalization of both σ and ς is Σ. It turns out that we have to edit the virtual property list file and correct the dimensions of the character c (which corresponds to the Greek final sigma ς; see Section 10.4) to have the dimensions of the character Σ and also map ς to Σ. The entry for c (which corresponds to ς) looks like

```
(CHARACTER C c
     (CHARWD R 496)
     (CHARHT R 447)
     (CHARDP R 208)
     )
```

and it should change to

```
(CHARACTER C c
     (CHARWD R 550)
     (CHARHT R 630)
     (MAP
        (SETCHAR O 123)
        )
     )
```

where the new dimensions are those of the capital sigma scaled to 80% or whatever is the scaling factor for the small capitals, and the glyph for c is the character with code point 83 (octal 123), which is the capital sigma.

The next application is more complex. We want to construct an underlined font. Underlining is not good for text work but may be useful in other applications such as posters. We saw how to underline with the package ulem on page 42. However, this was a bad underlining since the position of the underline depends on the depth of the glyphs. A good underlining should stay at the same position throughout the underlined text and should break nicely at all places where the glyphs extend below the baseline like this:

Quit the joy of gambling!

For this task, we should add underlines to all of the glyphs in our virtual fonts and take special care for those that extend well below the baseline.

Assume that we already have a font font.pfb for which we have prepared all necessary files such as font8a.vpl, font8a.vf, and so on. We repeat the same AFM2TFM command, but now we change the last two arguments to fontu8a and fontu8r (see Section 12.2.3). Obviously, the contents of the resulting fontu8a.vpl will be the same as that of font8a.vpl. Now, we edit the file fontu8a.vpl. The first step is to add a MAPFONT command so that fontu8a can refer to font8a. This is done by adding after the (MAPFONT D 0 (... the code

```
(MAPFONT D 1 (FONTNAME font8a))
```

Now, for each character that does not extend below the baseline, we add an underline of length equal to its width (CHARWD). For example, if the statement for the letter 'A' is

```
(CHARACTER C A
    (CHARWD R 777)
    (CHARHT R 663)
    (CHARDP R 29)
    )
```

we change it to

```
(CHARACTER C A
    (CHARWD R 777)
    (CHARHT R 663)
    (CHARDP R 29)
    (MAP
        (PUSH)
        (MOVEDOWN R 131)
        (SETRULE R 59 R 777)
        (POP)
        (SELECTFONT D 1)
        (SETCHAR C A)
        )
    )
```

The last modification says that we should move down 131 units and draw a line of height 59 units and length equal to the length of the letter 'A' (here 777). Then, at the same position ((PUSH) and (POP) make sure that we do not move forward), we typeset the character 'A' from the font 1; that is, the font font8a. Note that we have to link to another font (font8a) since referring to the current font would lead to a recursive font definition!

Let us see now the character 'j' that extends below the baseline. If the information for 'j' in the fontu8a.vpl file is

```
(CHARACTER C j
    (CHARWD R 233)
    (CHARHT R 683)
    (CHARDP R 280)
    )
```

we change it to

```
(CHARACTER C j
    (CHARWD R 233)
    (CHARHT R 683)
    (CHARDP R 280)
    (MAP
        (PUSH)
        (MOVEDOWN R 131)
        (SETRULE R 59 R 233)
        (POP)
        (SPECIAL ps: /SaveGray
         currentgray def 1 setgray)
        (PUSH)
        (MOVELEFT R 40)
        (SELECTFONT D 1)
        (SETCHAR C j)
        (POP)(PUSH)
        (MOVERIGHT R 40)
        (SELECTFONT D 1)
        (SETCHAR C j)
        (POP)
        (SPECIAL ps: SaveGray setgray)
        (SELECTFONT D 1)
        (SETCHAR C j)
        )
    )
```

As before, we first draw the underline rule. After that, we change the PostScript color to gray 100% (i.e., white). Then, with the white ink, we draw the character 'j' from the font 1, shifted to the left (MOVELEFT) and to the right (MOVERIGHT) by 40 units. This way we essentially erase the underline around the character parts that extend well below the baseline. We restore the black color and finally print the letter 'j'. There are cases where we need to shift the character up and down in addition to left and right. This is the case, for example, for the letter Q in in the ligature Qu.

After these modifications, we create our virtual fonts with VPTOVF.

12.4 Creating Support Packages and Font Definition Files

LATEX needs what we call *font definition* files if it is expected to properly handle size-changing commands such as \large, \tiny, or series- and shape-changing commands such as \textit and \textsc. A font definition file specifies the sizes and shapes available for a particular font family. An important characteristic of any font family is its font encoding. For any font family xxx that follows the font encoding yyy, we need the font definition file yyyxxx.fd in order to be able to use the fonts of the family. For any font encoding, we need to define a font encoding definition file. Given a font encoding called yyy, its font encoding definition file will be named yyyenc.def. If we have a font family that consists of fonts that contain supersets of the glyphs found in more than one font encoding, it is possible to create many font definition files. We will now discuss how to create a font definition file and a support package. We start with the font definition files.

First of all, we need to know the font family name. Next, we need to know the font encoding. In most cases, it is very easy to deduce the name of the font encoding. For example, for fonts usable for European languages, the possible choices are the OT1 and the T1 encodings. Once we have this information, we can create our font definition file. Suppose that we have a Greek font family. This means that we will use the LGR font encoding. The first line of a font definition file announces the font family and its encoding:

\DeclareFontFamily{LGR}{ptm}{}

Note that we specify the font encoding in capital letters. The empty argument should be used to specify loading settings, which we discuss below. Next, we specify a series of commands such as the following:

\DeclareFontShape{LGR}{ptm}{m}{n}{<-> font8a}{}

This says that whenever LATEX is asked to use the normal (n) shape of the medium (m) series of the of the ptm family encoded according to the LGR encoding, then the font to be used is font8a.

If we have other shapes of the fonts, such as an italic shape with name fonti8a, then in the same file we would add the following

\DeclareFontShape{LGR}{ptm}{m}{it}{<-> fonti8a}{}

Other alternatives for the series are usually b and bx for bold and extra bold and for the shape it, ui, sl, and sc for italics, upright italics, slanted (or oblique), and small caps.

The size specification <-> says that LATEX should use the same font (font8a) for creating all sizes in the output. This is an important function for METAFONT fonts, as they come in different files for different sizes. Alternatively, the specification <8> says that we should use the font at 8 pt only. The specifications <-8>, <8-16>, and 14.4- say that we should use the font that follows with all sizes up to 8 pt, all sizes between 8 and 16 pt, and all sizes greater than or equal to 14.4 pt. If we specify a list of size specifications, then this means that the specific font is available only at these sizes. An important addition to the above is the scaling function. Usually, we will want to use more than one font in a document. For example, one may want to use a special font in a document, but it may be the case that the document requires some math symbols available only from the default fonts. The problem is that the two fonts may have a different design size, and one of them may look much larger than the other. Typographically, it is certainly correct to completely avoid such situations by using only fonts with similar designs. However, this is usually not possible. For these cases, we can ask LATEX to scale one of the fonts on the fly. For example, if we want to match the design size for the default math symbols, we would like to scale the text fonts that we want to use so that they have the same height with the default fonts. If the scale factor is 0.9, then we modify the line in the fd file above to look like

```
\DeclareFontShape{LGR}{ptm}{m}{n}{<-> s * [0.9] font8a}{}.
```

Of course, we should run several tests to discover the correct scaling factor. We usually try to equalize the x-height of the fonts. A file to use for tests can look like this:

```
\documentclass{article}
\begin{document}
\Huge
x\fontfamily{ptm}\selectfont x
\end{document}
```

Now, if our font family does not contain a particular shape, we can fool LATEX with a declaration such as the following one

```
\DeclareFontShape{LGR}{ptm}{m}{sc}{
    <-> ssub * ptm/m/n}{}
```

The declaration above specifies that if we request the small caps shape, we should actually use the normal font of the medium series. For more information concerning the font definition files, the reader should consult [27].

Let us now go into the details of the loading-settings, which are instructions to LATEX on how to load a particular font. The commands here are executed immediately after loading this particular font shape. An example is the setting of the hyphenation character. This is set by the command

```
\hyphenchar\font=number
```

where *number* represents the position of this character in the encoding vector. The default value is 45. If set to   1, the hyphenation will be suppressed for this font. Other important commands that can go in this last argument are commands of the form

```
\fontdimennumber\font=dimension
```

The *number* can have at least one of the following seven values:

number **is 1** specifies the slant per point of the characters. Upright fonts have zero slant.

number **is 2** the interword space.

number **is 3** how much the interword space can stretch.

number **is 4** how much the interword space can shrink.

number **is 5** x-height of the font (i.e., the length of 1 ex).

number **is 6** length of the 1 em for this font.

number **is 7** the amount of extra space added after the end-of-sentence period (provided \nofrenchspacing is enabled).

Thus, the code

```
\DeclareFontShape{LGR}{ptm}{m}{n}{<-> s * [0.9] font8a}%
                               {\fontdimen2\font=.7em}
```

loads the font font8a scaled to 90% and adjusts the interword space to be equal to 0.7 em. It is better though (especially for the interword spacing) to use a factor of the original value since the size of the space should be allowed to adjust when we change font sizes for this font. Thus, the above will work better if it is set like this:

```
\DeclareFontShape{LGR}{ptm}{m}{n}{<-> s * [0.9] font8a}%
             {\fontdimen2\font=.7\fontdimen2\font}
```

Now, we multiply by 0.7 the default value \fontdimen2\font of this font.

Once we have completed the construction of the font definition file, we need to let LATEX know how to use the "new" font(s). As we have already explained, there are three kinds of families: the serifed or Roman, the sans serif and the typewriter. To change a default font family, we need to redefine the following commands: \rmdefault, \sfdefault, and \ttdefault; these commands produce the corresponding family name according to LATEX's conventions. Now, such redefinitions can be part of a new package or can appear in the preamble of a document. For example, the times actually contains the following code:

```
\renewcommand{\sfdefault}{phv}
\renewcommand{\rmdefault}{ptm}
\renewcommand{\ttdefault}{pcr}
```

Of course, the commands above assume that we do not actually change the font encoding that is actually in use. Otherwise, we must change the font encoding. We will not discuss how we can create font encoding definition files, as this task is quite complex and is best left to real experts.

The user should be warned that DVIPS may fail to create a PostScript file, and many times it exits with a message such as "Second number not found in Char string of '/FontName'" or something similar. In these cases, what fails is the partial download of the font that DVIPS attempts. In other words, DVIPS tries to include in the PostScript file only the glyphs of the font that are actually needed by the document, and it fails due to some bug, usually in the font itself. The way to overcome this is to run DVIPS turning off the partial glyph download using the switch -j0 or asking DVIPS to create bitmaps with the option -V; that is, for the file file.dvi, use

```
dvips -j0 file.dvi  or  dvips -V file.dvi
```

If one insists on correcting the bug, then one can use the programs T1DISASM and T1ASM (by Lee Hetherington and Eddie Kohler), which will convert the font to a human-readable form and back to the binary format, respectively. It seems that a simple pass through these programs solves the problem above. The programs are available in modern TeX installations. There is also a program named type1fix (by Péter Szabó). TYPE1FIX can be used to fix this problem (among many different problems) as well. It is available from http://www.inf.bme.hu/~pts/type1fix_pl-latest.tar.gz.

12.5 Systemwide Installation of Prepared Fonts

For systemwide installation of a font, we need to have the PFB (or PFA) files, the TFM, the VF, psfonts.map, the font encoding, the font definition and the package files, if available. Some LaTeX installations require in addition the afm files (they use them for their DVI previewer). The PFB, AFM, TFM, and VF files, respectively, go in directories inside

```
texmf/fonts/type1/
texmf/fonts/afm/   and   texmf/fonts/vf/
texmf/fonts/tfm/
```

The font definition files (.fd files) and packages (.sty files) go in a directory in texmf/tex/latex/. It is very important to *copy* the contents of the psfonts.map file in the system's psfonts.map, usually found in texmf/dvips/base/. At the same place, we put the enc files. We run mktexlsr or texhash on Unix or Refresh Filename Database in Windows installations, and we are ready to use our fonts by loading the appropriate package or by redefining the default font families as described above.

12.6 Installing Scalable Fonts for pdfLaTeX

When using pdfTeX, it is better to use scalable fonts (i.e., Type 1 and TrueType fonts). The main reason is that PDF files that use bitmapped fonts are poorly rendered. When we have a new font, we need to make pdfTeX aware of this new font. If this font is a Type 1 font, then we simply follow the procedure described above and add a line such as

```
tnr8r Times-NewRoman " myenc ReEncodeFont " <myenc.enc <tnr.pfb
```

to the file that is specified in the configuration file pdftex.cfg. Usually, this file is called psfonts.map or standard.map. To use a TrueType font, we must first get an appropriate font metric. Since pdfTEX suffers from the limitations of TEX, we need to use an encoding vector to create a TFM file with metric date for at most 256 glyphs. We can extract the metric data with a command such as the following:

$ ttf2tfm ArialRegular.ttf -p 8a.enc

This command will generate the file ArialRegular.tfm. Now, we must add the following line to the file psfonts.map (or standard.map):

ArialRegular ArialRegular <8a.enc <ArialRegular.ttf

Of course, if we wanted to have an alternative name for the font, say tar, then we should create the TFM file with the following command:

$ ttf2tfm ArialRegular.ttf -p 8a.enc tar.tfm

The program TTF2TFM can be used also to produce virtual property list files, fake small caps, and so on. The available options are shown in Table 12.1.

If we have a Japanese, a Korean, or a Chinese TrueType font, we cannot use it with pdfLATEX, as it usually contains several thousand glyphs; to use such a font with pdfLATEX, we need to split the font into several subfonts. Unfortunately, this limitation applies to Λ too, as there is no program to directly generate Ω font metrics. Such a utility can be easily constructed by modifying the source code of TTF2TFM. Now, we can generate subfonts by using a predefined subfont definition file. To make things clear, we give the command that we had to enter in order to create the necessary font metrics for the Japanese font that we used in Section 10.21:

$ ttf2tfm kochi-mincho ommincho@/path/to/Unicode@

Note that we need to specify the full path (surrounded by the symbol @) to the location where the subfont definition file Unicode.sfd resides. Once we have extracted the font metrics from a TrueType font, we can easily make it available to pdfLATEX (and LATEX, of course) without converting the font. All we have to do is to add the last line that TTF2TFM prints on the screen to file ttfonts.map. For instance, for our example, we have to add the following line:

ArialRegular   ArialRegular.ttf Encoding=8a.enc

Now, every time that we view or convert to PostScript a DVI file that uses this font, the program TTF2PK will generate the necessary bitmaps. To make the font accessible to Λ, we need an Ω virtual property list file, but we will come back to this issue in the next section.

If we do not like the idea of generating PostScript files with bitmapped fonts, we can try to generate Type 1 fonts for each subfont using a font conversion utility such as TEXTRACE. Just create a Type 1 font that for each TFM file you have.

Table 12.1: Options of the TTF2TFM program.

Option	Description
-c *real*	Use *real* for height of small caps made with -V 0.8
-e *real*	Widen (extend) characters by a factor of *real* (default value is 1.0)
-E *int*	Select *int* as the TrueType font encoding identification (default value is 1)
-f *int*	Select *int* as the font index in a TrueType collection (default value is 0)
-l	Create 1st/2nd byte ligatures in subfonts
-n	Use PostScript names of TrueType font
-N	Use only PostScript names and no cmap
-O	Use octal for all character codes in the vpl file
-p ENCFILE[.enc]	Read ENCFILE for the TrueType to raw TEX mapping
-P *int*	Select *int* as the TrueType file platform ID (default is 3)
-q	Suppress verbose output
-r *old new*	Replace glyph name *old* with *new*
-R RPLFILE[.rpl]	Read RPLFILE containing glyph replacement names
-s *real*	Oblique (slant) characters by *real*, usually much smaller than 1
-t ENCFILE[.enc]	Read ENCFILE for the encoding of the virtual property list file
-T ENCFILE[.enc]	Equivalent to -p ENCFILE -t ENCFILE
-u	Output only characters from encodings, nothing extra
-v FILE[.vpl]	Make a virtual property list file for conversion to a virtual font
-V SCFILE[.vpl]	Like -v, but synthesize small caps as lowercase
-x	Rotate subfont glyphs by 90 degrees
-y *real*	Move rotated glyphs down by a factor of *real* (default value is 0.25)

12.7 Installing Scalable Fonts for Λ

Since Ω is a TEX successor capable of handling Unicode input streams, it is quite natural to expect that Ω can handle Unicode fonts. Indeed, this is the case, but, unfortunately, as of this writing there are no publicly available tools that can produce Ω font metrics from Adobe font metrics or from TrueType Files. One way out is to use an Ω virtual property list file and generate subfonts that the Ω virtual font will use. For example, the Omega-j system provides an Ω virtual property list that is used to create all of the necessary font metric files. If we do not have a Unicode font, then we can either write an Ω virtual property list file or create a number of ΩTPs that will map the input stream to a Unicode stream and, after processing, the resulting stream to a stream that can be mapped to glyphs of the fonts. Here, we will briefly present the first case. The following is part of the OCherokee Ω virtual property list developed to support the preparation of Cherokee language documents with Λ:

```
(FAMILY CHEROKEE)
(CODINGSCHEME SHIFTED CHEROKEE TEX)
(DESIGNSIZE R 10.0)
(COMMENT DESIGNSIZE IS IN POINTS)
(COMMENT OTHER SIZES ARE MULTIPLES OF DESIGNSIZE)
(FONTDIMEN
    (SLANT R 0.0)
    (SPACE R 0.5)
    (STRETCH R 0.3)
    (SHRINK R 0.1)
    (XHEIGHT R 0.8)
    (QUAD R 1.0)
    )
(MAPFONT D 0
    (FONTNAME Cherokee)
    (FONTDSIZE R 10.0)
    )
(CHARACTER H 13B1 (COMMENT Unicode code point)
    (CHARWD R 0.665)
    (CHARHT R 0.7995)
    (MAP
        (SELECTFONT D 0)
        (SETCHAR O 41)
        )
    )
```

We see that the ΩVP file maps Unicode characters to glyphs of an existing font. Note that a hexadecimal number is used to identify the Unicode character. Files such as these are usually based on (virtual) property list files. To create the file above, we issued the following commands:

```
$ afm2tfm Cherokee.afm Cherokee.tfm
$ tftopl Cherokee.tfm > Cherokee.pl
```

The program tftopl generates a property list file from a TFM file. Then, we used the file Cherokee.pl to manually create the file OCherokee.ovp. This file was then processed with ovptoovf to get the Ω virtual font and the Ω font metric:

```
$ ovptoovf OCherokee.ovp OCherokee.ovf OCherokee.ofm
```

The program ovftoovp transforms an Ω virtual font into an Ω virtual property list. Finally, the programs ofm2opl and opl2ofm transforms an Ω font metric to an Ω property list and vice versa.

Although the current version of Ω virtual property list supports a number of new lists, these still are not recognized by Ω itself. We suppose that this is due to the fact that Ω is still an experimental system.

12.8 OpenType Fonts

OpenType is a new font format created jointly by Adobe Systems and Microsoft. An OpenType font can contain information about glyphs in both TrueType and Type 1 formats. A TrueType font is always an OpenType font. However, OpenType extends the capabilities of the Type 1 format. The Type 1 fonts cannot contain a big number of glyphs. This is a limitation that the OpenType removes. In addition, the OpenType format can contain additional information, such as kerning, that in a Type 1 font is provided separately.

This new font format will immediately be available for use with LaTeX and its friends the moment that the PostScript driver (for example, DVIPS) becomes capable of embedding such a font in a PostScript file and a tool that extracts the metric information (such as AFM2TFM) becomes available. For the moment, this is not the case, and the only way to use an OpenType font is to convert it to Type 1. This can be done with a tool such as the PFAEDIT font editor.

12.9 Installing Math Fonts for LaTeX

In math mode, we can either select a particular alphabet, which will be used to typeset a letter of a word, or we can select a particular symbol. So, when we install new math fonts, we need to make LaTeX aware of the new math alphabets and the "new" symbols. The standard font selection commands are:

Alphabet	Description	Example
\mathnormal	default	*abcAbc*
\mathrm	roman	abcAbc
\mathbf	bold roman	**abcAbc**
\mathsf	sans serif	abcAbc
\mathit	text italic	*abcAbc*
\mathtt	typewriter	abcAbc
\mathcal	calligraphic	*ABC*

In addition, the standard math symbol fonts are:

Symbol font	Description	Example
operators	symbols from \mathrm	$[+]$
letters	symbols from \mathnormal	$< * >$
symbols	most LATEX symbols	$\leq \star \geq$
largesymbols	large symbols	$\sum \prod$

After this necessary "reminder," we proceed with the presentation of the various commands. The first thing that we might be interested in is to define a new math version. This can be done easily with the following command:

$$\texttt{\textbackslash DeclareMathVersion}\{\textit{version-name}\}$$

In order to give meaning to this command, we first have to define a new math alphabet:

$$\texttt{\textbackslash DeclareMathAlphabet}\{\textit{alphabet}\}\{\textit{encoding}\}\{\textit{family}\}\{\textit{series}\}\{\textit{shape}\}$$

For example, the command

$$\texttt{\textbackslash DeclareMathAlphabet}\{\texttt{\textbackslash mathit}\}\{\texttt{OT1}\}\{\texttt{ppl}\}\{\texttt{m}\}\{\texttt{it}\}$$

declares that for the italic math alphabet LATEX should use the italic font of the Palatino font family. Now that we have defined a math alphabet, we can change or set a particular font to be used with a specified math version:

$$\texttt{\textbackslash SetMathAlphabet}\{\textit{alphabet}\}\{\textit{version}\}\{\textit{encoding}\}\{\textit{family}\}\{\textit{series}\}\{\textit{shape}\}$$

Following the example above, here is how we can redefine an existing math version:

$$\texttt{\textbackslash SetMathAlphabet}\{\texttt{\textbackslash mathit}\}\{\texttt{bold}\}\{\texttt{OT1}\}\{\texttt{ppl}\}\{\texttt{b}\}\{\texttt{it}\}$$

Now, it is time to see how we can declare symbol fonts. This can be accomplished with the following command:

$$\texttt{\textbackslash DeclareSymbolFont}\{\textit{sym-font-name}\}\{\textit{encoding}\}\{\textit{family}\}\{\textit{series}\}\{\textit{shape}\}$$

Of course, we can use this command to declare new symbol fonts (see page 110 for an example). Here are some example declarations:

```
\DeclareSymbolFont{operators}    {OT1}{zplm}{m}{n}
\DeclareSymbolFont{symbols}      {OMS}{zplm}{m}{n}
\DeclareSymbolFont{largesymbols}{OMX}{zplm}{m}{n}
```

These commands specify that, for example, the operators will be drawn from a font with font encoding OT1 that belongs to the font family zplm and has medium series and normal shape. Now, if we want to change the symbol font used for a particular math version, we have to use the following command:

$$\texttt{\textbackslash SetSymbolFont}\{\textit{sym-font}\}\{\textit{version}\}\{\textit{encoding}\}\{\textit{family}\}\{\textit{series}\}\{\textit{shape}\}$$

For example, the following commands are used to declare the bold version of the symbol fonts above. Note that all that changes is the *series* specification:

```
\SetSymbolFont{operators}    {bold}{OT1}{zplm}{b}{n}
\SetSymbolFont{symbols}      {bold}{OMS}{zplm}{b}{n}
\SetSymbolFont{largesymbols}{bold}{OMX}{zplm}{m}{n}
```

Another interesting command is the following one:

```
\DeclareSymbolFontAlphabet{alphabet}{sym-font}
```

This command allows the previously declared symbol font *sym-font* to be also the math alphabet *alphabet*. Here is an example:

```
\DeclareSymbolFontAlphabet{\mathrm}  {operators}
\DeclareSymbolFontAlphabet{\mathnormal}{letters}
\DeclareSymbolFontAlphabet{\mathcal}   {symbols}
```

So, for example, the operators symbol font will be used to draw the letters for the mathematical alphabet used for \mathrm. Now that we have finished with the math alphabets and the various other fonts, we proceed with the commands that declare the various symbols. With the following command, we can declare a new math symbol:

```
\DeclareMathSymbol{\symbol}{type}{sym-font-name}{slot}
```

This command defines a new command \symbol that when used will print the glyph of the symbol font *sym-font-name* that resides at *slot*. The *type* specifies how TEX will treat this symbol. The possible *types* are as follows:

Type	Meaning	Example
0 or \mathord	Ordinary	α
1 or \mathop	Large operator	$\sum$
2 or \mathbin	Binary operator	$\times$
3 or \mathrel	Relation	$\leq$
4 or \mathopen	Opening	$($
5 or \mathclose	Closing	$)$
6 or \mathpunct	Punctuation	$;$
7 or \mathalpha	Alphabetic character	A

For more information on the meaning of the first seven types, see page 110. In general the \mathalpha type behaves exactly like \mathord, except that commands that declare math alphabets will make \mathalpha pick up symbols from the newly declared math alphabet. Below, we give a couple of declarations:

```
\DeclareMathSymbol{\Gamma}{\mathalpha}{letters}{"00}
\DeclareMathSymbol{\hbar} {\mathord}  {AMSb}   {"7E}
```

The following command can be used to define a math delimiter:

```
\DeclareMathDelimiter{\cmd}{type}{font-1}{slot-1}{font-2}{slot-2}
```

With this command, we define \cmd to be a math delimiter whose small variant is at *slot-1* of the symbol *font-1* and whose large variant is at *slot-2* of the symbol *font-2*:

```
\DeclareMathDelimiter{\ulcorner}{\mathopen}{AMSa}{"70}{AMSa}{"70}
```

If we also want to define math accents also, we have to use the following command:

```
\DeclareMathAccent{\cmd}{type}{sym-font-name}{slot}
```

As is expected, the \cmd will be a command that will place the symbol at *slot* of the font *sym-font-name* above a symbol or a letter. The *type* can be either \mathord or \mathalpha; in the latter case, the accent character changes font when used in a math alphabet. Here are two examples:

```
\DeclareMathAccent{\widehat}{\mathord}{largesymbols}{"62}
\DeclareMathAccent{\mathring}{\mathalpha}{operators}{"17}
```

To define a new radical, we should use the following command:

```
\DeclareMathRadical{\cmd}{font-1}{slot-1}{font-2}{slot-2}
```

Here, \cmd is the new radical. The small variant of it is at *slot-1* of *font-1*, and the large variant of it is at *slot-2* of *font-2*. Here is the only available example:

```
\DeclareMathRadical{\sqrtsign}{symbols}{"70}{largesymbols}{"70}
```

The last thing that we must consider is the declaration of math font sizes. The standard command to declare math sizes is

```
\DeclareMathSizes{t-size}{mt-size}{s-size}{ss-size}
```

With this command, we declare that *mt-size* is the main math text size, *s-size* is the "script" size, and *ss-size* is the "scriptscript" size to be used in math (see page 108) when *t-size* is the current text size. Normally, *mt-size* and *t-size* will be identical. Here are a few examples:

```
\DeclareMathSizes{10}    {10}    {7.6}  {6}
\DeclareMathSizes{10.95}{10.95}{8}      {6}
\DeclareMathSizes{12}    {12}    {9}      {7}
```

In some cases, we need to specify the default ratio for math sizes. The default ratio for math sizes is 1 to \defaultscriptratio to \defaultscriptscriptratio. By default, this is 1 to 0.7 to 0.5. Here is how we may redefine these parameters:

```
\renewcommand{\defaultscriptratio}{.76}
\renewcommand{\defaultscriptscriptratio}{.6}
```

Now, we have all of the knowledge to build a new package that will provide support for some new math font.

When we install a new `symbols` or `largesymbols` font, we must make sure that the font is proper. This means that TeX will typeset math formulas containing glyphs from these fonts only if they have at least 22 and 13 font dimensions, respectively. Thus, if we want to install a new scalable math font (e.g., a PostScript math font), we need to add these parameters manually. To do this, we have to create a (virtual) property list file and modify its `FONTDIMEN` section. In particular, if we define a `symbols` font, we must add entries for the following font dimensions: `NUM1`, `NUM2`, `NUM3`, `DENOM1`, `DENOM2`, `SUP1`, `SUP2`, `SUP3`, `SUB1`, `SUB2`, `SUPDROP`, `SUBDROP`, `DELIM1`, `DELIM2`, and `AXISHEIGHT`. On the other hand, if we define a `largesymbols` font, we must add entries for the following font dimensions: `DEFAULTRULETHICKNESS`, `BIGOPSPACING1`, `BIGOPSPACING2`, `BIGOPSPACING3`, `BIGOPSPACING4`, and `BIGOPSPACING5`. The following is an example of a typical `FONTDIMEN` section for a `symbols` math font:

```
(FONTDIMEN
    (SLANT R 0.249977)

    . . . . . . . . . . . . . . . .
    (EXTRASPACE D 111)
    (NUM1 R 0.676508)
    (NUM2 R 0.393732)
    (NUM3 R 0.443731)
    (DENOM1 R 0.685951)
    (DENOM2 R 0.344841)

    . . . . . . . . . . . . . . . .
    (AXISHEIGHT R 0.25)
    )
```

Let us now explain the meaning of these extra font dimensions. `DEFAULTRULE-THICKNESS` is the thickness of the rule drawn above radicals, underlines, or overlines. The five `BIGOPSPACING` dimensions are used when TeX is typesetting an operator that has `\limits`. In particular, `BIGOPSPACING1` and `BIGOPSPACING3` are used to adjust the position of the box that encloses the upper limit; `BIGOPSPACING2` and `BIGOPSPACING4` are used to adjust the position of the box that encloses the lower limit; `BIGOPSPACING5` is some additional space placed under the subscript. When TeX is typesetting a "fraction," the three `NUM` and the two `DENOM` dimensions are taken into account. The remaining dimensions are used in the typesetting of superscripts and subscripts. For a complete description of all of these dimensions, the reader should consult Appendix G of [19].

12.10   Installing Math Fonts for Λ

If for some reason we want to use alternative math fonts and are using Λ, we must prepare a new package or use an existing one and, in addition, we must create the necessary MathML encoding files. This step is absolutely necessary because when we use a new math font, Ω expects to find the corresponding MathML encoding file. Of course, if we do not define this file, then it is not sure whether Ω will be able to process our input file.

In any MathML encoding file, we define to which MathML entity corresponds to each glyph of a particular math font. This correspondence is specified with the following command:

   \SGMLFontEntity{*math-font*}{*slot*}{*entity-name*}{*type*}{*attribute*}

Here, *math-font* is the actual font name without the size specification (e.g., we write cmr instead of cmr10), the *slot* of the symbol is a hexadecimal number, and *type* can be one of mi (math identifier), mn (math digits), or mo (math operators). The *attribute* parameter should be used to specify some extra font attributes. We will now give an example that we hope will make things clear.

Suppose that we want to typeset a mathematical document using alternative Greek letters. The first thing is to create a little package that will declare the new math font:

```
%package greekmath
\input{lgrenc.def}
\DeclareSymbolFont{grletters}{LGR}{cmr}{m}{n}
\DeclareSymbolFontAlphabet{\mathord}{grletters}
\DeclareMathSymbol{\alpha}{\mathord}{grletters}{"61}
.................................................
      many many lines omitted
.................................................
\endinput
```

The command \endinput is used to explicitly denote the end of an input file. Note that we must load the file lgrenc.def just because this file defines the LGR font encoding. Now, it is time to prepare the MathML encoding file grmn.onm. The name of the file derives from the name of the font used without the size specifier. Here are the first few lines of this file:

```
\newcommand{\SGMLname}[1]{\SGMLampersand#1;}
\SGMLFontEntity{grmn}{"61}{\SGMLname{alpha}}{mi}{}
.................................................
      many many lines omitted
.................................................
\endinput
```

The new command \SGMLname is used to define the *entity-name* name. Here, we use the form α. Alternatively, we could use the form &#HHHH;, where HHHH is a hexadecimal number denoting the Unicode code point of the corresponding symbol. The authors of Ω suggest the use of the following command for the second case:

```
\newcommand{\SGMLno}[1]{\SGMLampersand\SGMLhash#1;}
```

Now, it is time to test our work. In Figure 12.1 we give the code of an input file and the MathML code that is generated.

As we have already explained, the *attribute* should be used to provide information regarding the font attributes (e.g., is it a boldface font, and so on). The following example shows how we can pass this extra information:

```
\newcommand{\SGMLbold}{\SGMLattribute{fontweight}{bold}}
\SGMLFontEntity{eusb}{"00}{-}{mo}{\SGMLbold}
```

The command \SGMLattribute has two required arguments: the attribute name and its value. Note that in this example we specified the glyph name in a third way by simply typing it! This is possible for all ASCII characters and some combinations of them.

When in math mode, TEX can use up to sixteen font families (numbered from 0 to 15). Each font family consists of three fonts, which are declared with the following commands: \textfont, \scriptfont, and \scriptscriptfont. For example, to declare font family 6, one should use the following commands:

```
\textfont6=\mytextfont
\scriptfont6=\myscriptfont
\scriptscriptfont6=\mysscriptfont
```

Figure 12.1: An input file that uses the new math package and the resulting MathML output.

```
<mtext>
  <inlinemath>
    <math>                              \documentclass{article}
      <mrow>                            \usepackage{greekmath}
        <msup>                          \begin{document}
          <mi> &alpha; </mi>            \MMLmode%
          <mn> 2 </mn>                  \MMLstarttext%
        </msup>                         $\alpha^{2}=4$
        <mo> = </mo>                    \MMLendtext%
        <mn> 4 </mn>                    \noMMLmode
      </mrow>                           $$\alpha^{2}=4$$
    </math>                             \end{document}
  </inlinemath>
</mtext>
```

Now, each font can have up to 256 characters. This means that TₑX can access up to 12,288 characters in any formula. On the other hand, Ω increases TₑX's capabilities considerably and allows 256 font families, where each font may consist of up to 65,536 characters. This means that Ω can access up to 50,331,648 characters in any formula! However, it is not possible to access the additional characters with TₑX's primitives, so there are new primitives to assist package developers. For this reason, Ω reimplements the font selection commands so that now they can access up to 256 fonts. In addition, it provides new primitive commands that may give access to the additional characters that Ω can deal with.

Each character in math mode has an associated math code that can be assigned with the \mathcode command. By assigning a math code to a character, we can refer indirectly to any glyph in any family, by a simple keystroke. Typically, the math code that we assign to a character is a hexadecimal number that consists of four digits—the first from the left denotes the math type (e.g., binary operator), the second and the third the code point of the character in the font, and the last digit the family. This command is still valid when using Ω, but the system also provides the command \omathcode. The introduction of this command was dictated by the fact that Ω supports 65,536 math codes while TₑX supports only 256. Consequently, now the math code is a hexadecimal number that has seven digits—the first from the left denotes the math type, the next four denote the code point of the character, and the last two denote the family. Note that in case the math code is the number "0008000 (or "8000000), the character can be "programmed." Of course, we can program any character by setting its character code to 13. However, the programming of a character with this particular math code is not a straightforward task. We need to locally change the category code of the character to 13 and then define the new command. To do this, we need some advanced features of TₑX macro programming. We have to write:

```
\omathcode'\C="8000000
{\catcode'\C=13 \gdef\C{code}
```

where C is a character. Note that if we define a command in a local scope, TₑX will "forget" it once we leave the local scope. The command \gdef is used to globally define a new macro, so we actually fool Ω here! For more information on the \gdef command, the reader should consult the TₑXbook.

Instead of assigning a math code to some character, we can define a new command that will expand to some math code. The command \mathchar is analogous to the \symbol command. In addition, Ω provides the \omathchar command, which can be used when dealing with large fonts. Moreover, the commands \mathchardef and \omathchardef can be used instead of the commands \mathchar and \omathchar, so the definition

```
\newcommand{\sum}{\mathchar"1350}
```

is actually a complicated way to say

```
\mathchardef\sum="1350
```

To make a character act as a delimiter, we need to set its \delcode. A negative delimiter code means that the character does not behave like a delimiter. Any number less than "1000000 can be used to specify a delimiter code—the first three digits from the left specify that its small variant belongs to the family that the first digit from the left specifies and has the code point that the second and third digits specify; its large variant belongs to the family that the fourth digit specifies and has the code point that the fifth and sixth digits specify. In addition, Ω provides the command \odelcode. The Ω delimiter code is a hexadecimal number that has fourteen digits and the first seven digits from the left specify the small variant and the last seven the large variant. In both cases, if we want TEX and/or Ω to ignore a variant, we simply write zeroes instead of a number. Here are some simple examples:

```
\delcode'\(="028300 \odelcode'\.=0
```

Of course, the number zero means that the character has no variants. The command \radical is followed by a delimiter code and the command \mathaccent by a math code. Currently, the only radical is the √ symbol, but Ω provides the \oradical command, which is followed by an Ω delimiter code! Similarly, the \omathaccent is followed by an Ω math code. Note that actually all of these commands are followed by a delimiter code or a math code, respectively, and a character, a symbol, or an expression that is placed under the radical or the accent.

APPENDIX A

USING DVIPS

The DVIPS program is the most widely used PostScript driver for documents prepared with LaTeX (and ODVIPS for Λ). To make good use of the program's capabilities, the user should know a few things about its configuration and options. These are described in this appendix.

There are two important configuration files for DVIPS. One is called config.ps and although its filename extension is .ps it is a plain text file. The other important file is the psfonts.map file, which contains the information about fonts and how they are to be downloaded in our final PostScript file. The structure of the latter file has been described in the font installation chapter (see Chapter 12). Let us now describe the config.ps file.

Open this file in a text editor. The first thing we set in this file is the memory available to DVIPS. To figure this out for your system, save the following lines in a file (let us call it memory.ps):

```
%! Hey, we're PostScript
    /Times-Roman findfont 30 scalefont setfont 144 432 moveto
    vmstatus exch sub 40 string cvs show pop showpage
```

Now, open the file memory.ps with a PostScript viewer or print it. In any case, you will see a page with a number on it. Use this number for the parameter m in config.ps. For example, our system says:

```
                        m 1281145
```

Next is usually the "way to print." In most modern systems, config.ps contains a line saying "o |lpr" (without the quotes). This says that when DVIPS is run by the user, send the output to the printer. This is not so in non-Unix installations. If you do not want this to happen but prefer to write the output of DVIPS to a file, comment out this line; that is, change it to "%o |lpr".

After this, the default resolution is set. These are printer-dependent variables, so you need to know the resolution of your printer. The default settings are usually 600 (that is, 600 dots per inch), so we have the declarations

```
D   600
X   600
Y   600
```

for the default (D) resolution, the resolution in the horizontal direction (X), and the resolution in the vertical direction (Y).

Now, we have to specify the printer. This is very important for the bitmap generation procedure when we run dvips. The default setting is M ljfour. This says that our printer is the Hewlett-Packard Laser-Jet 4. One should change this variable depending on the available printer. To find the printer's name that will be understood by dvips and the utilities that generate the bitmaps, we should check the file modes.mf. This file contains a list of all printers supported and gives the proper resolution values for them (to be used in the resolution settings above). If your printer is not listed, use a close match. This variable is very important to set correctly when doing commercial-grade work using METAFONT fonts. For example, when we prepare a book to be published, we must contact the publisher and get the information about the printer that will be used for the final printing. Then, we copy the config.ps file in the directory in which we work and change the M variable to the publisher's printer. For example, a common printer is the Linotype Linotronic 300 that prints at 2540 dots per inch. The modes.mf file calls this printer linotzzh. Thus, in the config.ps| file, we set the resolution to 2540 dots per inch and the variable M to linotzzh.

Next is the partial download capability. When using PostScript fonts, dvips can download into the PostScript output only the glyphs actually used in the file and not the whole font. This practice produces smaller PostScript files in the output. The default is to have this feature on (option j). However, some fonts do not work properly. In such cases, one may disable this feature by changing j to j0 (this is a zero, not a capital O).

What follows now are the offset variables for the printer, whether the bitmaps will be compressed or not, additional font maps, and, finally, paper dimensions. Offset variables can be checked by running the testpage.tex file through LaTeX and printing it after converting it to PostScript with dvips. This file is available in every installation. If the variable Z is set the bitmaps generated will be compressed. This is the default, and all modern hardware can compress very fast.

In addition, the config.ps file can be used to specify the available paper sizes and to instruct dvips to search a number of additional files for information about PostScript fonts. This last feature is extremely important, as there is no need to update the psfonts.map file everytime we add new PostScript fonts in our installation. Here is how we specify a sample paper size and the names of three files that contain information about additional PostScript fonts:

```
@ Springer 7in 9.5in
@+ %%PaperSize: Springer
@+ ! %%DocumentPaperSizes: Springer
p +textrace.map
p +inuit.map
```

```
p +cherokee.map
```

Creating Encapsulated PostScript with DVIPS

Encapsulated PostScript is PostScript with the additional information of the bounding box (i.e., the dimensions of the "content"). This information is used when the "content" is to be embedded in another file. For example, most of the pictures in this book have been prepared as Encapsulated PostScript files and then imported using the mechanisms of the graphicx package.

The procedure is to create a LaTeX file as follows:

```
\documentclass{article}
\pagestyle{empty}
other necessary preamble commands
\begin{document}
commands creating our ''picture''
\end{document}
```

We need to use the empty page style since otherwise LaTeX will include page numbers, which will affect the dimensions of the resulting file.

Now, we run this file through LaTeX, and after a successful run we use DVIPS to create the PostScript file. However, we want the resulting PostScript file to contain dimension information (bounding box). For this task, we use the –E option of DVIPS:

```
dvips -E file.dvi
```

Note that it is not necessary to specify the filename extension. If we now open the file.ps with GV or GHOSTVIEW, we will see that the window that opens has the dimensions of the file.

DVIPS tries to do its best to compute the bounding box, but sometimes it fails. In these cases, we need to manually edit the PostScript file and adjust the bounding box information. GV helps a lot with this, as it always displays the coordinates of the location of the pointer (mouse). Figure A.1 is self-explanatory nf how to determine the bounding box. We get the coordinates of the two pointer locations (lower left and upper right) whose locations set the bounding box of the picture. Then, if DVIPS was run with the –E option, the PostScript file contains a line of the form

```
%%BoundingBox: 104 553 300 747
```

We open the PostScript file with a text editor and adjust the numbers of this line. The first two numbers are the coordinates of the lower-left corner of the bounding box, and the last two are the coordinates of the upper-right corner as presented by the GV program. Thus, in order to get the Figure 9.10, we adjusted the bounding box above to

```
%%BoundingBox: 127 600 270 700
```

The first two numbers (127 600) are shown (inside the ellipse) in Figure A.1.

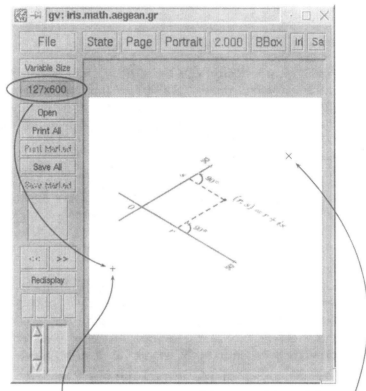

Mouse position of the lower left corner of the correct bounding box.

Mouse position of the upper right corner of the correct bounding box.

Figure A.1: Determining the bounding box manually.

Command-Line Options of DVIPS

We give here most options of DVIPS from its documentation as a handy reference.

Many of the parameterless options listed here can be turned off by suffixing the option with a zero (0); for instance, to turn off page reversal, use -r0. Such options are marked with a trailing *.

'-' Read additional options from standard input after processing the command line.

'--help' Print a usage message and exit.

'--version' Print the version number and exit.

'-a*' Conserve memory by making three passes over the DVI file instead of two and only loading those characters actually used.

'-A' Print only the odd pages. This option uses TEX page numbers, not physical page numbers.

'-b *num*' Generate *num* copies of each page, but duplicating the page body. This can be useful for color tricks, but we do not cover this here.

'-B' Print only the even pages. This option uses TEX page numbers, not physical page numbers.

'-c *num*' Generate *num* consecutive copies of every page (i.e., the output is uncollated).

'-C *num*' Generate *num* copies but collated (by replicating the data in the PostScript file). Slower than the -c option, but easier on the hands, and faster than resubmitting the same PostScript file multiple times.

'-d *num*' Set the debug flags, showing what dvips (thinks it) is doing.

'-D *num*' Set both the horizontal and vertical resolution to *num*, given in dpi (dots per inch). This affects the choice of bitmap fonts that are loaded and also the positioning of letters in resident PostScript fonts. It must be between 10 and 10000. This affects both the horizontal and vertical resolution.

'-e *num*' Maximum drift in pixels of each character from its "true" resolution-independent position on the page. The default value of this parameter is resolution-dependent (it is the number of entries in the list [100, 200, 300, 400, 500, 600, 800, 1000, 1200, 1600, 2000, 2400, 2800, 3200, ...] that are less than or equal to the resolution in dots per inch). Allowing individual characters to "drift" from their correctly rounded positions by a few pixels, while regaining the true position at the beginning of each new word, improves the spacing of letters in words.

'-E*' Generate an EPS file (Encapsulated PostScript File) with a tight bounding box. This only looks at marks made by characters and rules, not by any included graphics. In addition, it gets the glyph metrics from the TFM file, so characters that print outside their enclosing TFM box may confuse it. In addition, the bounding box might be a bit too loose if the character glyph has significant left or right side bearings. Nonetheless, this option works well enough for creating small EPS files for equations or tables or the like. (Of course, dvips output, especially when using bitmap fonts, is resolution-dependent and thus does not make very good EPS files, especially if the images are to be scaled; use these EPS files with care.) For multiple page input files, also specify -i to get each page as a separate EPS file; otherwise, all of the pages are overlaid in the single output file.

'-f*' Run as a filter. Read the dvi file from standard input and write the PostScript to standard output. The standard input must be seek-able (i.e., we can perform the seek system call), so it cannot be a pipe. If your input must be a pipe, write a shell script that copies the pipe output to a temporary file and then points dvips at this file.

'-F*' Write control-D (ASCII code point 4) as the very last character of the PostScript file. This is useful when dvips is driving the printer directly instead of working through a spooler, as is common on personal systems. On systems shared by more than one person, this is not recommended.

'-i*' Make each section be a separate file; a "section" is a part of the document processed independently, most often created to avoid memory overflow. The filenames
are created by replacing the suffix of the supplied output file name with a three-
digit sequence number. This option is most often used in conjunction with the -S
option, which sets the maximum section length in pages; if -i is specified and -S
is not, each page is output as a separate file. For instance, some phototypesetters
cannot print more than ten or so consecutive pages before running out of steam;
these options can be used to automatically split a book into ten-page sections, each
to its own file.

'-j*' Download only needed characters from Type 1 fonts. This is usually the default.

'-k*' Print crop marks. This option increases the paper size (which should be specified
either with a paper size special or with the -T option) by a half inch in each
dimension. It translates each page by a quarter inch and draws cross-style crop
marks. It is most useful with typesetters that can set the page size automatically.

'-l [=] num' The last page printed will be the first one numbered num. Default is the
last page in the document. If num is prefixed by an equals sign, then it (and the
argument to the -p option, if specified) is treated as a physical (absolute) page
number rather than a value to compare with the TEX \count0 values stored in the
dvi file. Thus, using -l =9 will end with the ninth page of the document, no matter
what the pages are actually numbered.

'-m*' Specify manual feed if supported by the output device.

'-mode mode' Use mode as the METAFONT device name for path searching and font
generation. This overrides any value of the M variable in the config.ps file.

'-M*' Turns off automatic font generation.

'-n num' Print at most num pages. Default is 100,000.

'-o name' Send output to the file name. If -o is specified without name, the default
is file.ps, where the input DVI file was file.dvi. If -o is not given at all, the
configuration file default is used. If name is -, output goes to standard output. If the
first character of name is ! or |, then the remainder will be used as an argument to
popen; thus, specifying |lpr as the output file will automatically queue the file for
printing as usual. (The DOS/Windows version will print to the local printer device
PRN when name is |lpr and a program by that name cannot be found.)

'-O x-offset,y-offset' Move the origin by x-offset,y-offset, a comma-separated pair of dimensions such as .1 in, .3 cm. The origin of the page is shifted from the
default position (of one inch down, one inch to the right from the upper-left corner
of the paper) by this amount. This is usually best specified in the printer-specific
configuration file. This is useful for a printer that consistently offsets output pages
by a certain amount. You can use the file testpage.tex to determine the correct
value for your printer. Be sure to do several runs with the same O value —some
printers vary widely from run to run.

'-p [=] num' The first page printed will be the first one numbered num. Default is
the first page in the document. If num is prefixed by an equals sign, then it (and

the argument to the -l option, if specified) is treated as a physical (absolute) page number rather than a value to compare with the TEX \count0 values stored in the DVI file. Thus, using -p =3 will start with the third page of the document, no matter what the pages are actually numbered.

'-pp *first-last*' Print pages *first* through *last*; equivalent to -p *first* -l *last*, except that multiple -pp options accumulate, unlike -p and -l. The - separator can also be :.

'-P *printer*' Read the configuration file config.*printer*, which can set the output name (most likely +o |lpr -P*printer*), resolution, METAFONT mode, and perhaps font paths and other printer-specific defaults. It works best to put sitewide defaults in the one master config.ps file and only things that vary from printer to printer in the config.*printer* files; config.ps is read before config.*printer*.

'-q*' Run quietly. Do not chatter about pages converted to standard output and so on; report no warnings (only errors) to standard error.

'-r*' Output pages in reverse order. By default, page 1 is output first.

'-R' Run securely. This disables shell command execution in \special[1] (via ') and config files (via the E option), pipes as output files, and opening of any absolute filenames.

'-s*' Enclose the output in a global save/restore pair. This causes the file to not be truly conformant, and is thus not recommended, but is useful if you are driving a deficient printer directly and thus do not care too much about the portability of the output to other environments.

'-S *num*' Set the maximum number of pages in each section. This option is most commonly used with the -i option; see its description above for more information.

'-t *papertype*' Set the paper type to *papertype*, usually defined in one of the configuration files, along with the appropriate PostScript code to select it. You can also specify a *papertype* of landscape, which rotates a document by 90 degrees. To rotate a document whose paper type is not the default, you can use the -t option twice, once for the paper type and once for landscape.

'-T *hsize,vsize*' Set the paper size to (*hsize,vsize*), a comma-separated pair of dimensions such as .1 in, .3 cm. It overrides any paper size special in the DVI file.

'-U*' Disable a PostScript virtual memory-saving optimization that stores the character metric information in the same string that is used to store the bitmap information. This is only necessary when driving the Xerox 4045 PostScript interpreter, which has a bug that puts garbage on the bottom of each character. Not recommended unless you must drive this printer.

'-V*' Download nonresident PostScript fonts as bitmaps.

'-x *num*' Set the x-magnification ratio to *num*/1000. Overrides the magnification specified in the DVI file. Must be between 10 and 100,000. It is recommended that you use standard magstep values (1095, 1200, 1440, 1728, 2074, 2488, 2986, and so on)

---

1. This command is used to directly include PostScript code in the generated DVI file. The code shoudl be surrounded by braces and must immediately follow the \special command.

to help reduce the total number of PK files generated. *num* may be a real number, not an integer, for increased precision.

'-X *num*' Set the horizontal resolution in dots per inch to *num*.

'-y *num*' Set the y-magnification ratio to *num*/1000. See -x above.

'-Y *num*' Set the vertical resolution in dots per inch to *num*.

'-z*' Pass html hyperdvi specials through to the output for eventual distillation into PDF. This is not enabled by default to avoid including the header files unnecessarily and use of temporary files in creating the output.

'-Z*' Compress bitmap fonts in the output file, thereby reducing the size of what gets downloaded. Especially useful at high resolutions or when very large fonts are used. May slow down printing, especially on early 68000-based PostScript printers. Generally recommended today and can be enabled in the configuration file.

VISUAL EDITING

A considerable number of LaTeX users have complained about the lack of a tool that would allow them to perform *visual editing* of their documents. In other words, users wanted a tool that would allow them to view their DVI file and at the same time be able to see the part of the input file to which the formatted output corresponds. The srcltx package (designed initially by Aleksander Simonic) is such a tool and has been designed to work with the WinEdt Windows editor (also by Aleksander Simonic). The package has been further improved by Stefan Ulrich, and it can now be used with most DVI viewers, such as the YAP Windows DVI viewer and of course the xdvi[1] DVI viewer. When loaded in a LaTeX file, the package forces the typesetting engine to place a number of "hooks" (*specials*, in TeX parlance) in the resulting DVI file. Here is how this works: When we view the DVI file with some DVI viewer and position the mouse on a paragraph or a math equation and then press (at least when using xdvi) the Ctrl button and the left button of the mouse,[2] the cursor on the editor screen is positioned at the beginning of the code that corresponds to the beginning of the paragraph or math equation on which we had positioned the mouse. Before we describe how to set up our favorite editor, we must stress that with the latest releases of the TeX typesetting engine,[3] we can get the effect of the srcltx package by issuing the command

$$\text{\$ latex -src-specials } input\text{-}file$$

---

1. Note that the features described in this appendix work with releases of xdvi that have a version number greater than 22.39.

2. When using YAP, just double-click somewhere on the DVI view; this causes YAP to bring the editor window to the front, moving the text cursor directly to the line that corresponds to the view location.

3. A program can be called TeX only if it can process a plain TeX file called `trip.tex`. This test is known as the "trip" test. Any program that by default enables features described in this appendix fails to pass the "trip" test and thus cannot be called TeX. Since these programs are actually "supersets" of TeX, TeX purists believe they should be called something else (e.g., MINT for *MINT is not TeX*). In our opinion, only the banner should change once these extra features are enabled. Of course, this "rhetoric" does not apply to the other typesetting engines described in this book, as their authors do not seem to be so strict about what is, say, ε-TeX or not.

By default, TEX places a hook at the beginning of a paragraph, but we can force it to place hooks in other places too. To do this, we simply use the following command to process our input file:

$$\text{\$ -src-specials=} hooks\ input\text{-}file$$

where *hooks* is a comma-separated list of the following keywords: display, math, and par. The meaning of these keywords is that a hook is placed at the beginning of a display math text, in-line math, or a paragraph, respectively. We will now present what we have to do in order to be able to communicate with the input file while we browse the DVI file.

We show how one can use the EMACS editor (with version number greater than 20.3) and the GVIM editor. If we use EMACS, we have to add the following line in our .emacs file:

$$\text{(server-start)}$$

If we are using XEMACS, then we should place the following line in the .xemacs file:

$$\text{(gnuserver-start)}$$

Now, if we use EMACS, we should start XDVI with the command

```
$ xdvi -editor "emacsclient --no-wait +%1 %f" sample
```

provided of course that we have already opened the file sample.tex with the EMACS editor. For XEMACS, one should start XDVI with the following command:

```
$ xdvi -editor "gnuclient -q +%1 %f" sample
```

To use the GVIM editor use the following command:

```
$ xdvi -editor "gvim +%1 %f" test
```

Note that each time we execute the command above we actually create a new instance of the editor so there is no need to have our input file opened with GVIM.

If we are using EMACS (XEMACS), then there is a tool that allows users to see the parts of the formatted output to which a particular piece of input "code" corresponds. The facilities described in this paragraph and those described above form a very powerful toolbox that really facilitates the work of novice as well as advanced users. To enable this new feature, we have to append the following lines into our .emacs file:

```
(require 'xdvi-search)
(add-hook 'tex-mode-hook (lambda ()
        (local-set-key "\C-x\C-j" 'xdvi-jump-to-line)))
```

These lines will make EMACS load the file xdvi-search.el and make the key sequence Ctrl-x and Ctrl-j a shorthand. By pressing this shorthand, XDVI will start and will place a frame around the formatted text that corresponds to the input "code" where the EMACS cursor is located. This assumes that we have installed the STARXDVI batch file:

```
#!/bin/sh
### name of xdvi executable
XDVI_EXEC="/usr/local/teTeX/bin/i386-pc-solaris2.8/xdvi"
### set this to /dev/null if you do not want to see the xdvi output:
XDVI_LOG="/tmp/XDVI.log"
### end of customizable variables

XDVI_CALL="$XDVI_EXEC -name xdvi -sourceposition"
echo "calling $XDVI_CALL \"$1\" $2" > $XDVI_LOG
$XDVI_CALL "$1" $2 >> $XDVI_LOG 2>&1 &
exit 0
```

If we are using the AUCTEX editing system,[4] then we must append the following lines into our .emacs file:

```
(require 'xdvi-search)
(add-hook 'LaTeX-mode-hook (lambda ()
        (local-set-key "\C-x\C-j" 'xdvi-jump-to-line)))
```

Preview-LATEX is another tool that facilitates visual editing of LATEX documents. The tool has been designed by David Kastrup, Alan Shutko, Jan-Åke Larsson, and Nick Alcock and consists of a LATEX package and an EMACS Lisp module. Preview-LATEX makes use of the facilities provided by quite recent versions of EMACS to render equations and other stuff inside your EMACS window. This way, one avoids the continuous switching between the DVI-viewer's window and the editor's window. The EMACS module needs AUCTEX and can be activated by adding the following lines into our .emacs file:

```
(require 'tex-site) ;;Activates AUCTeX
(add-hook 'LaTeX-mode-hook #'LaTeX-preview-setup)
(autoload 'LaTeX-preview-setup "preview")
```

The preview package should be included in our LATEX file if we want to be able to make use of the features of Preview-LATEX. The package has a number of arguments, which correspond to the parts of the input file that will be processed. For example, the options floats, sections, displaymath and textmath should be used to "preview" floats, section headings, display math, and in-line math, respectively. Figure B.1 is a screen shot of an EMACS window with Preview-LATEX enabled. One of the weak points of Preview-LATEX is that it cannot work properly with the inputenc package. In addition, we do not expect the tool to work with Λ.

The YaTEX package by Yuuji Hirose (also known as "Wild Bird") integrates typesetting and previewing of LATEX files and offers automated completion of commands such

---

4. AUCTEX is a set of EMACS Lisp macros (a collection of commands that adds user-friendly functionality to EMACS) that gives shortcuts for frequently used LATEX commands as well as a drop-down menu called "Command" that enables you to process input files and to generate PostScript files within EMACS. AUCTEX was designed by Kresten Krab Thorup and is maintained by Per Abrahamsen.

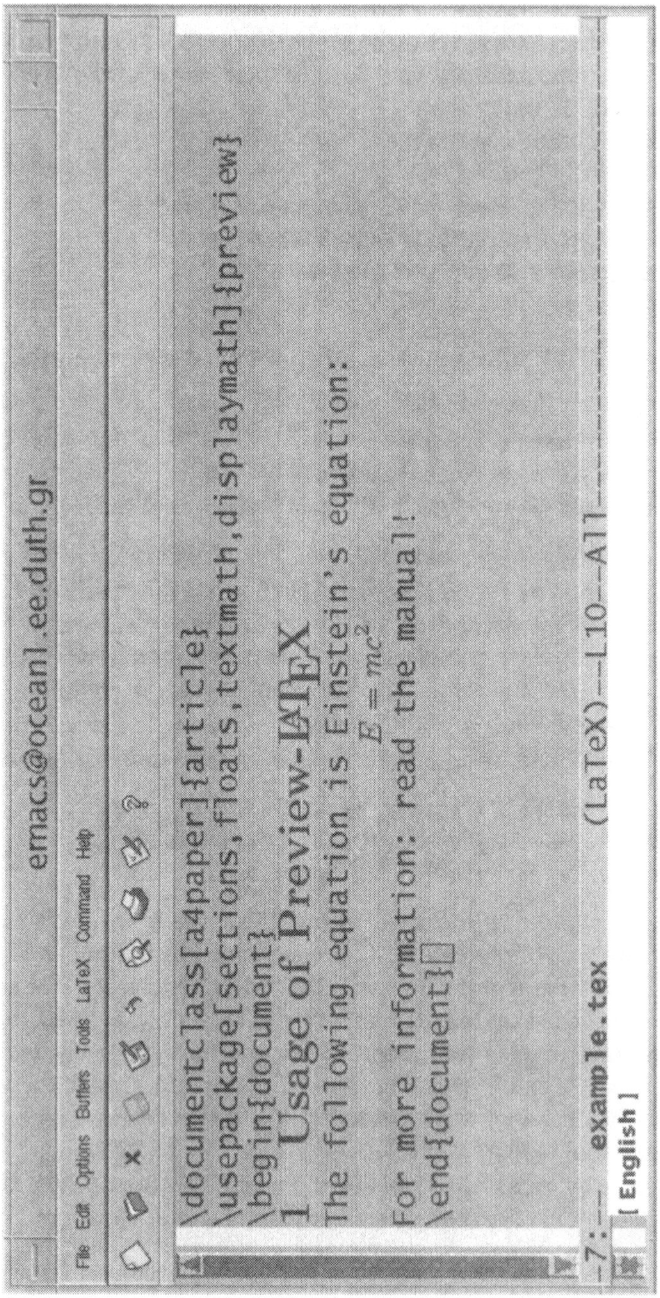

Figure B.1: Preview-LaTeX in action.

as \begin{}* ... \end{}. It jumps directly to an error line in the input file, learns un-known and new LATEX commands for their next completion, reads in the arguments of complex commands, blankets commenting out of text regions, offers keyboard short-cuts for accents, math-mode commands, and Greek symbols, and has a hierarchical browser for document files.

Many of YaTEX's shortcuts involve the use of a *prefix* keypress; for example, the default is Ctrl-c, followed by a short sequence of letters. Thus, [Ctrl-c t j invokes the LATEX typesetting engine, and Ctrl-c t b calls BIBTEX and Ctrl-c t p starts up the preview program. Some of the environment completion commands are Ctrl-c b c for the center environment, Ctrl-c b q for the quotation environment, and CTRL-c b l for the flushleft environment. However, a number of the keystroke combinations are very similar and require some memorization before their usage is straightforward; for example, to get a tabbing environment we use Ctrl-c b t, while Ctrl-c b T produces a tabular environment; and Ctrl-c b^T gives us a table environment. When in a math environment, the prefix ; allows shortcuts for mathematical signs and the prefix ; , followed by the mapped Latin character speeds up the typing in of Greek mathematical symbols. As may be realized by now, the YaTEX editor is not for the casual user but may prove useful to users who like to control the editing process from the keyboard rather than with a mouse and graphical user interface.

TYPESETTING XML

This appendix assumes familiarity with XML, so it can be skipped by readers not familiar with XML. We have already briefly explained why XML is a very important document format. But since XML is so important, it would make sense to be able to typeset XML documents using TEX. Indeed, David Carlisle has created XMLTEX [5], a LATEX extension capable of parsing XML content and typesetting it according to some user-specified formatting conventions. Although every XML document must be structured according to some Document Type Definition, XMLTEX itself does not need to know anything about any particular DTD. However, XMLTEX must be aware of a file that contains the necessary formatting commands. This file is called an *xmt* file in XML-TEX parlance. The file xmltex.cfg is the place where we associate DTDs with xmt files. This file may contain five different kinds of entries:

$$\NAMESPACE\{URI\}\{xmt\text{-}file\}$$
$$\PUBLIC\{FPI\}\{xmt\text{-}file\}$$
$$\SYSTEM\{URI\}\{xmt\text{-}file\}$$
$$\NAME\{element\text{-}name\}\{xmt\text{-}file\}$$
$$\XMLNS\{element\text{-}name\}\{URI\}$$

The first command should be used to associate a namespace specified with a *URI*. For example, if we have the following tag in an XML file,

```
<scientist xmlns="http://ocean1.ee.duth.gr/sciperson">
```

then we must insert a line such as the following one into xmltex.cfg:

```
\NAMESPACE{http://ocean1.ee.duth.gr/sciperson}{person.xmt}
```

If we use a predefined DTD as in the example

```
<!DOCTYPE scientists SYSTEM
    "http://ocean1.ee.duth.gr/dtds/scientists.dtd">
```

then we should place the following entry into the configuration file:

```
\SYSTEM{http://ocean1.ee.duth.gr/dtds/scientists.dtd}
                                         {scientists.xmt}
```

If we specify an internal DTD subset such as the following one,

```
<!DOCTYPE pupil [
<!ENTITY name (#PCDATA)
<!ENTITY behavior SYSTEM "behavior.xml">
..................................]>
```

then we should add the following line to the configuration file:

$$\text{\NAME\{pupil\}\{pupil.xmt\}}$$

Now, we must create the xmt file to be able to typeset specific XML content. XMLTEX provides the following commands:

\FileEncoding{*encoding*} With this command, we can specify the input encoding. The default is UTF-8.

\DeclareNamespace{*prefix*}{*URI*} This command should be used to declare *prefix* to be used for referring to elements in the specified namespace. Any empty *prefix* implies the use of the default namespace.

\XMLelement{*element*}{*attribute*}{*bcode*}{*ecode*} With this command, we declare that the codes *bcode* and *ecode* must be executed at the beginning and end of each instance of *element*. This is actually like the declaration of a new environment. The *attribute* should be used to declare a list of attributes (see the description of the \XMLattribute command below).

\XMLelement{*element*}{*attribute*}{\xmlgrab}{*ecode*} This version of the \XMLelement command should be used when we want to process the *element* content with the command(s) of *ecode*.

\XMLattribute{*attribute*}{*command*}{*default*} This command may only be used as an argument to \XMLelement. The first argument specifies the name of the attribute (using any namespace prefixes "active" for this file). The second argument is the name of the LATEX command that will be used to access the value of this attribute in the *bcode* and *ecode* for the *element*. The third argument provides a default value that will be used if the attribute is not used on an instance of this element.

\XMLnamespaceattribute{*prefix*}{*attribute*}{*command*}{*default*} This is a command similar to \XMLattribute but it is used "globally," (i.e., this way we avoid specifying the same thing for each \XMLelement).

\XMLentity{*name*}{*code*} This command should be used to declare an (internal parsed) entity. This is actually equivalent to a <!ENTITY ...> declaration, except that the preplacement text is specified in LATEX syntax.

\XMLname{*name*}{*command*} Declare the LATEX command to hold the normalized, internal form of the XML name given in the first argument.

\XMLstring{*command*}<>*XML Data*</> With this command, we can actually save an XML fragment as a LATEX *command*.

Now that we have presented all of the necessary information, we will present a trivial, though complete working example. The following is the code of a real DTD file:

```
<!-- file: scientists.dtd -->
<!ELEMENT first_name (#PCDATA)>
<!ELEMENT last_name  (#PCDATA)>
<!ELEMENT profession (#PCDATA)>
<!ELEMENT name       (first_name, last_name)>
<!ELEMENT person     (name, profession*)>
<!ELEMENT scientists (person*)>
```

This DTD is used to construct the following (validated) XML file:

```
<!-- file: greatsci.xml -->
<?xml version="1.0" standalone="no"?>
<!DOCTYPE scientists SYSTEM
    "http://ocean1.ee.duth.gr/dtds/scientists.dtd">
<scientists>
  <person>
    <name>
      <first_name>Alan</first_name>
      <last_name>Turing</last_name>
    </name>
    <profession>Computer Scientist</profession>
    <profession>Mathematician</profession>
    <profession>Cryptographer</profession>
  </person>
  <person>
    <name>
      <first_name>Leslie</first_name>
      <last_name>Lamport</last_name>
    </name>
    <profession>Computer Scientist</profession>
  </person>
</scientists>
```

Here is the file that will be used by XMLTEX to typeset our XML document:

```
% file : scientists.xmt
%
\XMLelement{scientists}
{}
  {\documentclass[a4paper,12pt]{article}
```

```
  \begin{document}\begin{itemize}}
  {\end{itemize}\end{document}}
\XMLelement{person}
{}
  {\item}
  {}

\XMLelement{name}
{}
  {}
  {:}

\XMLelement{first_name}
{}
  {\xmlgrab}
  {#1}

\XMLelement{last_name}
{}
  {\xmlgrab}
  {\textsc{#1}}

\XMLelement{profession}
{}
  {\xmlgrab}
  {#1\ }
```

The line that should appear in file `xmltex.cfg` has been given above. To typeset the XML file above, we have to enter the following command:

```
$ xmltex greatsci.xml
This is TeX, Version 3.14159 (Web2C 7.3.3.1)
(./greatsci.xml
LaTeX2e <2000/06/01>
Babel <v3.7h> and hyphenation patterns for
american, english, greek, loaded.
xmltex version: 2000-03-08 v0.14 DPC
(./xmltex.cfg)
No File: greatsci.cfg
Document Element: scientists
Doctype Public:
Doctype System:
http://ocean1.ee.duth.gr/dtds/scientists.dtd
  <0:scientists (./scientists.xmt)  >
```

```
(/usr/local/teTeX/share/texmf/tex/latex/base/article.cls
Document Class: article 2000/05/19 v1.4b Standard LaTeX
document class
(/usr/local/teTeX/share/texmf/tex/latex/base/size12.clo))
(./greatsci.aux)

  <0:person  >
(/usr/local/teTeX/share/texmf/tex/latex/fd/omscmr.fd)

  <0:name  >
  <0:first_name  >
  </0:first_name>
.............................................
........ many many lines omitted .....
.............................................
  Grabbed content
  End Grabbed content
  </0:person>
  </0:scientists> [1] (./greatsci.aux) )
Output written on greatsci.dvi (1 page, 464 bytes).
Transcript written on greatsci.log.
```

Here is the typeset result:

- Alan TURING : Computer Scientist Mathematician Cryptographer
- Leslie LAMPORT : Computer Scientist

WEB PUBLISHING

The term "Web Publishing" refers to our capability to publish documents on the World Wide Web, often just called the Web. Documents published on the Web are usually marked up in HTML or XHTML, which is HTML writen in XML. There are two ways to publish LaTeX documents on the Web: either by providing the source file of our document or by transforming the LaTeX file to HTML or some other markup such as DocBook, TEI, and others. Naturally, giving away the LaTeX file is of no real interest, as Web navigators usually prefer to have a quick look at some document before they actually start reading it, so we must definitely have tools that will transform our LaTeX source files to a "Web-aware" format. Nowadays, there are two categories of tools that differ in the way they parse the input file. Tools that belong to the first category do not use LaTeX to parse the input file, while tools that belong to the second category do exactly the opposite. In this appendix, we present two tools—LaTeX2HTML and tex4ht. The former belongs to the first category and the latter to the second one. We describe their basic functionality and show the output that we get from the same input file.

D.1 LaTeX2HTML

LaTeX2HTML is a Perl script originally written by Nikos Drakos and now maintained by Ross Moore. Before we actually start processing our LaTeX file with LaTeX2HTML, we must feed it twice to LaTeX. Then, we feed it to LaTeX2HTML. The program creates a directory that contains all of the generated files and has the same name as the input file; that is, if the LaTeX file is named example.tex, LaTeX2HTML will create the directory example. This directory will contain at least two HTML files: index.html and example.html. These files will have identical contents and are the files we have to open in order to view the output of LaTeX2HTML. Before we discuss the most common command-line switches, we give you the contents of a LaTeX file that was transformed to HTML by LaTeX2HTML:

```
\documentclass[a4paper,11pt]{article}
\begin{document}
```

```
            Here is an interesting equation:
            \begin{displaymath}
              n!=\int_{0}^{\infty} e^{-t}t^{n}\,dt
            \end{displaymath}
            And here is a table:
            \begin{center}
              \begin{tabular}{|l|r|c|}\hline
                111 & 222  & 333\\ \hline
                1   & 2    & 3 \\ \hline
              \end{tabular}
            \end{center}
          \end{document}
```

The output generated is "shown" in Figure D.1 on page 448.

When LATEX2HTML encounters a mathematical expression, it transforms it to some graphics file. If you wonder why LATEX2HTML does not convert math formulas to MATHML, the answer is that at the time LATEX2HTML was written, MATHML simply did not exist! If we have some simple mathematical formulas such as $x + y = 2$, then it is better to use LATEX2HTML with the -no_math command-line option. This option prevents LATEX2HTML from creating image files for simple formulas. When using the latest version of LATEX2HTML, we must also specify the format of the graphics files that the program will generate. We can do this by using the command-line switch

-image_type image_type

where <i>image_type</i> can be either gif or png.

The option -local_icons should be used if we want LATEX2HTML to include all necessary graphics files in the new directory. Note that the generated HTML files use some graphics files that are part of the LATEX2HTML distribution, so this option forces LATEX2HTML to actually copy these extra graphics files to the new directory. Different versions of HTML provide different features, which are not always implemented by all Web browsers. Consequently, it is wise to be able to choose the version of the resulting HTML files. This can be specified with the -html_version switch, which must be followed by the version number (e.g., -html_version 4). This option can be used also to specify the input encoding of the resulting HTML files. Just place a comma immediately after the version number and then the input encoding; for example,

-html_version 4,latin2

The supported input encodings are latin1, latin2, latin3, latin4, latin5, latin6, and unicode (partially supported). If we want some other input encoding, we can change the generated HTML files with a script such as ch_enc, which is available from the Web site of this book.

The program automatically splits chapters, sections, and so on, to separate HTML files and creates a table of contents with links to these files. In addition, it uses a number of standard words to name chapters, sections, and so on, just like LATEX does. LATEX2HTML achieves this by using predefined "modules" for each language. If your language is not supported but it is supported by the babel package, it is easy to create a new "module" for your language—just copy the file that contains an existing "module" to a new file that will be called *lang*.perl, where *lang* is the name of your language, and modify the new file accordingly. To enable the use of a particular language "module," set the environment variable LANGUAGE_TITLES to the name of the language.

D.2 tex4ht

The tex4ht system (by Eitan M. Gurari) can be used to transform any LATEX file to a wide range of output formats that include HTML, XHTML, TEI, EBook, and DocBook. In addition, one can choose whether the mathematical content will be transformed to images or to MATHML content. Instead of a large number of command line switches, the system provides a large number of little programs that are actually used to transform a LATEX file to any of the output formats supported. The following table shows the available programs.

Output Format	Bitmap Math	Bitmap Math & Unicode	MATHML	Mozilla
HTML	htlatex	uhtlatex		
XHTML	xhtlatex	uxhtlatex	xhmlatex	mzlatex
TEI		teilatex	teimlatex	
EBook		eblatex	ebmlatex	
DocBook		dblatex	dbmlatex	
HTML for MS Word	wlatex			
XHTML for MS Word	wlatex			

The programs above do not create a directory where all output files are placed, but instead they are placed in the current directory, so it is a wise idea to create a directory, place the LATEX input file there, and then proceed with the transformation. Figure D.2 on page 449 shows the output generated by htlatex when fed with the LATEX file above.

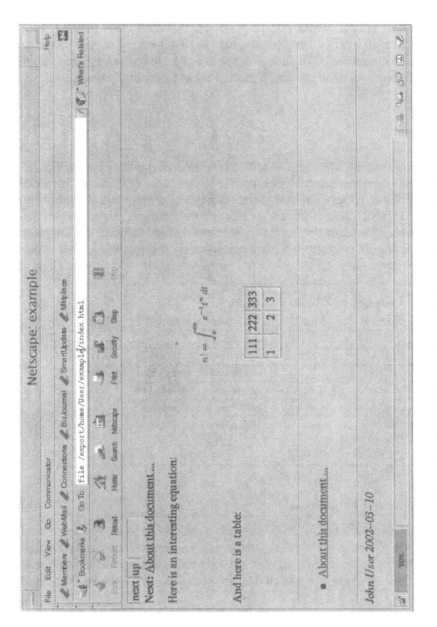

Figure D.1: LaTeX2HTML output opened with our favorite Web browser.

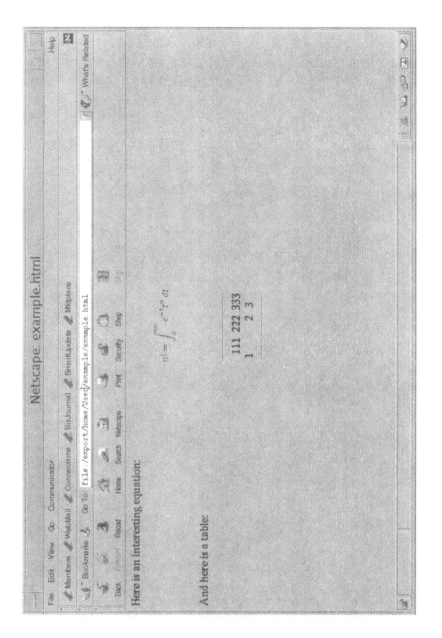

Figure D.2: tex4ht output opened with our favorite Web browser.

APPENDIX E

NEW FEATURES INTRODUCED TO Ω 1.23

In this book we describe Ω version 1.15, which is considered a rather stable release. Newer unstable versions of Ω introduce a number of new commands and features, which need testing before a new stable release will be available to the TEX community. In this appendix we summarize the changes that are introduced in version 1.23. In addition, we take this opportunity to present future directions in the development of Ω.

To make the Ω MathML engine more flexible a number of new commands have been introduced, while quite a few commands have been eliminated and replaced by new ones. The commands \MMLmode and \noMMLmode are no longer available. Instead, one should use the commands \SGMLmode and \noSGMLmode, respectively.

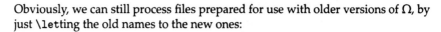

Obviously, we can still process files prepared for use with older versions of Ω, by just \letting the old names to the new ones:

```
\let\MMLmode\SGMLmode
\let\noMMLmode\noSGMLmode
```

The command \SGMLfilesuffix can now be used to specify the filename extension of the generated file that will contain the XML/SGML content. For example, the command

```
\SGMLfilesuffix{.html}
```

specifies that the filename extension of the generated file will be html. The new symbol-generating commands introduced to version 1.23 are presented below:

\SGMLquote	,	\SGMLbackquote	'
\SGMLdoublequote	"	\SGMLtilde	~
\SGMLexclamationmark	!	\SGMLquestionmark	?
\SGMLequals	=	\SGMLcolon	:
\SGMLsemicolon	;		

Although there are some other commands, we will not present them since they are completely experimental and their use is not recommended.

When we need to change the writing direction in a page or a paragraph, we must be careful to do the change in an empty page or outside a paragraph, respectively. Normally, the writing direction of mathematical text is identical to the writing direction of European languages. However, in many cases we use the advanced typesetting capabilities provided by Ω inside mathematical text to typeset ordinary text. For example, this happens when we typeset tables. The command \nextfakemath is used to typeset mathematical text as if it were ordinary text and thus employing the writing direction of the main text. Since the tabular environment is actually defined in math mode, we need this command to ensure that the text in the cells of a table will have the proper writing direction.

Patching Λ

A patch is a temporary addition to a piece of code, usually as a quick-and-dirty remedy to an existing bug or misfeature. LATEX was built so that the contents of the ltpatch.ltx file form a patch to the LATEX kernel. Since Λ is affected by the changes introduced to Ω, we need to fix LATEX with a patch. The following code is a patch that can be used in generating the Λ format with Ω 1.23:

```
\newcommand*\fmtversion@topatch{2002/06/01} %% LaTeX version the
\newcommand*\patch@level{1} %%Patch level    %% patch is applied
%%
\newcommand*\@tabular{\leavevmode \hbox \bgroup
    \nextfakemath$\let\@acol\@tabacol
    \let\@classz\@tabclassz
    \let\@classiv\@tabclassiv \let\\\@tabularcr\@tabarray}
%%%
%%% Comment out the above code if you are using Omega 1.15.
%%% Without the next two lines, we cannot use the command
%%% \verb and the verbatim environment.
%%%
\IfFileExists{ot1uctt.fd}{%
    \renewcommand*\ttdefault{uctt}}{}%
%%%
%%% Commands like \textit break the paragraph making algorithm
%%% of Omega. The following commands should be used only if your
%%% input file contains the \textit command; otherwise, you
%%% should put the code in a package file and use it accordingly.
%%%
\let\mydocument\document
\renewcommand*\document{\mydocument\textit{ }}
```

The Future of Ω

The development of Ω is far from being complete, and so we cannot really talk about a stable release. However, as Yannis Haralambous has explained in his foreword to this book, the kernel features of the system are somehow frozen. Since the developers of Ω and the people who support its development in various ways are quite enthusiastic about these exciting new technologies, it is inevitable that we will see the incorporation of these technologies into Ω. Here we will present the new features that are planned for future realeses of Ω as these have been communicated to us by John Plaice.

The Unicode standard specifies that a plain Unicode text file consists of 16-bit character sequences. Practically, this means that there are 65,536 different code points available. However, one can readily access 63,486 code points. The remaining 2,048 are used to represent an additional 1,048,544 characters through pairs of 16-bit sequences. These additional code points are usually called *surrogates*. Future releases of Ω will make it possible to access characters that have a code point less than or equal to 1,114,111 (or 10FFFF). So, all code points up to 65,535 will be accesible with the `^^^^hhhh` notation, while characters with code point above 65,535 will be accessible with the new `^^^^^^hhhhhh` notation. Although this may not seem a serious improvement, still it is very important, since many symbols (e.g., Western and Byzantine musical symbols) have code points above 65,535. So for example, we cannot create true Unicode-aware packages for musical text without this new feature. The capability to recognize surrogates will also affect the way Ω treats math codes and math delimiters.

The command `\charset={char-set}` will be introduced to allow users to easily specify the input encoding of a file. In addition, a special configuration file will allow the user to declare the default input encoding. Although the character ZERO WIDTH NO-BREAK SPACE (with code point FEFF) is used as a byte order mark, Ω still does not make any special use of this charcater, something we believe is necessary.

The writing direction commands will be modified as the current implementation fails in certain cases.

The big innovation will come the day Ω becomes a real XML-aware typesetting engine. This practically means that Ω will be able to generate XML output just as pdfTEX generates PDF output.

SOLUTIONS TO ALL EXERCISES

2.1 (page 15):
You have simply to type in the following characters:

```
You have a 30\% discount and so the price is \$13.
```

2.2 (page 15):
We use the complete example of the first chapter to create the complete LaTeX file:

```
\documentclass{article}
\begin{document}
The characters \{ and \} are special. They are used
to create a local scope.

Comments are introduced with the character \% and
extend to the end of the line.
\end{document}
```

2.3 (page 16):
At the end of the second paragraph, just type in the command \par.

2.4 (page 19):
We always have to consult the tables in order to write down the correct \docu-mentclass command. In our case, we want to prepare an article (i.e., we have to use the article document class). Moreover, we want to typeset our article at 11 pt for two-sided printing (i.e., the optional arguments must be 11pt and twosided), so the complete command is:

```
\documentclass[11pt,twoside]{article}
```

2.5 (page 21):
\section{Sectioning Commands}.

2.6 (page 24):
We first have to define the command that will put the asterisk in the table of contents and then the sectioning command:

```
\newcommand{\newaddcontentsline}[3]{%
  \addtocontents{#1}{%
  \protect\contentsline{#2}{*#3}{\thepage}}}
\let\oldaddcontentsline\addcontentsline
\newcommand\firstarg{}
\newcommand{\starredsection}[2][]{%
  \let\addcontentsline\newaddcontentsline%
  \renewcommand\firstarg{#1}%
  \if\firstarg\empty
     \relax\section{#2}%
  \else%
     \section[#1]{#2}%
  \fi%
  \let\addcontentsline\oldaddcontentsline}
```

Note that we are not allowed to use a local scope in the definition of the new sectioning command, so we have to play with the \let command.

2.7 (page 27):

If we put the abstract environment before the \maketitle command, then the abstract appears on a separate page. On the other hand, if we put the environment after the \maketitle command, the abstract appears just after the title of the document.

2.8 (page 28):

We only show the postal code part:

$$671\text{\textvisiblespace}00\text{\textvisiblespace}\text{\textvisiblespace}Xanthi$$

3.1 (page 45):

Note that in order to correctly typeset this text, one has to use both forms of the font selection commands:

```
Because of \textttt{typewriters} and the usually
\textbf{incomplete} fonts that come with
\textsc{word processors}, many people have learned
to \textsl{emphasize} by \textit{underlining}
the text. That is \textbf{really \textit{poor}} as
it disturbs the \textsl{balance} of the
''{\fontshape{ui}\selectfont page color}'' and
makes it look {\sffamily unprofessional}.
```

3.2 (page 45):

Here we used the \itshape command to denote natural numbers. We avoided math mode since it is not discussed up to this point:

```
\textbf{Theorem 1} Every natural number bigger than 1 has a
prime divisor.
```

```
\mbox{\letterspace to 3.5cm {Sketch of Proof:}} We use
```
induction. For the number 2, the result is correct since 2 is
a prime itself. Let `{\itshape n}` be a natural number. If
`{\itshape n}` is prime, then we are done. If not, then assume
that `{\itshape k}` is a divisor different from 1 and
`{\itshape n}`. Then, `{\itshape k}` is less than `{\itshape n}` and
consequently `(\textit{by induction})` it has a prime divisor,
say `{\itshape p}`. Then, `{\itshape p}` is a divisor of
`{\itshape n}` as well.

3.3 (page 46):
Note the font selection commands in the following solution:

```
This is {\fontfamily{pag}\selectfont Avant Garde},
{\fontfamily{pbk}\selectfont Bookman Old Style},
{\fontfamily{pzc}\fontshape{it}\selectfont Chancery
italic}. Also available in modern installations are
{\fontfamily{bch}\selectfont the Charter font},
{\fontfamily{ccr}\selectfont the Concrete font},
{\fontfamily{pcr}\selectfont the Courier font},
{\fontfamily{phv}\selectfont the Helvetica font},
{\fontfamily{pnc}\selectfont the New Century
Schoolbook Roman font}, {\fontfamily{ppl}\selectfont
the Palatino font}, {\fontfamily{panr}\selectfont the
Pandora font}, {\fontfamily{ptm}\selectfont the Times
font}, and {\fontfamily{put}\selectfont the Utopia
font}. If the \textsf{oldgerm} package is available,
then you can write with \textfrak{Fraktur},
\textgoth{Gotisch}, or \textswab{Schwab}.
```

3.4 (page 48):
The solution of this exercise demands the correct use of the various glyph selection
commands:

```
H\'an Th\^e Th\'anh, {\L}ata{\l}a, Livshi\v{c}, G\"odel,
Geometri\ae\ Dedicata, Na{\"\i}ve Set Theory, {\AA}ke
Lundgrem \"ar f\"odd 1951 i byn kusmark utanf\"or
Skellefle\aa, How many {\fontencoding{LGR}\selectfont
\euro}'s is a US\$, The University of the {\AE}gean, ?'!'
\copyright 1998 TV Espa\~na, Gro{\ss}en Planeten, \'a des
soci\'et\'es, H\"useyn a\u{g}abeyi ile birlilcte hal\^a
\c{c}al\i\c{s}{\i}yor.
```

3.5 (page 57):
We need to use the extended document classes:

```
\documentclass[8pt]{extbook}
\documentclass[14pt]{extarticle}
```

3.6 (page 58):

Note the combination of the various font selection commands:

```
\textttt{\large Starting with large typewriter typeface}
\textsc{\LARGE going on with large small capitals}
\textsf{\bfseries \small and ending with small sans
serif typeface.}
```

3.7 (page 58):

In order to typeset this text, we need to use the \font primitive command:

```
{\fontfamily{panr}\selectfont This is the Pandora
font family at} \font\pandoraone=pnr10 at 13.3pt
\font\pandoratwo=pnr10 at 21.5pt
\font\pandorathree=pnr10 at 34.8pt
{\pandoraone 13.3pt} {\pandoratwo 21.5pt}
{\pandorathree 34.8pt}
```

3.8 (page 60):

Since the table does not contain all possible symbols, first we define the missing symbols:

```
\newcommand{\rcl}{\raisebox{1ex}{\rc{~}}\hspace{-0.265em}l}
\newcommand{\actildeo}{\upaccent{\aboxshift{\'{}}}{\~o}}
\newcommand{\actildei}{\upaccent{\aboxshift{\'{}}}{\~{\i}}}
```

Now, we can use these new commands to create the phonetic transcriptions:

```
dl\actildeo\actildeo\glottal    (prairie dog)
t\esh\,'ah                      (hat)
\rcl itsx\uplett{w}o            (yellow-orange)
ts\actildei\actildei\rcl        (haste)
\glottal ak'os                  (neck)
xa\vari h                       (winter)
```

4.1 (page 62):

The lengths in ascending order follow:

1. 1.9 cc	2. 0.89 cm
3. 101.7 dd	4. 45 mm
5. 2.5 in	6. 33 pc
7. 536 bp	8. 674 pt
9. 123456789 sp	

4.2 (page 62):
They are exactly 7272 pt.

4.3 (page 64):
Certainly, one has to use the `verse` environment to typeset the poem. The interline space can be achieved by using the third form of the \\ command:

```
Soon we will plunge ourselves into cold shadows,\\[3,5pt]
```

4.4 (page 65):
A simple solution is to use the `twocolumn` option and the `description` environment.

4.5 (page 67):
The two commands must simply change the category code of the *at* character:

```
\newcommand{\makeatletter}{\catcode'\@=11}
\newcommand{\makeatother}{\catcode'\@=12}
```

4.6 (page 68):
We first define a new length

```
\newlength{\exlabel}
```

Then, we set this length equal to the width of a long enough label such as EX viii:

```
\settowidth{\exlabel}{EX\ viii}
```

Finally, we advance the value of \leftmarginii by this length:

```
\addtolength{\leftmarginii}{\exlabel}
```

That's all!

4.7 (page 71):
A simple solution is to create a command that will behave identically to \footnotesize:

4.8 (page 72):
The following redefinition achieves the desired effect:

```
\addtolength{\skip\footins}{-2pt}
\renewcommand{\footnoterule}{\vspace*{-5pt}
    \noindent\rule{\textwidth}{4pt}\vspace*{1pt}}
```

4.9 (page 82):
See how we fool LaTeX into correctly placing the horizontal and vertical lines:

```
\begin{tabular}{@{}ccc@{}}
\phantom{$A^\bot$}&\phantom{$A^{\bot}\otimes B$} &
\phantom{$A$}\\
\cline{1-3}\\[-12pt]
\multicolumn{1}{@{\vline}}{} & &
\multicolumn{1}{r@{\vline}}{}\\[4pt]
\multicolumn{1}{@{\vline}}{} & & $A$\\
```

```
\multicolumn{1}{@{\vline}}{} & & $\vdots$\\
$A^\bot$ & & $B$\\[3pt]
\hline\\[-6pt]  & $A^{\bot}\otimes B$
```

5.1 (page 110):
Just use the following piece of code:

```
$$\doubleint_{\{x\in X\,:\,\|x\|\leq 1\} }$$
```

5.2 (page 110):

```
\newcommand{\grcos}{\mathop{\mathgroup\symgroperators sun}\nolimits}
```

5.3 (page 111):
We have to use twice the \atop command:

```
\begin{displaymath}
\sum_{{0\le i\le r} \atop {%
0\le j\le s \atop 0\le k\le t}} C(i,j,k)
\end{displaymath}
```

5.4 (page 148):
Since the MathML translator is in a rather experimental stage, try to avoid placing a \SGMLlonetag somewhere between ordinary tags:

```
\documentclass{article}
\begin{document}
  \MMLmode
  \SGMLstarttexttag{html}%
    \SGMLstarttexttag{head}%
      \SGMLstarttexttag{title}%
      Planck's Equation
      \SGMLendtexttag{title}%
    \SGMLendtexttag{head}%
    \SGMLstarttexttag{body}%
      \SGMLstarttexttag{h2}%
        \SGMLstarttexttag{center}%
        Planck's Equation:
%\SGMLlonetag{br}
          \MMLstarttext
          $E=\hbar\nu$
          \MMLendtext%
        \SGMLendtexttag{center}%
      \SGMLendtexttag{h2}%
    \SGMLendtexttag{body}%
  \SGMLendtexttag{html}%
\end{document}
```

Uncomment the \SGMLlonetag to see what happens.

6.1 (page 166):

```
a\hspace{\stretch{5}}b\hspace{\stretch{2}}c
```

6.2 (page 167):

We have to use the \stretch command to get the desired effect:

a

```
\vspace{\stretch{2}}
```

b
```
\vspace{\stretch{1}}
```

c

6.3 (page 170):

Note that we must redefine even the \thepage command:

```
\renewcommand{\pagenumbering}[1]{%
  \setcounter{page}{1}%
  \renewcommand{\thepage}{%
    \thechapter-\csname @#1\endcsname{\value{page}}}}}
```

In addition, we must redefine the \chapter command so that it sets the page counter to one each time we start a new chapter.

6.4 (page 172):

All we have to do is to ignore the first argument of the command.

6.5 (page 183):

Here is a possible solution to George's problem:

```
\documentclass[a4paper,twoside]{article}
  \pagestyle{myheadings}
  \markboth{George Typesetter}{On the use of \TeX}
```

6.6 (page 192):

The obvious solution is to use the hyperref package. More specifically, we should use the hyperref package with no options loaded, and in each slide we should place some \Acrobatmenu commands that will allow readers to navigate in the slide show. Here is a skeleton file that can be used to prepare "hyper"-slides:

```
\documentclass[slideColor,total,pdf]{prosper}
\usepackage{hyperref}
\title{''Hyper''-slides}
\author{George Showman}
```

```
\slideCaption{Conference of ``Hyper''-shows}
\begin{document}
\maketitle
\begin{slide}[Box]{First Slide}
one one one one one one one one one one one
.........................................
one one one one one one one one one one one
.........................................
\Acrobatmenu{NextPage}{Next}
\end{slide}
\begin{slide}[Glitter]{Second Slide}
two two two two two two two two two two two
.........................................
two two two two two two two two two two two
.........................................
\Acrobatmenu{PrevPage}{Previous}
\Acrobatmenu{NextPage}{Next}
.........................................
\end{document}
```

6.7 (page 197):
We only give you the code that is responsible for the framed boxes:

$$\framebox[2\width][l]{...}$$
$$\framebox[2\width][r]{...}$$
$$\framebox[2\width][c]{...}$$
$$\framebox[2\width][s]{...}$$

6.8 (page 197):
We only give the La part of the logo:

```
L\hspace{-.36em}\raisebox{.2em}{\scriptsize A}\hspace{-.15em}
```

6.9 (page 197):
We first declare a new box variable and then define it using a strut:

```
\newsavebox{\tmpbox}
\savebox{\tmpbox}[3ex]{\rule{0pt}{4ex}}
\setlength{\fboxsep}{0pt}
\fbox{\usebox{\tmpbox}}
```

6.10 (page 200):
The following definition is rather nonstandard (i.e. we have to use \equation and \endequation instead of the expected opening and closing commands just because such commands are not allowed in a definition when the amsmath package is used). Naturally, if we do not use this package, we can use the "normal" commands.

```
\newlength{\mylen}
\setlength{\mylen}{\textwidth}
\addtolength{\mylen}{-2\fboxsep}
\addtolength{\mylen}{-2\fboxrule}%
\newenvironment{fequation}{%
   \Sbox\minipage{\mylen}%
   \setlength{\abovedisplayskip}{0pt}%
   \setlength{\belowdisplayskip}{0pt}%
   \equation}%
   {\endequation\endminipage%
    \endSbox\[\fbox{\TheSbox}\]}}
```

6.11 (page 203):
The definition is almost identical to the definition of the command \seq:

```
\newcommand{\seqk}{\{x_1,\ldots,x_k\}}
```

6.12 (page 204):
In case the argument of the command is just a mathematical variable, things are fine. However, if the argument is a mathematical expression, we need to have the argument in curly brackets to make sure that the whole expression will be a subscript to x. Change the definition and use the modified command with x^2 as an argument to see the difference.

6.13 (page 204):
The first step is to define a command that will simply print the required phrase. Then, we can define another command that calls this command one hundred times, or we can be smart and do the following:

```
\newcommand{\one}{\noindent I love to fly!\\}
\newcommand{\five}{\one\one\one\one\one}
\newcommand{\twentyfive}{\five\five\five\five\five}
\newcommand{\hundred}{\twentyfive\twentyfive\twentyfive
   \twentyfive}
```

Now, it is easy to define another command that prints one hundred times the phrase "I X to Y."

6.14 (page 208):
The frame is generated by a tabular environment. However, since the tabular environment adds some space, we remove this space from the line width.

```
\newlength{\Fquotewidth}
\newenvironment{Fquote}{\begin{quote}
\setlength{\Fquotewidth}{\linewidth}%
\addtolength{\Fquotewidth}{-2\tabcolsep}%
\addtolength{\Fquotewidth}{-2\arrayrulewidth}%
\begin{tabular}{|p{\Fquotewidth}|}\hline}{%
```

```
\\\hline\end{tabular}\end{quote}}
```

6.15 (page 210):

In what follows, we give only the redefinition of \makelabel:

```
\setlength{\labelwidth}{3cm}
\renewcommand{\makelabel}[1]{%
  \settowidth{\LabelWidth}{##1}%
  \ifthenelse{\lengthtest{\LabelWidth>\labelwidth}}{%
    \makebox[\LabelWidth][r]{##1}}{%
    \makebox[\labelwidth][r]{##1}}}
```

Note that we use a new length variable, namely \LabelWidth.

7.1 (page 223):

We give a sample of the code:

```
\documentclass[11pt]{article}
\usepackage{currvita}
\begin{document}
\def\today{January 4, 310}
\frenchspacing
\begin{cv}{Curriculum vitae of Steve Worker}
\begin{cvlist}{Personal Information}
\item Steve Worker\\
      University of Sparta\\
      Department of Fine Arts\\
      Sparta
\item Email: \texttt{sworker@fn.sparta.edu}
\item Born in Thebes
\end{cvlist}

\begin{cvlist}{Studies}
\item[June, 300] B.Sc. in Mathematics...
\item[September, 305] Ph.D. in....
Title: \textit{Symmetrizations and ....}
Thesis advisor: Euclid.
\end{cvlist}

\begin{cvlist}{Research Interests}
\item Geometry of Convex Bodies, ....
\end{cvlist}

\begin{cvlist}{Publications List}
\item
\begin{enumerate}
```

```
\item \textit{On a Conjecture by Aristarchus}
{Annals of the Athenean Mathematical Society}
60:187--206, 306.
```

```
\item \textit{Is Hippasus's Theorem Correct?}
 (to appear).
\end{enumerate}
\end{cvlist}
\thispagestyle{empty}
\end{cv}
\end{document}
```

8.1 (page 231):

Since our redefinition uses the @ character, we must make it an ordinary letter with the \makeatletter command. After the redefinition is finished, we have to make @ an "other" character with the \makeatother command.

8.2 (page 244):

Here is the code that prints the index header and makes the corresponding entry in the table of contents

```
\newpage \twocolumn[{\Large\bfseries #2 \vspace{4ex}}]
\markright{\uppercase{#2}}
\addcontentsline{toc}{section}{#2}
```

Now, if we replace the above code with the following piece of code, we get the required behavior.

```
\markboth{\uppercase{#2}}{\uppercase{#2}}
\chapter*{#2}
\addcontentsline{toc}{chapter}{#2}
```

8.3 (page 250):

We have to set headings_flag to some positive value so that each group starts with a capital letter. Moreover, we have opted to center this letter on the line. Here is the complete code:

```
heading_prefix "{\\hfil "
heading_suffix " \\hfil}\\nopagebreak\n"
headings_flag 1
delim_0 ", {\\scshape "
delim_1 ", {\\scshape "
delim_t "}"
```

9.1 (page 255):

The following code does exactly what we want:

```
\begin{picture}(0,0)
\put(10,20){\framebox(10,10)[c]{$\heartsuit$}}
\end{picture}
```

9.2 (page 268):

The following sketch assumes that we use the fancyhdr package:

```
\newsavebox{\image}
\savebox{\image}{\includegraphics[height=70pt,%
                    width=360pt]{image.eps}}
\setlength{\topmargin}{0pt}
\setlength{\voffset}{-0.5in}
\setlength{\headheight}{70pt} \fancyhead{}
\fancyhead[LE,LO]{} \fancyhead[RE,RO]{}
\fancyhead[CO,CE]{\usebox{\image}\\
\hfill\thepage\hfill}
\renewcommand{\headrulewidth}{0pt}
```

9.3 (page 269):

A\reflectbox{B}BA.

9.4 (page 280):

In the code that follows, we do not include all points, for obvious reasons.

```
\setcoordinatesystem units <10pt,10pt>
\setplotarea x from 0 to 25, y from 65 to 92
\axis bottom label {Time in hours since midnight}
    ticks numbered from 0 to 25 by 5 /
\axis left label {Temp $\mbox{}^\circ\mathrm{F}$}
    ticks numbered from 65 to 90 by 5 /
\put{+} at 1 65   \put{+} at 2 69   \put{+} at 3 74
```

Note that we do not have to transform the data, but instead we "enter" them as they are.

9.5 (page 283):

\setdashpattern <5pt,5pt>.

9.6 (page 292):

In the following code, we reexpressed the data in a linear form:

```
verbatimtex \documentclass[a4paper]{article}
\begin{document} etex
input graph;
beginfig(1); draw begingraph(200pt,100pt);
setrange(10,1,210,"1e4"); setcoords(linear,log);
glabel.lft(btex
\shortstack{l\\o\\g\\a\\r\\i\\t\\h\\m\\i\\c} etex,
  OUT);  glabel.bot(btex etex, OUT);
```

```
gdraw "dummy.data" plot btex $\circ$ etex;
endgraph; endfig; end
```

9.7 (page 295):

One easy way is to use the following pdfLATEX file:

```
\documentclass[a4paper]{article}
\usepackage[pdftex]{color}
\definecolor{X}{rgb}{1.0,1.0,1.0}
\definecolor{Y}{cmyk}{1.0,1.0,1.0,1.0}
\begin{document}
\Huge \textcolor{X}{THIS} \textcolor{Y}{THAT}
\end{document}
```

9.8 (page 299):

It is better to use a predefined length variable to avoid filling TEX's memory with new variables:

```
\setlength{\@tempdima}{\paperwidth}
\setlength{\paperwidth}{\paperheight}
\setlength{\textheight}{\@tempdima}
```

10.1 (page 305):

To embed English text in an Arabic document, we should use the command {\textdir TLT *text*}. The opposite is achieved with the command {\textdir TRT *text*}. Of course, this solution assumes that we have prepared our documents with a Unicode editor and that we are using a complete Unicode font.

10.2 (page 310):

We assume that the following ΩTP will be applied to Unicode input.

```
input: 2; output: 2;
aliases:
LETTER  = (@"03AC-@"03D1 | @"03D5 | @"03D6 | @"03F0-@"03F3 |
           @"1F00-@"1FFF) ;
expressions:
^({LETTER})@"03B8({LETTER} | @"0027) => \1 @"3D1 \3;
. => \1;
```

10.3 (page 310):

Suppose that we are using a font that provides the ligature C for the characters A and B. To prevent the formation of the ligature, Knuth in [19] suggests the use of the construct A{\kern0pt}B. The \kern command is used to adjust space between glyphs in typeset text. Based on this suggestion, we can easily write the code for the ΩTP:

```
input: 1; output: 1;
expressions:
'f' 'i' => \1 "{\kern0pt}" \2;
```

10.4 (page 312):

It is on A4 paper! The reason, of course, is that Ω has this as the default paper size.

10.5 (page 324):

The main problem is that the stroke of the letter t is not symmetrical, so we have to redraw this stroke:

```
\newcommand{\tx}{t\kern-0.3em\rule[0.82ex]{0.3em}{0.1ex}%
            \kern-0.3em\rule[0.41ex]{0.3em}{0.1ex}}
\newcommand{\Tx}{T\hspace{-.55em}\raisebox{0.4ex}{--}%
            \hspace{0.05em}}
```

10.6 (page 328):

We could use a local scope, but this is not necessary.

```
\newenvironment{Tibetan}{\tbfamily%
         \pushocplist\TibetanInputOcpList}{}
```

10.7 (page 335):

`\pagenumbering{gana}`.

10.8 (page 342):

The `\noarmtext` command will actually be an application of `\aroff`:

```
\newcommand{\noarmtext}[1]{{\aroff #1}}
```

Note that we need to create a local scope or else, once we use this command, TeX will continue typesetting using a Latin font.

10.9 (page 346):

Naturally, the most difficult task is to find which letter we are using:

```
$\sin\ethmath{ki}^2 +\cos\ethmath{ki}^2=1$
```

Bibliography

[1] Beccari, C., and Syropoulos, A. New Greek fonts and the greek option of the babel package. *TUGboat 19*, 4 (1998), 419–425.

[2] Berdnikov, A. S. Package accentbx: Some problems with accents in TeX and how to solve them. *Εὔτυπον*, 5 (2000), 25–50.

[3] Berry, K. Filenames for fonts. *TUGboat 11*, 4 (1990), 517–520.

[4] Bowman, J. *Greek Printing Types in Britain: From the Late Eighteenth to the Early Twentieth Century*. Typophilia, Thessaloniki, Greece, 1998.

[5] Carlisle, D. xmltex: A nonvalidating (and not 100% conforming) namespace aware XML parser implemented in TeX. *TUGboat 21*, 3 (2000), 193–199.

[6] Francis, B., Green, M., and Payne, C. *GLIM 4: The Statistical System for Generalized Linear Interactive Modelling*, second ed. Oxford University Press, Oxford, U.K., 1993.

[7] Girard, J.-Y. Linear logic: Its syntax and semantics. In *Advances in Linear Logic*, J.-Y. Girard, Y. Lafont, and L. Regnier, Eds. Cambridge University Press, Cambridge, U.K., 1995, pp. 1–42.

[8] Goossens, M., Mittelbach, F., and Samarin, A. *The LATEX Companion*. Addison Wesley Publ. Co., 1994.

[9] Haralambous, Y., and Plaice, J. Τὸ Ω καὶ τὰ Ἑλληνικά, ἤ «ὁ ἄυλος αὐλός». *Εὔτυπον*, 2 (1999), 1–15.

[10] Haralambous, Y., and Plaice, J. Produire du MathML et autres *ML à partir d'Ω: Ω se généralise. *Cahiers GUTenber*, 33–34 (2001), 173–182.

[11] Haralambous, Y., and Plaice, J. Traitement automatique des langues et compositions sous Omega. *Cahiers GUTenberg*, 39–40 (2001), 139–166.

[12] Harold, E. R., and Means, W. S. *XML in a Nutshell*. O'Reilly, 2001.

[13] Hobby, J. Introduction to METAPOST. *Εὔτυπον*, 2 (1999), 39–53.

[14] Hutcheson, G. D., Baxter, J., Telfer, K., and Warden, D. Child witness statement quality: general questions and errors of omission. *Law and Human Behaviour 19* (1995), 631–648.

[15] Knuth, D. E. *Computer Modern Typefaces*, vol. E of *Computers and Typesetting*. Addison–Wesley Publ. Co., Reading, MA, USA, 1986.

[16] Knuth, D. E. The errors of TeX. *Software— Practice & Experience 19, 7* (July 1989), 607–685.

[17] Knuth, D. E. *Literate Programming*. Center for the Study of Language and Information, Stanford University, Stanford, CA, USA, 1992.

[18] Knuth, D. E. *The METAFONT book*, vol. C of *Computers and Typesetting*. Addison–Wesley Publ. Co., Reading, MA, USA, 1992.

[19] Knuth, D. E. *The TeX book*, vol. A of *Computers and Typesetting*. Addison–Wesley Publ. Co., Reading, MA, USA, 1993.

[20] Lamport, L. *LaTeX: A Document Preperation System*, 2nd ed. Addison–Wesley Publ. Co., Reading, MA, USA, 1994.

[21] Ryle, G. *The Concept of Mind*. Penguin Books, 1949.

[22] Sofka, M. Color book production with TeX. *TUGboat 15, 3* (1994), 228–238.

[23] Syropoulos, A. *LaTeX*. "Paratiritis" Editions, Thessaloniki, Greece, 1998. ISBN: 960-260-990-7.

[24] Syropoulos, A., and Tsolomitis, A. Γραμματοσειρές TrueType και LaTeX2ε. *Εὔτυπον, 2* (1999), 17–22.

[25] N𝒯S Team, and Breittenlohner, P. The ε-TeX manual, Version 2. MAPS, 20 (1998), 248–263.

[26] Thành, H. T., Rahtz, S., and Hagen, H. The pdfTeX users manual. MAPS, 22 (1999), 94–114.

[27] The LaTeX3 Project Team. LaTeX2e font selection. LaTeX document that is distributed with every release of LaTeX, 2001.

[28] Tukey, J. W. *Exploratory Data Analysis*. Addison–Wesley Publishing Company, Reading, MA, USA, 1977.

[29] Wall, L., Christiansen, T., and Orwart, J. *Programming Perl*, third ed. O'Reilly, Sebastopol, CA, USA, 2000.

[30] Wichura, M. J. *The PiCTeX Manual*. No. 6 in TeXniques. 1992. Available from www.pctex.com.

[31] Zubrinić, D. The exotic Croatian glagolitic script. *TUGboat 13, 4* (1992), 470–471.

[32] Zubrinić, D. Croatian Fonts. *TUGboat 17, 1* (1996), 29–33.

Name Index

SUBJECT INDEX

◊

The
book "Digital
Typography Using
LATEX" was typeset us-
ing LATEX 2$_\varepsilon$. Some parts were
typeset by Λ, ε-LATEX and pdfLATEX.
The working platforms were Solaris 8 x86
and Linux 2.4 x86. The final PostScript file was
generated with DVIPS. Some parts that used Λ were
generated with ODVIPS. The bibliography was generated with BIBTEX8 and the indices with MAKEIN-
DEX. Screen shots were captured with GIMP
and transformed to EPS with AU-
TOTRACE. Editing was entirely
done with GNU Emacs.
The size of the final
DVI file was
1.6 MB.

◊